W0036353

Durable Resistance in Crops

NATO Advanced Science Institutes Series

A series of edited volumes comprising multifaceted studies of contemporary scientific issues by some of the best scientific minds in the world, assembled in cooperation with NATO Scientific Affairs Division.

This series is published by an international board of publishers in conjunction with NATO Scientific Affairs Division

A	**Life Sciences**	Plenum Publishing Corporation
B	**Physics**	New York and London
C	**Mathematical and Physical Sciences**	D. Reidel Publishing Company Dordrecht, Boston, and London
D	**Behavioral and Social Sciences**	Martinus Nijhoff Publishers The Hague, Boston, and London
E	**Applied Sciences**	
F	**Computer and Systems Sciences**	Springer Verlag Heidelberg, Berlin, and New York
G	**Ecological Sciences**	

Durable Resistance in Crops

Edited by

F. Lamberti

Institute of Agricultural Nematology of the CNR
Bari, Italy

J. M. Waller

Commonwealth Mycological Institute
Kew, Surrey, England

and

N. A. Van der Graaff

Food and Agricultural Organization of the United Nations
Rome, Italy

Plenum Press
New York and London
Published in cooperation with NATO Scientific Affairs Division

Proceedings of a NATO Advanced Study Institute
held September 30–October 11, 1981,
in Martina Franca, Italy

Library of Congress Cataloging in Publication Data

Main entry under title:

Durable resistance in crops.

(NATO advanced science institutes series. Series A, Life sciences; v. 55)
  "Proceedings of a NATO advanced study institute held September 30–October 11, 1981, in Martina Franca, Italy"—T.p. verso.
  "Published in cooperation with NATO Scientific Affairs Division."
  Includes bibliographical references and index.
  1. Plants—Disease and pest resistance—Congresses. 2. Plant—Disease and pest resistance—Genetic aspects—Congresses. 3. Plant-breeding—Congresses. I. Lamberti, F. (Franco), 1937–   . II. Waller, J. M. III. Van der Graaff, N. A., 1945–   IV. North Atlantic Treaty Organization. Scientific Affairs Division. V. Series.
SB750.D87 1982   632'.3   82-18980
ISBN 978-1-4615-9307-2   ISBN 978-1-4615-9305-8 (eBook)
DOI 10.1007/978-1-4615-9305-8

©1983 Plenum Press, New York
Softcover reprint of the hardcover 1st edition 1983
A Division of Plenum Publishing Corporation
233 Spring Street, New York, N.Y. 10013

All rights reserved. No part of this book may be reproduced, stored in a retrieval system, or transmitted in any form or by any means, electronic, mechanical, photocopying, microfilming, recording, or otherwise, without written permission from the Publisher

PREFACE

    Plant diseases and pests are a major constraint to agricultural
production despite the various measures used to control them.
Chemical control, although often effective, may pose environmental
hazards and is relatively expensive, especially in developing
countries where it may be completely uneconomic. Control through
genetically mediated resistance to diseases and pests, is both cheap
and environmentally safe and at present most diseases and pests of
staple food crops are controlled through some form of resistance.

    One of the basic problems in the use of resistance is its fre-
quent lack of durability; very often a type of resistance is used
that 'breaks down' after a certain period. The temporary nature of
this resistance, due to the development of new strains of pest or
pathogen able to overcome it, has seriously hindered the improvement
of the yield potential of many crops as a continuing effort is needed
to replace old cultivars who resistance has failed, with new ones.
Following Vanderplank's now classical publications (1963, 1968)
which differentiated horizontal and vertical resistance, studies on
several host-parasite systems have shown that different types of
resistance can be distinguished genetically and epidemiologically,
and on the ability of the pests or pathogens to adapt to them. A
knowledge of how resistance operates at the population level has also
opened up possibilities of 'managing' relatively simple resistance
types in such a way that a stable host-pathogen system can be produced with a minimum of crop loss.

    Several research and breeding programs have been working to-
wards these objectives both at national and international level.
The Food and Agricultural Organisation of the United Nations began
its International Programme on Horizontal Resistance in 1975 with
the objective of accumulating high levels of supposedly durable
resistance. Considerable interest and scientific debate was generated by these programs which necessitated the meeting and interaction of those concerned. Discussion between FAO and the Special
Projects Committee of the International Society for Plant Pathology
concluded that a NATO Advanced Study Institute would be a suitable

v

forum at which recent results of these programs could be communi-
cated and the various strategies which had been developed could be
adequately discussed. The Institute was held in Martina Franca,
Italy from September 30 - October 11, 1981.

The lectures and contributions reviewed many theoretical sub-
jects, research results and practical breeding programs. The
presentations often generated extensive discussions much of which
are included in the Proceedings as we consider them an integral
part of these. As in any relatively new field of science, ideas
are many and diverse, but the practical selection, breeding and
management strategies in which they result differ much less. We
believe that the Institute achieved its purpose very well and that
these Proceedings will provide a useful reference point in the quest
for durable resistance in all crops.

We would like to thank all those who made this NATO Advanced
Study Institute possible and particularly Dr. L. Chiarappa for much
help and encouragement in planning the program, the Mayor of
Martina Franca for granting use of the Municipal Hall for our meeting,
Mrs. Angela Tosi, secretary of the Institute, Mr. F. Elia and Mr.
G. Zaccheo for their assistance before and during the meeting and to
Mrs. M. Rainbow and Mrs. B. Ritchie for help in preparing these
Proceedings for publication. On behalf of all those who participated
we gratefully acknowledge the Scientific Affairs Division of NATO
without whose approval and financial support this Advanced Study
Institute would not have been possible.

F. Lamberti
J.M. Waller
N.A. Van der Graaff

March 1982

CONTENTS

CONTENTS

DURABLE RESISTANCE AND AGRICULTURAL DEVELOPMENT

L. Chiarappa

Food and Agriculture Organization of the United Nations
Rome, Italy

I have great pleasure in bringing my personal greetings and those of the Organization I represent to this important gathering of scientists. I should congratulate the NATO Advanced Study Institute Program, the International Society of Plant Pathology and, above all, Prof. Franco Lamberti on convening the meeting.

Although great technological advances have been made in recent years, the problem of crop failures and of food shortages still remains one of primary concern. In fact, crop failures still occur in many countries, even where these have reached a high degree of development and are fully capable to respond promptly to large scale crop parasite attacks. This means that there is still much to be gained by looking in greater depth into the various causes of crop failure and to identify those factors that can be controlled by man.

I do not wish to recall here some of the classic examples of crop failures due to parasites. These are too often cited in plant pathology text books. Nor do I wish to repeat here some of the broad reaching statements, so often made at scientific meetings, which emphasize the importance that we have, as plant pathologists and plant breeders, in serving society and world agriculture. We all know that our field of science is extremely important. But this importance must be unequivocally demonstrated in practice by applying scientific knowledge in reducing or eliminating crop failures.

The needs of the world's increasing population call for a determined scientific effort to develop adequate understanding of the causes underlying crop failures and to promote practical actions to overcome them. This international meeting is called here specifically to examine one of the important causes of crop failure: the

breakdown of genetic resistance to crop parasites. Many and quite
different theories have been put forward to explain this phenomenon.
We must recognize that during the past twenty years considerable
experimental work has been conducted to confirm or to disprove a
number of these scientific hypotheses. While in many respects we
are still debating on theoretical grounds, we can also say that
some valuable progress has been achieved.

The primary scope of this meeting is to analyze some of the
theories underlying durable resistance and to examine also some of
the practical resistance breeding programs which have been based
on these theories. This should allow us to discover where we have
failed in the past and in which direction we should re-address our
thinking and future practice. The eminence and the diversity of
scientists convened here should make this exercise quite useful.

I wish to mention at this point why FAO is so much interested
in this meeting.

Until not long ago durable resistance was only a subject for
scientific debate. Did this resistance really exist? Was there
any resistance that could be considered really stable? Was horizontal resistance something entirely different from vertical resistance? and so on. All these questions were raised for the sake of
scientific discussion but, I should say, with little effort to do
anything about this matter in a practical way.

In 1970 the World Food Congress recognized the fragile basis
of the "green revolution" and put forward a recommendation for the
"scientific elimination" of its instability. Again, in 1974 the
World Food Conference emphasized the need for a continuing research program into the mechanism of resistance especially in tropical areas. It was at this point that FAO came forward with an
International Program on Horizontal Resistance (IPHR). One year
later this program became operational having as major objectives
the following ones:

(a) Provide convincing demonstrations of the validity of
     horizontal/durable resistence;

(b) Develop methods for accumulating such resistance in crops;

(c) Assist national institutes of developing countries in
     producing new high-yielding, good quality local cultivars
     with comprehensive horizontal resistance to all locally
     important crop parasites.

What FAO was looking for in launching this type of program
was a breeding strategy more adaptable to the developing world and
responding better to the immediate needs of the small farmers.

These, as we all know, do not have easy access to the technology or economics of pesticide usage, nor to expensive improved-seed distribution schemes. To test its strategy FAO selected a number of situations in Africa and Latin America where breeding work could be conducted with a minimum of infrastructures and manpower. A number of crops and disease situations were selected in which vertical resistance had clearly been demonstrated to be of no practical value. The outcome of some of these projects will be illustrated during this meeting when some of the scientists who have participated in FAO's IPHR will present the results of their work. Although these projects have been in existence for only a short time, and much more time will probably be needed to confirm the durability of the resistances so far accumulated, we expect that the results now available can already be analyzed and some trends identified.

I should also add that from this meeting FAO wishes to derive much fresh knowledge which will also serve to guide future government policies in national plant breeding programs. We, in fact, foresee that a meeting of government representatives of FAO member countries will be organized by FAO next year to present in a suitable form the knowledge gained here.

Another point I wish to call to your attention has to do with organizational matters. I am referring to the unexpected absence of two eminent scientists who recently had to cancel their participation because of health problems: they are Dr. J.E. Van der Plank and Dr. J.S. Niederhauser. In planning this meeting, one of the major objectives of the program was to ensure the participation of these two men not only to honor their pioneering work and outstanding ideas on durable resistance, but to permit a unique opportunity to the younger scientists for free and ample discussions.

In closing, I wish again to express my appreciation to the organizers and to give a warm welcome to all participants.

GENETIC BACKGROUND OF DURABLE RESISTANCE

R. Johnson

Plant Breeding Institute
Trumpington
Cambridge CB2 2LQ, England

SUMMARY

Durable disease resistance is defined as resistance that has remained effective while a cultivar possessing it has been widely cultivated in an environment favoring the disease. This characteristic of resistance is recognised retrospectively, as are all other characteristics of interactions between hosts and pathogens. Resistance that has been shown to be durable should be considered as a valuable resource for further breeding and for investigation of the causes of durability. Durable resistance is often achieved against diseases that show little or no special-isation into races pathogenic to particular cultivars. For such diseases, the control of resistance may be by single genes or by several genes, each of small effect, and the resistance may be either complete or incomplete (quantitative). For pathogens that possess pathogenicity specific for particular cultivars, failures of resistance can often be shown to be due to race-specificity of resistance genes and the evolution and spread of pathogen races with matching pathogenicity. These race-specific genes may have large (major) or small (minor) effects, giving either complete or quantitative resistance, and may operate at any stage of host development. Despite failures of resistance due to race-specificity in some cultivars, other cultivars remain resistant for relatively long periods, though this does not prove that their resistance will be permanent. For some diseases such durable resistance may be controlled by one or few genes of large effect. More usually a number of genes, not necessarily a large number, giving cumulative and often quantitative resistance, may appear to control durable resistance. Breeding programs should be modified to increase the chance of incorporating genetic components that

appear to be related to durable resistance. However, breeding
involves the selection of new genotypes, and the achievement of
durable resistance in new cultivars, whatever the breeding method,
will only be proved by a widespread and prolonged test.

INTRODUCTION

In 1971 and 1972 there was a yellow rust epidemic on the
winter wheat cultivar Joss Cambier in Britain (Walker and Roberts,
1974). Prior to 1971 its resistance had been adequate but
incomplete (quantitative) and thus of a type that reduced the rate
of disease development. This was a criterion for the identification
of horizontal resistance (equally effective against all races of the
pathogen) according to Vanderplank (1968, pp 164, 169). Events of
1971 and 1972 proved, however, that the resistance of Joss Cambier
was highly race-specific (Johnson and Taylor, 1972). I tried to
explain the initial hopes and subsequent disappointment with the
resistance of Joss Cambier to a research scientist* in the field
of human pathology. I made little progress in this explanation
using such terms as horizontal, vertical or general resistance
and slow-rusting because each required elaboration and definition
or was inadequate or subject to exceptions. I compared the
resistance to yellow rust that had failed in Joss Cambier with
resistance in Cappelle-Desprez that had remained effective for many
years. I pointed out that although it could be suggested that
resistance in Cappelle-Desprez was not race-specific and would be
permanent, this would be unacceptable to many people because they
would doubt the fixed relationship it implied between a host and its
pathogen.

It was apparent that these technical terms and the contro-
versies surrounding their use were not ideal for presenting the
simple notion I wished to convey. I decided that it was best
expressed by saying that the resistance of Cappelle-Desprez had been
durable, due to unknown causes, whereas that of Joss Cambier had
been quickly overcome by a previously unknown race of *Puccinia
striformis*. This was an accurate statement of our information and
in 1972 I therefore proposed to my colleagues at the Plant Breeding
Institute (PBI) that the term 'durable resistance' would be useful
to describe resistance that had remained effective for a long time.
In this chapter I hope to show its value as an aid to the
description of an aspect of the genetic control of pest and disease
resistance in plants that is of vital concern to plant breeders and
growers of crop cultivars.

---

*Dr. D.E. Bowyer, Department of Pathology, University of Cambridge

DURABLE RESISTANCE

Durable resistance is resistance that has remained effective in a cultivar during its widespread cultivation for a long sequence of generations or period of time, in an environment favorable to a disease or pest (Johnson and Law, 1973; Johnson 1979, 1981). It is a description based on observation and does not assume or explain any underlying cause for the durability. It does not imply that resistance will remain permanently effective (Johnson, 1981; Johnson and Law, 1975; Scott et al,. 1980). It does, however, focus attention on a particular attribute of resistance (continued effectiveness despite widespread and prolonged exposure to the pathogen) that is of great practical importance. Often no underlying cause is known for such consistency and the description durable is then more appropriate than many terms that imply a particular explanation, such as a mechanism or system of genetical control.

It is evident that the ability of a cultivar in commercial use to remain resistant can be influenced by several factors which include:

1) The inherent durability of its resistance.

2) The inherent variability in pathogenicity of the pathogen.

3) The life-cycle of the host and pathogen.

4) The manner and extent of deployment of the cultivar and of other resistant and susceptible host cultivars or host species.

5) Epidemiological factors affecting the size of the pathogen population. These include (a) climate and weather. A cultivar may be more susceptible in an environment highly favorable to the development of disease than in a less favorable environment (b) the level of susceptibility of the cultivar and other hosts.

TERMS DESCRIBING SPECIFICITY IN INTERACTIONS BETWEEN HOSTS AND PATHOGENS

The ability of a host to hinder a pathogen and reduce disease is called resistance and the ability of a pathogen to attack a host and promote disease is called pathogenicity (Robinson, 1969). Resistance that is more effective against some races of a pathogen than against others may be described as race-specific. Where resistance is apparently equal to all races it may be described as

race-non-specific. However, the description of resistance as
race-non-specific appears to imply that it will be non-specific also
to races of the pathogen with which it has not been tested. This
assumption may not be correct and resistance that is not known to
show race-specificity can only be described as apparently race-non-
specific (Scott et al., 1980).

    Race-specific resistance, as defined here, is synonymous with
vertical resistance (Vanderplank, 1963). Race-non-specific
resistance is synonymous with horizontal resistance defined as
resistance spread evenly against all races (Vanderplank, 1963).
It is clear that horizontal resistance thus defined is mainly of
interest because it implies resistance that will be durable.
Robinson (1976) has suggested that the terms vertical and horizontal
have the advantage of being conceptual or abstract and therefore
precise and appropriate in many contexts. Unfortunately, perhaps
because of this abstract quality, resistance has sometimes been
defined as horizontal on criteria other than its apparent race-non-
specificity, such as epidemiological or genetic characteristics.
In part these alternative definitions are due to statements by
Vanderplank (1963, 1968, 1975) who evidently assumed that certain
features observed in the interaction of potatoes with *Phytophthora
infestans* would be common to other host-pathogen interactions. Thus
it has been assumed that quantitative or incomplete resistance is
horizontal resistance. To define horizontal resistance thus
(Nelson, 1978) is misleading because quantitative resistance is not
necessarily either race-non-specific or durable. It is also often
assumed that horizontal resistance implies polygenic resistance.
The over-emphasis of this supposed relationship (Vanderplank, 1975
p. 173) has tended to obscure the fact that, in many examples of
durable resistance to pathogens and pests, resistance is apparently
due to a single or very few genes. These are of special interest to
plant breeders because of the ease with which they can be
manipulated.

    Because I do not wish my discussion of specificity in
resistance to be confused by the alternative interpretations of
horizontal resistance I prefer the descriptive terms race-specific
and race-non-specific resistance to the abstract terms vertical and
horizontal resistance.

    Matching terms applied to pathogenicity are host-specific and
host-non-specific (Scott et al., 1980). In the context of this
chapter host-specificity refers to specific pathogenicity towards
individual host cultivars or resistance genes. Host-specific
pathogenicity is synonymous with virulence as defined by Vanderplank
(1963) and vertical pathogenicity as defined by Robinson (1969).
Host-non-specific pathogenicity is synonymous with aggressiveness as
defined by Vanderplank (1963) and horizontal pathogenicity as
defined by Robinson (1969).

THE GENE-FOR-GENE HYPOTHESIS

From genetic studies of the flax rust pathogen *Melampsora lini* and its host, Flor (1955) deduced that for each race-specific resistance gene in flax there was a corresponding specific gene for pathogenicity in *M. lini*. If a cultivar possessed two race-specific resistance genes, the pathogen must carry both of the corresponding genes for matching pathogenicity in order to attack the cultivar (Person, 1959). For many diseases, even where comparative genetic studies in the host and pathogen have not been carried out, evidence for a gene-for-gene relationship has been based on differential interactions between host cultivars and pathogen strains (Day, 1974)

The existence of a gene-for-gene relationship between a host and its pathogen has an important implication for the plant breeder. It suggests that if he produces resistance by combining together two or more race-specific genes, the resistance will remain effective only so long as the pathogen is unable to produce and multiply a single race containing the corresponding combination of matching genes for pathogenicity.

The gene-for-gene theory describes a particular type of genetical interaction between hosts and their pathogens but is not necessarily applicable to all genetically controlled interactions between hosts and pathogens.

DURABLE RESISTANCE TO PATHOGENS THAT DO NOT DISPLAY HOST-SPECIFIC PATHOGENICITY

Many pathogens, including some that are important, show little or no tendency to develop host-specific pathogenicity. This is worth emphasizing because some authors forget that such pathogens exist and write as though virtually all interactions between hosts and pathogens show specificity and operate according to a gene-for-gene pattern (e.g. Ellingboe, 1981). Resistance towards pathogens that do not show host-specific pathogenicity is likely to be durable, provided that selection for such resistance is carried out using a pathogen isolate of adequate pathogenicity, under an environment favoring disease.

Milo disease of Sorghum has been adequately controlled, apparently by a single resistance gene, described as dominant or partially dominant (Russell, 1978). There has been no evidence to indicate the evolution of any race of the pathogen *Periconia circinata* overcoming this resistance. Another single gene, causing a hypersensitive reaction and complete resistance to *Helminthosporium carbonum* (the *Drechslera* state of *Cochliobolus carbonus*), was said to have provided resistance of maize to leaf spot for more than 25 years in the USA (Caldwell, 1968). Numerous

other examples are listed by Walker (1966), Eenink (1977), Russell
(1978) and elsewhere in this book. In many of these cases there is
little doubt that the durable resistance is controlled by the
recognised individual genes, since their absence gives a high level
of susceptibility. However, one note of caution should be sounded.
If a cultivar displays durable resistance to disease and a single
gene for resistance is identified this does not necessarily indicate
that the cultivar possesses no other genes for resistance that might
be contributing to the durability.

Perhaps more frequently, resistance to pathogens that do not
appear to display host-specific pathogenicity is controlled by
several genes with cumulative effects. The two diseases of wheat,
caused by *Septoria nodorum* and *Pseudocercosporella herpotrichoides,*
appear to be of this type (Scott *et al.*, 1982; Law *et al.*, 1975).
A further example may be resistance of coffee to the coffee berry
disease caused by *Colletotrichum coffeanum* (Van der Graaff, 1981).
In each of these cases resistance is apparently under the control of
several genes and there is no evidence of a gene-for-gene relationship.

These examples illustrate that when a breeder is facing a new
disease he should not automatically assume that he will have to deal
with a high degree of race-specificity in resistance. Where there
is no race-specificity in resistance, more rapid progress in
breeding for resistance may be achieved by using resistant cultivars
as sources of resistance than by trying to accumulate resistance by
intercrossing rather susceptible cultivars and selecting for
increased resistance among their progeny, as suggested by Robinson
(1973). However, vigilance should always be exercised when
attempting to breed for resistance to pathogens about which little
is known, in case specificity should occur.

DURABLE RESISTANCE TO PATHOGENS THAT DISPLAY HOST-SPECIFIC
PATHOGENICITY

In the following paragraphs some examples are given of
interactions between hosts and pathogens in which certain genetic
components of resistance have been shown to be race-specific.
In such cases resistance has often failed to be durable in new
cultivars because it has depended on one or a few race-specific
genes. Pathogens have often shown their ability to evolve or
recombine and spread rapidly the matching genes for pathogenicity.
In many cases where such specificity is found, cultivars sometimes
occur with resistance that is not rapidly overcome and can be
classified as durable. The contribution of recognised race-specific
components to the durability of resistance and the ease with which
these can be distinguished from components in which no specificity

has been observed, differ from case to case as illustrated in the examples.

Mosaic Virus in Tomatoes, Virus X in Potatoes and Anthracnose on Beans

Three genes for resistance to tomato mosaic virus have been named $Tm-1$, $Tm-2$ and $Tm-2^2$ (Hall, 1980). All have been used in the development of tomato cultivars resistant to the disease. When cultivars with $Tm-1$ or $Tm-2$ were introduced, strains of the virus that overcame the resistance rapidly emerged. By contrast, the gene $Tm-2^2$ which conditions a hypersensitive reaction to the virus, has now been widely used in new tomato cultivars in Western Europe. Strains pathogenic to $Tm-2^2$ were found shortly after the introduction of cultivars homozygous for this gene but they have not spread and Hall (1980) suggested that these strains may lack transmissibility. The gene $Tm-2^2$ has undoubtedly provided durable resistance to tomato mosaic virus and there is some prospect, though not certainty, that it will continue to be effective. It cannot be classified as a race-non-specific gene, but it is nevertheless of great value to tomato breeders. I am grateful to J.T. Fletcher of the Agricultural Development and Advisory Service for England and Wales (ADAS) for bringing this example to my attention.

Resistance to potato virus X has been controlled for many years by the dominant resistance gene Nx which causes a hypersensitive reaction to infection giving complete resistance. Virus strains able to overcome Nx are known to exist but have never caused any problem (see Howard, 1978).

Another example, described by Bannerot and Pochard (1972) suggests that the 'Cornell' gene gave more durable resistance of green beans to *Colletotrichum lindemuthianum* (anthracnose) than did previously used genes, towards which the pathogen had displayed rapidly evolving specific pathogenicity. However, I do not know whether the early promise of this gene has been maintained.

These examples suggest that initial evidence of specificity in resistance and failure of the first genes used should not necessarily be taken to indicate certain failure of all subsequent attempts to introduce single genes.

Late Blight on Potatoes and Brown Rust on Barley

Several individually recognised, highly race-specific genes, causing resistance reactions that have been described as hypersensitive, are known in barley conditioning resistance to brown rust caused by *Puccinia hordei* (Roane and Starling, 1967) and in potatoes conditioning resistance to late blight caused by *Phytophthora infestus* (Black *et al.*, 1953). In both cases these resistance genes give a high level of resistance and apparently interact according to a gene-for-gene pattern with specific genes for pathogenicity in the respective pathogens. Even where several such resistance genes have been combined together in potato cultivars they have been quickly overcome by *P. infestans* (Mastenbroek, 1966). It seems likely that a similar fate would await attempts to exploit combinations of the known race-specific genes that give resistance to brown rust in barley.

It has also been observed in both hosts that quantitative resistances to the respective diseases have remained effective for many years (Thurston, 1971; Habgood and Clifford, 1981). In both cases it has been suggested that the quantitative resistances are under the control of a number of genes each of small but cumulative effect (Thurston, 1971; Parlevliet, 1976).

As a result of observing that quantitative resistance to late blight in potato cultivars did not display differential interactions with various races of *P. infestans*, classified according to their interaction with recognised race-specific genes, Vanderplank (1963) concluded that this quantitative resistance was, and always would be, equal to all races. I do not believe it was justifiable to assume that resistance that had not displayed differential interactions with a limited number of races would never display such interactions. It was a hypothesis that could not be proved. In fact small degrees of race-specificity have been demonstrated in quantitative resistance both to brown rust in barley (Clifford and Clothier, 1974; Parlevliet, 1978) and to late blight in potatoes (Caten, 1974). However, these demonstrations have not altered the important observation that quantitative resistances to both diseases have often been durable, and this suggests that the respective pathogens have, as yet, been unable to evolve greater pathogenicity towards certain components of the resistance.

It is logical to use these cultivars with durable, quantitative resistance in further breeding and this will increase the probability that resistance will be durable in the resulting cultivars. It is fortunate that, in these two cases and probably some others, it has often been easy to distinguish the quantitative and durable resistance from the high-level race-specific resistance that has not been durable. However, it seems unwise, even with these diseases, to assume that all quantitative resistance will be durable, and for breeding purposes it would still seem preferable, where possible, to utilise resistance that has been shown by previous use to be durable. Unfortunately, for some other important diseases, to which some components of resistance are race-specific, there is no simple way to identify the genetic components that provide durable resistance and distinguish them from components that contribute little or nothing to durability.

Yellow Rust on Wheat

Race-specifity of resistance. Several genes conditioning resistance to yellow rust in wheat seedlngs have been identified and named (Macer 1972). Of these, all except *Yr-5* have been overcome somewhere in the world by races of *Puccinia striiformis* with specific matching pathogenicity. As far as I am aware the gene *Yr-5* has never been used in a widely grown cultivar so it is not known whether it would provide durable resistance. Some of the named genes, such as *Yr-6* and *Yr-7*, provide less than complete resistance and allow some sporulation.

Combinations of some of these genes provided resistance to all races of *P. striiformis* available at the PBI in the mid-1960s, and were used in some new wheat cultivars. These combinations of genes were rapidly overcome by previously unknown races that had matching genes for pathogenicity. In some cases the cultivars were too susceptible for further use and were more susceptible than the durably resistant ancestral cultivar Cappelle-Desprez (Table 1). This indicated that certain resistance factors from Cappelle-Desprez were lost during selection of the cultivars, due to the epistatic expression of the combination of race-specific genes they contained. In other cases, however, the cultivars still retained adequate resistance for commercial use (Table 1).

After the failures of resistance in cultivars possessing combinations of race-specific genes that were effective at the seedling stage it appeared possible that cultivars that were susceptible to yellow rust as seedlings and became more resistant as they matured, might possess more durable resistance. However, as pointed out in the Introduction, it was soon found that in some cases such resistance was highly race-specific and the resistance was not durable, as illustrated by the wheat cultivar Joss Cambier. In fact, Manners (1950) and Zadoks (1961) had both previously

Table 1   Percent leaf area infected with yellow rust in field trials (1970) of wheat cultivars with Cappelle-Desprez in their pedigree after their initially effective combinations of race-specific genes were overcome

Cultivar	Race-specific *Yr* genes	Date 18 June	Date 10 July
Cappelle-Desprez	3a,4a	6	45
Maris Envoy	1,3a,4a	45	75
Maris Ranger	6,3a,4a	16	74
Maris Beacon	2,3b,4b	12	70
Maris Nimrod	2,3a,4a	5	50
Maris Huntsman	2,3a,4a	1	23

demonstrated race-specificity in quantitative resistance to yellow rust in wheat cultivars, but sometimes it seems that personal experience is necessary to reinforce information gained from reading the literature. In the years following 1971 the lesson was impressed upon me frequently. Although the specificity of resistance in Joss Cambier was detected by measuring sporulation of *P. striiformis* on seedlings (Johnson and Taylor, 1972; Johnson and Bowyer, 1974), other cases were detected from field trials and in some cases their specificity could not be detected in seedlings.

The type of field trial used was based upon the method developed by Zadoks (1961) in which isolated nurseries contained numerous cultivars to be tested and a susceptible cultivar which was inoculated with a single isolate of *P. striiformis*. In the Netherlands, nurseries with different rust races were separated by wide barriers of a non-cereal crop on newly reclaimed land. At the PBI, shortage of land prevented use of the wide barrier to separate races. Instead the different nurseries, each inoculated with a single race of *P. striiformis* were separated by a barrier only 1 m wide of a tall resistant cultivar. This barrier did not prevent cross-contamination between rust races, and it was necessary to take account of this in interpreting the results. However, when a cultivar became infected due to cross-contamination, in a nursery inoculated with a race to which it was known to be resistant, it was almost invariably infected much less than in a nursery inoculated with a race to which it was known to be susceptible.

In most commercial fields the cultivar Cappelle-Desprez developed very little yellow rust but in these inoculated nurseries it became moderately infected due to the high level of inoculum. It has been found that other cultivars with no more yellow rust than Cappelle-Desprez in these nurseries also have adequate resistance for commercial use.

In 1970 two new semi-dwarf wheat cultivars, Maris Bilbo and Hobbit, were being developed at the PBI. Both had quantitative resistance but were more resistant than Cappelle-Desprez (Table 2). In 1972 severe foci of rust infection were found in a multiplication stock of Maris Bilbo. An isolate collected from these infections caused increased infections on Maris Bilbo and Hobbit so that both were now more susceptible than Cappelle-Desprez as shown in the data for 1977 under race 104 E137(3) (Table 2). To demonstrate the specificity of this effect a comparison with race 41 E136(3) and two cultivars, Maris Huntsman and Cappelle-Desprez, is included. Race 41 E136(3) was obtained in 1974 from crops of Maris Huntsman that had higher levels of infection than had been seen previously (Johnson and Taylor, 1980) and its specific pathogenicity on Maris Huntsman can be observed in Table 2. Another feature demonstrating the specificity of pathogenicity in these races is that race 104 E137(3) infects Cappelle-Desprez less than do the other races.

Table 2   Percent leaf area infected with yellow rust in field
            trials of wheat cultivars inoculated with several races of
            Puccinia striiformis

Date	10.7.70	19.7.77			23.6.81		
Race	41 E136(1)	41 E136(1)	41 E136(3)	104 E137(3)	41 E136(3)	104 E137(3)	41 E136(4

Cultivar

Cultivar	10.7.70	19.7.77			23.6.81		
Bilbo	23	63*	= 53*	< 88	70*	< 87 <	95
Hobbit	10	25*	= 25*	< 67	20*	< 45 <	76
Huntsman	23	23	< 40	> 18	35	> 10 =	17
Cappelle	45	43	= 42	> 28	40	> 33 <	43

= Resistance thought to be equal to two races (taking account of
  these and previous data)

> Resistance thought to be differential to two races (taking account
< of these and previous data)

* Some infection due to cross-contamination between races. See
  text.

     Maris Bilbo was discarded by the PBI; Hobbit narrowly avoided
the same fate and, despite its rather high susceptibility to yellow
rust, was grown commercially with the help of fungicides, achieving
very high yields and reaching about 17 per cent of the total wheat
area in Britain in 1977 and 1978.   In 1978 a further isolate of
P. striiformis was collected from an infected crop of Hobbit and
classified as race 41 E136(4).   It gave even higher levels of
infection on Hobbit and Maris Bilbo than race 104 E137(3) as shown
in Table 2.

     These results confirm that quantitative resistance of wheat to
yellow rust can be highly race-specific and may not be durable.
This probably applies to other hosts in their interactions with
pathogens, including wheat to other rust diseases. Perhaps this
type of specificity in quantitative resistance has been less often
observed in some other host-pathogen interactions than with yellow
rust on wheat because assessments for resistance in field trials
have often been carried out with mixtures of races or with only one
race, where such specificity cannot be observed. A further
implication from these data is that the pathogen may show more than
one step in increasing its pathogenicity towards a cultivar. Other
conclusions are that race-specificity can be detected having both
large (major) effects and small (minor) effects, and towards
resistance that operates at any stage in the development of the

host. Thus minor genes for resistance to yellow rust obtained from
exotic wheat cultivars or from other closely related cultivars
cannot be assumed to be race-non-specific.

Durable resistance to yellow rust. Despite the wide range of degrees
of race-specificity shown by wheat in its resistance to yellow rust,
and the failure of both complete and quantitative resistance to be
durable in some cultivars, others have remained adequately resistant
while being widely cultivated for many years. These include
Cappelle-Desprez, which occupied more than 80% of the total wheat
area in Britain for more than 10 years and was important for 20
years (Johnson, 1978). Many other cultivars in North-Western Europe
have had durable resistance to yellow rust, sometimes at higher
levels than in Cappelle-Desprez (Johnson and Taylor, 1980). Genetic
analyses of resistance in some durably resistant cultivars have
indicated that several genes on different chromosomes may be
involved (Law et al., 1978; Labrum, 1980). Nevertheless, the
transfer of adequate levels of resistance from durably resistant
cultivars into new cultivars has been achieved (Lupton and Johnson,
1970; Bingham, 1981).

     In all cases so far noted durable resistance to yellow rust
has been incomplete or quantitative, not complete. It is evident
that, with present information, the only way of distinguishing
quantitative resistances that are durable from those that are not is
by an adequate test in agriculture. Because of the many factors
that affect the rate of evolution of the pathogen and the complex
range of specificity in resistance it cannot be assumed that
resistance that has been durable will be permanent. Nevertheless
cultivars that have displayed durable resistance are considered to
be a valuable resource and it is our policy at the PBI to utilise
them to try to incorporate durable resistance to yellow rust into
new cultivars.

Breeding for durable resistance. Because combinations of race-
specific genes giving complete resistance are epistatic and prevent
selection of the quantitative resistance from the cultivars with
durable resistance, such combinations are avoided by selecting
against them if they occur. The new PBI wheat cultivar Bounty was
produced by this method from a cross between the two cultivars,
Ploughman and Durin. Each parent carried recognised genes for race-
specific resistance expressed in seedlings. When these genes were
overcome by appropriate races of *P. striiformis* Durin was rather
susceptible to yellow rust but Ploughman had a satisfactory level of
quantitative resistance thought to be derived mainly from Maris
Widgeon, a cultivar with durable resistance. The race-specific
genes *Yr-3b* and *Yr-4b* from Ploughman gave complete resistance to the
race of *P. striiformis* to which Durin was susceptible, but lines
carrying these genes were discarded. Selection was then carried out

among the remaining progeny using a race of *P. striiformis* that gave
maximum yellow rust on Durin so that the resistance from Ploughman,
without *Yr-3b* and *Yr-4b* could be detected. A satisfactory level of
resistance to yellow rust was achieved in the cultivar Bounty and
because the breeding procedure made it likely that this was derived
mainly from Maris Widgeon we hope the resistance will be durable.

The procedure for producing Bounty would have been simpler if
Ploughman had not possessed the genes *Yr-3b* and *Yr-4b* or if a race
pathogenic for the combination of *Yr-1* (from Durin) with *Yr-3b* and
Yr-4b had been available. Selection could then have been carried
out using a race to which Durin was highly susceptible and to which
Ploughman displayed its quantitative and probably durable
resistance, thus providing the opportunity to select the resistance
from Ploughman. This procedure bears some similarity to the
proposal of Robinson (1973) to select for resistance using a race of
the pathogen that overcomes all the race-specific components of
resistance in the parents. He suggested accumulating resistance
from recurrent polycrosses between several cultivars all of which
were susceptible to a single race of the pathogen. It was assumed
that the susceptible cultivars would not possess any unmatched
race-specific components of resistance so that any accumulated
resistance would be race-non-specific. However, as indicated by the
data given here, any degree of resistance, no matter how small, can
be race-specific. The use of several cultivars in the recurrent
polycross technique might increase the chance of selecting race-
specific minor genes.

By contrast, the method used at the PBI employs a source of
resistance already shown to be durable and selection is carried out
using a race of the pathogen that matches the known race-specific
components of resistance or by eliminating race-specific components
for which this cannot be achieved. It seems more likely that
durable resistance will be achieved in new cultivars by this method
than by attempting to accumulate resistance from sources that are
susceptible and also less intensively investigated. However, no
method can eliminate the possibility of further evolution in the
pathogen and resistance achieved, either by the method of Robinson
(1973) or by the method used at the PBI, will only be shown to be
durable after widespread and prolonged use in agriculture.

Black Rust on Wheat

Present evidence indicates that combinations of genes already
known to be highly race-specific have not resulted in durable
resistance to yellow rust in wheat in Britain. This provides a
contrast with some apparent evidence for the control of black (stem)
rust on wheat using race-specific genes for resistance.

        Genes for resistance to *Puccinia graminis* have been described
at more than 30 loci in wheat, with allelic series at some loci such
as *Sr-9* (McIntosh, 1973, 1980). All appear to be race-specific and
most are expressed at the seedling stage though many do not provide
complete resistance. Various combinations of them have been used to
control black rust and their use has often necessitated replacement
of cultivars for which *P. graminis* developed matching pathogenicity
(e.g. Watson and Luig, 1963). In some cases, however, resistance
has remained effective for long periods before being overcome (Green
and Campbell, 1979). An interesting example of resistance that has
remained effective for longer than expected is provided by the wheat
cultivar Timgalen in Australia (R.A. McIntosh, pers. comm.). A gene
SrTt-1 derived from *Triticum timopheevi*, was used in two wheat
cultivars Mendos and Mengavi, but their resistance was rapidly
overcome. It was also used in the cultivar Timgalen which has
remained resistant for about 14 years. During the last seven years
it seems that the only remaining effective gene providing resistance
in Timgalen has been *SrTt-1*, races having been found with
pathogenicity matching the other known resistance genes in the
cultivar. The durability of this resistance is difficult to
explain.

        In such cases of unexpected persistence of resistance, it is
possible that other unidentified genes for resistance are present.
There is evidence that some resistance to black rust is controlled
by genes other than the identified race-specific genes, (e.g. Knott,
1977). In some cases such genes seem to be associated with
increased durability of resistance as in the two wheat cultivars,
Kenya Page and Africa Mayo, which have shown unusually durable
resistance to black rust in Kenya, having been grown commercially
for 12 and 15 years respectively compared with about four years for
many cultivars in commercial use. (De Pauw, 1978; De Pauw and
Buchannon, 1975).

        Several potential sources of durable resistance to stem rust
such as Hope, H-44 and Thatcher have been identified (e.g. Hare and
McIntosh, 1979; Knott, 1977) though the contribution of different
genetic components in such cultivars to durability has been
difficult to assess. This is not surprising in view of the apparent
potential for certain combinations of race-specific genes ·to provide
resistance lasting many years. Analysis of the causes of durable
resistance of the Canadian wheat cultivar Selkirk to black rust
provides an example of this difficulty.

        It has recently been shown that Selkirk possesses six
recognisable race-specific genes *Sr-6*, *Sr-7b*, *Sr-9d*, *Sr-17*, *Sr-23*
(Green and Campbell, 1979) and *Sr-2* (Hare and McIntosh, 1979). The
gene *Sr-2* conditions resistance in adult plants and was first
identified in Hope and H-44 the latter being a parent of Selkirk
(Knott, 1968). In surveys of populations of *P. graminis* in Canada,

races able to overcome all the five genes for seedling resistance in Selkirk were rarely found. Green and Campbell (1979) suggested that this indicated that such races had reduced aggressiveness, but presented no direct evidence to support this statement. As Selkirk possesses *Sr-2* which was not noted by Green and Campbell (1979), it is likely that those rare races with pathogenicity matching all the other five genes still could not attack Selkirk to a high level, and this could have helped to prevent rapid multiplication of such races (Johnson, 1981). It is also possible that Selkirk possesses still further unidentified resistance genes.

Vanderplank (1978) analysed the frequency of occurrence and association of pathogenicity for *Sr-6* and *Sr-9d* in populations of *P. graminis* in North America. Surveys indicated that combined pathogenicity for these two genes was common in Mexico and Southern Texas but rare further north and in Canada. He interpreted this as showing that there was strong dissociation of the pathogenicity genes matching *Sr-6* and *Sr-9d* as *P. graminis* travelled northwards each year towards Canada and suggested that this might have occurred because races with this combination of pathogenicity were poorly adapted to the higher temperatures encountered as the season advanced. No direct evidence supporting this hypothesis was presented. It was also suggested that it was the dissociation of these two genes for pathogenicity that protected Selkirk from black rust. The analysis on which this conclusion was based treated the population of *P. graminis* in North America as a single random-mating population, which was not a valid assumption (Wolfe and Knott, 1982). However, even if the analysis is accepted it is clear, from the gene-for-gene hypothesis, that possession of pathogenicity for both *Sr-6* and *Sr-9d* would not, of itself, enable a race to attack Selkirk. An explanation for the low frequency of combined pathogenicity matching *Sr-6* and *Sr-9d* in Canada when Selkirk was widely grown must involve analysis of the association of these two genes for pathogenicity with all the other four (perhaps more) genes essential to attack Selkirk. It must also involve an assessment of the influence of the resistance genes in other cultivars on the pathway of *P. graminis* in North America (Wolfe and Knott, 1982). No such analysis has been presented and the question of why resistance to black rust in Selkirk was so durable remains open.

This discussion of resistance to black rust in wheat indicates that it has been relatively easy to identify race-specific resistance genes and to recognise the combinations of them that have been overcome by *P. graminis*. When resistance has been durable it has been much more difficult to identify the precise genetic causes of the durability. Apparently the successful use of cultivars possessing certain combinations of recognised race-specific genes has reduced interest in the possibility of utilising genetic components of resistance that might have greater inherent durability.

CONCLUSION

Durable resistance can be complete or quantitative and therefore cannot be defined in epidemiological terms. It is recognised retrospectively as are all the other characteristics of the interaction between any host and pathogen. It is only on the basis of their past performance that we can identify pathogens that display little or no host-specificity. Likewise it is only on the basis of past performance and experiment that we can assess the number of genes controlling resistance and identify the race-specificity of some of them.

When we identify genes that, due to their race-specificity fail, either singly or in simple combinations to provide durable resistance, it is logical to cease to combine such genes together as the sole means of achieving resistance. If some genetic components of resistance have failed to provide durable resistance while others have succeeded, it is logical to utilise the latter as far as possible. However, as illustrated in this chapter, it is often difficult to distinguish two such groups of genes. The diversity of genetic systems apparently controlling durable resistance is sufficient warning against basing our action on theoretical models rather than on actual experience. The numerous factors that can influence the rate of evolution of pathogens should make us cautious of assuming that resistance that has been effective for a long period will certainly continue to be effective into the future.

Plant breeding involves the construction and selection of new assortments of genes. We cannot be certain, especially where resistance is under the control of several genes, that a newly constructed genome will perform exactly as we would predict. Therefore it is not possible to state, in advance, that a new cultivar will possess durable resistance, but only to try to increase the probability that it will do so. Only a widespread test will indicate whether or not resistance is durable in a new cultivar, no matter by what method the cultivar has been produced.

Because of these uncertainties it will always be wise, where it is possible, to maintain genetic diversity between cultivars for their resistance characters, and to exploit them in ways that will reduce the possibilities of major epidemics due to the rapid spread of matching pathogen races.

ACKNOWLEDGEMENTS

I am grateful to P.R. Scott and M.S. Wolfe for help with this chapter and on many previous occasions, to my support staff especially A.J. Taylor and G.M.B. Smith and to the PBI cereals breeders with whose work my own is closely integrated.

REFERENCES

Bannerot, H.E. and Pochard, E., 1972, Four cases of 'non-specific' resistance in vegetables, Pages 109-125 in "The Way Ahead in Plant Breeding", F.G.H. Lupton, G. Jenkins and R. Johnson, eds. *Proc. Sixth Eucarpia Congress 1971*, Plant Breeding Institute, Cambridge.

Bingham, J., 1981, Breeding wheat for disease resistance, Pages 3-14 in "Strategies for the Control of Cereal Diseases", J.F. Jenkyn and R.T. Plumb, eds. Blackwell Scientific Publications, Oxford.

Black, W.C., Mastenbroek, W.R., Mills, W.R. and Peterson, L.C., 1953, A proposal for an international nomenclature of races of *Phytophthora infestans* and of genes controlling immunity in *Solanum demissum* derivatives, *Euphytica*, 2:173-179.

Caldwell, R.M., 1968, Breeding for general and/or specific plant disease resistance, Pages 207-216 in *Proc. Third Int. Wheat Genetics Symp. Australian Acad. Sci. Canberra.*

Caten, C.E., 1974, Intra-racial variation in *Phytophthora infestans* and adaptation for field resistance to potato blight, *Ann. appl. Biol.*, 77:259-270.

Clifford, B.C. and Clothier, R.B., 1974, Physiological specialisation of *Puccinia hordei* on barley hosts with non-hypersensitive resistance, *Trans. Br. mycol. Soc.*, 63:421-430.

Day, P.R., 1974, "Genetics of Host-Parasite Interactions", W.H. Freeman and Company, San Francisco.

De Pauw, R.M., 1978, Breeding for post-seedling resistance to wheat stem rust, *Cereal Res. Communications*, 6:249-253.

De Pauw, R.M. and Buchannon, K.R., 1975, Post-seedling response of wheat to stem rust, *Can. J. Pl. Sci.*, 55:385-390.

Eenink, A.H., 1977, Genetics of host-parasite relationships and the stability of resistance, Pages 47-56 in "Induced Mutations Against Plant Diseases", IAEA Vienna.

Ellingboe, A.H., 1981, Changing concepts in host-pathogen genetics, *Ann. Rev. Phytopathol.*, 19:125-143.

Flor, H.H., 1955, Host-parasite interactions in flax rust - its genetic and other implications, *Phytopathology*, 45:680-685.

Green, G.J. and Campbell, A.B., 1979, Wheat cultivars resistant to *Puccinia graminis tritici* in western Canada: their development, performance and economic value, *Can. J. Plant Pathol.*, 1:3-11.

Habgood, R.M. and Clifford, B.C., 1981, Breeding barley for disease resistance: the essence of compromise. Pages 15-25 in "Strategies for the Control of Cereal Disease", J.F. Jenkyn and R.T. Plumb, eds. Blackwell Scientific Publications, Oxford.

Hall, T.J., 1980, Resistance at the Tm-2 locus in the tomato to tomato mosaic virus, *Euphytica*, 29:189-197.

Hare, R.A. and McIntosh, R.A., 1979, Genetic and cytogenetic studies of durable adult-plant resistances in 'Hope' and related cultivars to wheat rusts, *Z. Pflanzenzuchtg.*, 83:350-367.

Howard, H.W., 1978, The production of new varieties, Pages 607-646, in "The Potato Crop: the Scientific Basis for Improvement",

P.M. Harris ed. Chapman and Hall, London.

Johnson, R., 1978, Practical breeding for durable resistance to rust diseases in self-pollinating cereals, *Euphytica*, 27:529-540.

Johnson, R., 1979, The concept of durable resistance, *Phytopathology*, 69:198-199.

Johnson, R., 1981, Durable resistance: definition of, genetic control, and attainment in plant breeding, *Phytopathology*, 71: 567-568.

Johnson, R. and Bowyer, D.E., 1974, A rapid method for measuring production of yellow rust spores on single seedlings to assess differential interactions of wheat cultivars with *Puccinia striiformis*,   *Ann. appl. Biol.*, 77:251-258.

Johnson, R. and Law, C.N., 1973, Cytogenetic studies on the resistance of the wheat variety Bersee to *Puccinia striiformis*. *Cereal Rusts Bull.*, 1:38-43.

Johnson, R. and Law, C.N., 1975, Genetic control of durable resistance to yellow rust (*Puccinia striiformis*) in the wheat cultivar Hybride de Bersee, *Ann. appl. Biol.*, 81:385-391.

Johnson, R. and Taylor, A.J., 1972, Isolates of *Puccinia striiformis* collected in England from the wheat varieties Maris Beacon and Joss Cambier, *Nature, Lond.*, 238:105-106.

Johnson, R. and Taylor, A.J., 1980, Pathogenic variation in *Puccinia striiformis* in relation to the durability of yellow rust resistance in wheat, *Ann. appl. Biol.*, 94:283-286.

Knott, D.R., 1968, The inheritance of resistance to stem rust races 56 and 15B-1L (Can) in the wheat varieties Hope and H-44, *Can. J. Genet. Cytol.*, 10:311-320.

Knott, D.R., 1977, Studies on general resistance to stem rust in wheat. Pages 81-86 in "Induced Mutations Against Plant Diseases", IAEA, Vienna.

Labrum, K.E., 1980, Investigations into the genetic control and development of resistance to yellow rust in wheat, Ph.D. Thesis, University of Cambridge.

Law, C.N., Scott, P.R., Worland, A.J. and Hollins, T.W., 1975, The inheritance of resistance to eyespot (*Cercosporella herpotrichoides*) in wheat, *Genet. Res. Camb.*, 25:73-79.

Law, C.N., Gaines, R.C., Johnson, R. and Worland, A.J., 1978, The application of aneuploid techniques to a study of stripe rust resistance in wheat, Pages 427-436 in *Proc. Fifth Int. Wheat Genetics Symposium*, IARI, New Delhi.

Lupton, F.G.H. and Johnson, R., 1970, Breeding for mature-plant resistance to yellow rust in wheat, *Ann. appl. Biol.*, 66: 137-143.

Macer, R.C.F., 1972, The resistance of cereals to yellow rust and its exploitation by plant breeding, *Proc. R. Soc. Lond. B.*, 181:281-301.

McIntosh, R.A., 1973, A catalogue of gene symbols for wheat, Pages 893-937. *Proc. Fourth Int. Wheat Genetics Symposium*, Missouri Agr. Exp. Sta., Columbia.

McIntosh, R.A., 1980, Catalogue of gene symbols for wheat, 1980 supplement. *Cereal Res. Communications*, 8:437-439.

Manners, J.G., 1950, Studies on the physiological specialisation of yellow rust (*Puccinia glumarum* (Schm.) Erikss. & Henn.) in Great Britain, *Ann. appl. Biol.*, 37:187-214.

Mastenbroek, C., 1966, Some major points from 22 years of experience in breeding potatoes for resistance to late blight (*Phytophthora infestans*), *Amer. Pot. J.*, 43:261-277.

Nelson, R.R., 1978, Genetics of horizontal resistance to plant diseases, *Ann. Rev. Phytopathol.*, 16:359-378.

Parlevliet, J.E., 1976, Partial resistance of barley to leaf rust, *Puccinia hordei* III. The inheritance of the host plant effect on latent period in four cultivars, *Euphytica*, 25:241-248.

Parlevliet, J.E., 1978, Race-specific aspects of polygenic resistance of barley to leaf rust, *Puccinia hordei*, *Netherlands J. Plant Path.*, 84:121-126.

Person, C.O., 1959, Gene-for-gene relationships in host: parasite systems, *Can. J. Bot.*, 37:1101-1130.

Roane, C.W. and Starling, T.M., 1967, Inheritance of reaction to *Puccinia hordei* in barley II. Gene symbols for loci in differential cultivars, *Phytopathology*, 57:66-68.

Robinson, R.A., 1969, Disease resistance terminology, *Rev. appl. Mycol.*, 48:593-606.

Robinson, R.A., 1973, Horizontal resistance, *Rev. Plant Path.*, 52: 483-501.

Robinson, R.A., 1976, "Plant Pathosystems", Springer-Verlag, Berlin.

Russell, G.E., 1978, "Plant Breeding for Pest and Disease Resistance", Butterworths, London.

Scott, P.R., Johnson, R., Wolfe, M.S., Lowe, H.J.B. and Bennett, F.G.A., 1980, Host-specificity in cereal parasites in relation to their control, *Appl. Biol.*, 5:349-393.

Scott, P.R., Benedikz, P.W. and Cox, C.J., 1982, A genetic study of the relationship between height, time of ear emergence and resistance to *Septoria nodorum* in wheat, *Plant Path.*, 31: In Press.

Thurston, H.D., 1971, Relationship of general resistance: late blight of potato, *Phytopathology*, 61:620-626.

Van der Graaff, N.A. 1981, Selection of arabica coffee types resistant to coffee berry disease in Ethiopia, *Meded. Landbouwhogeschool Wageningen*, 81-110 (1981).

Vanderplank, J.E., 1963, "Plant Diseases: Epidemics and Control", Academic, New York.

Vanderplank, J.E., 1968, "Disease Resistance in Plants", Academic, New York.

Vanderplank, J.E., 1975, "Principles of Plant Infection", Academic, New York.

Vanderplank, J.E., 1978, "Genetic and Molecular Basis of Plant Pathogenesis", Springer-Verlag, Berlin.

Walker, J.C., 1966, The role of pest resistance in new varieties, Pages 219-242 in "Plant Breeding", K.J. Frey ed. Iowa State University Press, Ames.

Walker, A.G. and Roberts, E.T., 1974, The wheat yellow rust
     (*Puccinia striiformis*) epidemic 1971-1972, Pages 159-162 in
     ADAS Science Arm Annual Report, 1972. HMSO, London.
Watson, I.A. and Luig, N.H., 1963, The classification of *Puccinia
     graminis* var. *tritici* in relation to breeding resistant
     varieties, *Proc. Linn. Soc. N.S.W.*, 88:235-258.
Wolfe, M.S. and Knott, D.R., 1982, Collection and analysis of data
     on pathogenicity in populations of plant pathogens, *Plant
     Path.*, 31: In Press.
Zadoks, J.C., 1961, Yellow rust of wheat: studies in epidemiology
     and physiologic specialisation, *Tijdschr. PlZiekt.*, 67:69-
     256.

DISCUSSION

PARLEVLIET: The 'Cornell' resistance gene in beans to anthracnose
has been overcome by a race reported first from the area around
Munich in West Germany. Have you any more information about this?

DE PONTI: This occurred about four years ago, but for some unknown
reason has not spread over Europe as was first thought. This is
similar to Johnson's example of $Tm-2^2$ in tomato.

PARLEVLIET: The Selkirk resistance to stem rust in North America
may also be due to the fact that stem rust is not prevalent any more
in recent years in the wheat belt of North America. New races are
less likely to evolve under such low population conditions.

JOHNSON: It is possible that both limited population size and the
epidemiological constraints on *Puccinia graminis* in North America
could have contributed to the durability of resistance of Selkirk.

SHARP: Could genes for lack of aggressiveness associated with genes
for virulence in the pathogen explain the apparent resistance of
Selkirk to stem rust, rather than R genes other than *Sr-6* and *Sr-9d*
in Selkirk?

JOHNSON: I think it would be very difficult to test the hypothesis
that genes in *P. graminis* affecting aggressiveness or non-specific
pathogenicity, associated with genes for specific pathogenicity
affected the resistance of Selkirk, so it remains a possibility.
However, the genetic analysis of Selkirk has now shown that it
contains at least six identified genes for resistance so a race
able to attack Selkirk would have to carry a combination of genes
pathogenic to *Sr-6* and *Sr-9d* plus genes able to counteract the other
four resistance genes present. Thus analysis of the frequency of
pathogenicity for *Sr-6* and *Sr-9d* in the population of *P. graminis*
without reference to the frequency of pathogenicity for the other
four genes in Selkirk cannot illuminate why it remained resistant

for so long. Also, it remains possible that the entire resistance
genotype of Selkirk has not yet been revealed.

CIRULLI: With retrospective judgement, how many years of effectiveness are needed before a resistance can be defined as 'durable'?

JOHNSON: The period of time and extent of use of a cultivar necessary
before its resistance can be classified as durable will depend on
several factors, including a judgement about the usefulness of a
given time of durability in relation to the performance of other
cultivars, the rate of turnover of cultivars and the length of their
life cycle. There will be a subjective element in this judgement,
but less subjective than to pronounce that the resistance was race-
non-specific and permanent.

ZADOKS: Durability of the host resistance may be a function of the
size of the pathogen population. The wheat cultivar Alba had
apparently durable resistance, as it stayed resistant (with a high
degree of partial resistance) for about 20 years. It survived the
severe 1955 yellow rust epidemic in the low countries but succumbed
to a new race in 1957. A possible explanation is that the rust
population could produce the new mutant Alba-compatible race only
when it was very large, i.e. during a really severe epidemic, either
in 1955 or 1957. If this explanation is true, the 'durability' was
due to lack of opportunity for the rust population to change.

JOHNSON: Thankyou for describing this example. It is clear that the
ability of a pathogen to evolve a biotype able to overcome resistance
will be affected by the size of the pathogen population. It will
also, of course, be affected by many other factors, including the
area occupied by the cultivar. This example illustrates the inadvis-
ability of leaping to the conclusion that resistance is race-non-
specific.

ROBINSON: Your examples all seem to involve vertical resistance
which is quantitative in its effect. Some of these resistances last
longer than others; they are durable. The term durable is useful
for retrospective judgement but not for looking forward. If we are
looking ahead, we must distinguish between horizontal and vertical
resistance because it is extremely dangerous to rely on the dubious
durability of vertical resistance.

JOHNSON: My examples did not all involve race-specific resistance
that is quantitative in effect. Some of the durable resistances
referred to gave complete resistance, some were durable and quantitative and, as yet no race has occurred to overcome these resistances.
In noting this fact it cannot be categorically asserted that no race
will ever occur that can overcome the resistance and I therefore
prefer to describe it as durable and not to claim that it is race
non-specific or horizontal both of which would imply that I was

confident that the resistance would be permanent. You suggest that
the term durable is only useful if I am looking backward but not
forward. I suggest to you that potato cultivars described by
Vanderplank as having horizontal resistance were first identified
from his evidence that the resistance had been durable; it was a
retrospective judgement. I believe that it is wise to use such
durable resistance as a resource in further breeding, as did
Vanderplank, but I do not share the conviction he had that such
resistance will be permanent, especially when transferred to a new
genetic background. Use of a durably resistant cultivar as a parent
may increase the probability that derived cultivars will also possess
durable resistance but not to the point of absolute certainty. In
this respect, resistance described as horizontal has no greater
chance of success, in my view, than resistance that has already had
a thorough test of its durability. Indeed if resistance is
designated as horizontal on any other criterion than its durability
it may have less chance than a thoroughly tested resistance of
being successful and durable when used in a new cultivar. I do not
accept the view that designation of a resistance by the abstract
definition 'horizontal' permits its performance in the future to be
confidently predicted.

WOLFE: To deal with the future, one can deal only with probabilities,
not with certainties. On the one hand, if a resistance is exposed
repeatedly on a large scale there may be an increasing probability
that it will be eventually overcome. On the other hand, if a variety
produced along the lines suggested by Johnson is exposed repeatedly
and remains resistant, it may exhibit a conditional probability;
each annual experience increases the confidence that the variety
does have durable resistance. The kernel of the problem of inherent
durability is therefore to be able to move from the first to the
second type of probability.

PRESENT KNOWLEDGE AND THEORIES

CONCERNING DURABLE RESISTANCE

C. Person, R. Fleming, L. Cargeeg and B. Christ

Department of Botany, University of British Columbia
Vancouver, B.C. V6T 2B1, Canada

SUMMARY

Examination of the additive and specific interaction models
proposed by Parlevliet and Zadoks (1977) led to the conclusion
that the comparison made of these models by these authors was not
valid, and that their conclusions were therefore not supported by
their analysis. Examination of the so-called Person model, which
assumes polygenes that interact nonspecifically, showed that this
model predicts constant ranking of both hosts and pathogens, and
that at least two factors, (viz. "phenotypic damping," and op-
position of selective forces operating in the two interacting
populations), would contribute to stability. Research on the
genetics of pathogenicity of *Ustilago hordei* toward cultivated
barley, also described in this paper, showed that for this system
pathogenicity is determined by major genes, by polygenes, and by
interaction between major genes and polygenes. It was concluded
that an understanding of durable resistance must be based on
detailed genetic knowledge of host resistance as well as of
pathogenicity, and in particular of how polygenes and major genes
of host and pathogen interact. It may be that no single model or
theory will turn out to be universally applicable.

INTRODUCTION

Johnson and Law (1975) have defined durable resistance as
resistance that has been maintained for a long time by a cultivar
that has been widely grown in the presence of the disease. The
genetic basis of durable resistance is not well understood. A
better understanding must await detailed genetic study, at the

populational level, of hosts and their parasites and of the interactions that they undergo. It may then be learned that no single
explanation or theory will apply to all examples of durable resistance.

An example of durable resistance is the "field resistance" of
potatoes to the late blight disease. Driver (1962), in outlining
the characteristics of this resistance, states that: "The varieties
can be rated in the same order and with the same degree of infection
even though some are more than one hundred years old. This is true
throughout the world, so that there is no evidence of potato blight
varying the balance between it and the field resistance of commercial
varieties over the past century." This statement contains three
postulates, namely: (i) that varieties with field resistance will
exhibit constant ranking; (ii) that the interactions are nonspecific
(clearly implied by constant ranking maintained throughout the world);
and (iii) that field resistance is stable.

The field resistance of potatoes to the late blight disease
was also extensively reviewed and discussed by Vanderplank (1963)
who gave it the epidemiologically-based label "horizontal
resistance" (HR). The main characteristics of HR, described in
books by Vanderplank (1963) and Robinson (1976) are constant ranking, nonspecificity, and stability.

The basis of the stability of field resistance (=HR) is not
known. Driver (1962) attributed it to polygenes, the explanation
being "that to overcome (polygenic) resistance several gene changes
may be necessary by the fungus." Vanderplank's explanation was
different. He attributed the stability to HR to stabilizing
selection (in its original meaning) acting independently in each of
the two interacting populations (Vanderplank, 1968). The explanation given by Robinson (1976) included a model, referred to by
him as the "Person" model, together with a brief description of it.
This model will be more fully described in this paper.

Although it appears that the majority of authors incline to
the view that polygenically-determined resistance will tend to be
more stable than major-gene resistance, there is no general agreement that polygenically-determined resistance is nonspecific.
There are known examples in which major genes for resistance continue to confer a low level of residual resistance after being
"matched" and, in doing so, they mimic the action of minor genes.
In fact Ellingboe (1976) states that all resistance that appears
to be governed by minor genes is attributable to major genes that
have been matched. In such a case low levels of resistance would
originate from specific interaction. The idea that minor genes
can engage in specific interaction is supported by studies of
resistance in which the analysis of variance reveals a component
of variability that is due to variety-race interaction (Parlevliet

& Zadoks, 1977). However, proof that this component of variability flows from gene-for-gene interaction *sensu* Flor is lacking
and, as Parlevliet has pointed out, may be unobtainable.

POLYGENE MODELS

     Parlevliet & Zadoks (1977) have attempted to assess the re-
lative stabilities of disease systems in which the minor-gene interactions are nonspecific, on the one hand, and specific, on the other.
To do this they formulated two models: (i) the "addition model", in
which the interactions were assumed to be nonspecific; and (ii) the
"interaction model" in which the minor genes were assumed to interact on a gene-for-gene basis and the interactions were therefore
assumed to be specific. For each of the two models it was assumed
that host and pathogen were both diploid and that each had five
genetic loci; alleles at these loci were designated either "plus"
or "minus".

     For the *addition model* each plus allele of the host would
decrease, and each plus allele of the pathogen would *increase*,
the level of disease by 10 percent, and where the numbers of plus
alleles in host and pathogen were equal there would be no change
in the level of disease. Based on these assumptions the "baseline" level of disease would be expressed whenever the numbers of
plus alleles in host and pathogen are equal. With all 10 host
alleles plus and all 10 pathogen alleles minus there would be a
100 percent reduction from the baseline level, and with all 10
host alleles minus and all 10 pathogen alleles plus there would be
a 100 percent increase over the baseline level of disease. The
range of disease expression should therefore be from no disease
(100 percent reduction) to twice the "baseline" level of disease
expression. However the method of analysis used by Parlevliet &
Zadoks, which involved subtraction of the numbers of plus alleles
in the pathogen from the numbers of plus alleles in host, gave
negative values when the former exceeded the latter. To avoid the
mathematical and statistical problems associated with negative
values, Parlevliet & Zadoks chose for their comparison only those
values that were positive. As a consequence the comparison
involved only that part of the range of variability for which
disease levels were below the "baseline" level. The upper half of
the range of variability of disease expression was not included as
part of the analysis.

     For the *interaction model* Parlevliet & Zadoks assumed 'gene-
for-gene' interaction, i.e. that a plus allele of the host would
interact only with its related plus allele in the pathogen. Other
characteristics (e.g. dominance and epistasis of R-genes) that
usually associate with gene-for-gene interaction were not assumed.
The scheme below represents a summary of possible interations
that is applicable to each of the five gene-for-gene relationships

that were assumed for the interaction model: it shows, for a pair
of interacting loci, that where the numbers of plus genes in host
and pathogen are equal there is no change in the level of disease
expression,

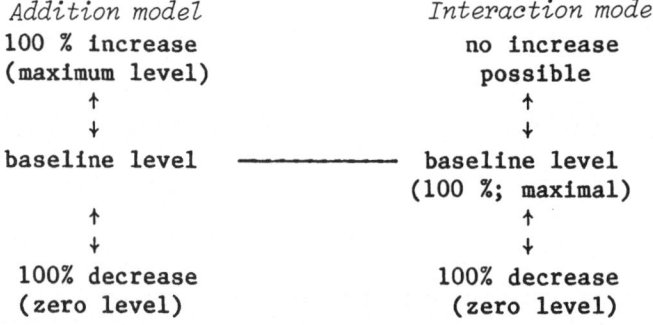

and that where one, or two plus genes of the host are not matched
the level of disease expression is reduced by 10, or 20 percent,
respectively. (We are grateful to Doctor Parlevliet for checking
the accuracy of this scheme). With the interactions shown in this
scheme operating at each of the five pairs of interacting loci,
and with each having the potential of reducing the level of
disease by 20 percent, it is evident that the range of disease
expression will vary from a minimum of zero to a maximum of 100
percent. It is also evident that for the interaction model the
"baseline" for disease expression is 100 percent, and that the
effect of plus genes is limited to decreasing the expression of
disease below this level.

      A comparison of the possible ranges of disease expression,
for the two models is shown below:

Addition model	Interaction model
100 % increase	no increase
(maximum level)	possible
↑	↑
↓	↓
baseline level ———————	baseline level
	(100 %; maximal)
↑	↑
↓	↓
100% decrease	100% decrease
(zero level)	(zero level)

      As mentioned earlier the comparison made by Parlevliet and
Zadoks (1977) involved only part of the potential variability of
the addition model. Their analytical procedure had the added
effect of equating only half of the range of potential variability of the addition model with the full range of the interaction
model. Their analysis led them to the conclusion that systems in
which there is specific interaction will be more stable than

systems in which there is nonspecific interactions. In our judgement there are weaknesses in their analysis that are so serious
that their conclusion is unfounded.

The "Person" model, referred to earlier was based on the
assumption that the disease interactions were polygenically-
determined and nonspecific. The polygenes in both host and pathogen were assumed to act equally and additively, to generate continuous variability of the kind that is described by the normal
curve. The objective was to try to determine whether a model based
on these simple assumptions would lead to the expectations of:
(i) constant ranking; and (ii) stability.

Two slightly different versions of the model are presented in
Figures 1a and 1b. The normal curves in the upper part of these
figures represent the polygenically-based potentials of host and
pathogen to enable the disease to develop. Each of the upper curves
is divided by one, two and three standard deviations from the mean
into six classes to which are assigned numbers (from zero to five,
inclusive) that represent the potentials of these classes for enabling the disease to develop. The numbers in the matrix represent
the levels of disease that are assumed to develop when the six
classes of host and the six classes of pathogen interact in all
possible combinations. For the additive model (Fig. 1a) these
"realized" levels of disease are obtained by adding, and for the
multiplicative model (Fig. 1b) by multiplying, the numbers that
had been assigned to the classes of hosts and pathogens. The curves
at the bottom of Figure 1 describe the "realized" levels of disease
for the two models.

From Figure 1 it is seen that both versions of the model lead
to the expectation of constant ranking. For the additive model it
has been shown mathematically (Fleming and Person, in press): (i)
that the "realized" levels of disease would also conform to a normal
frequency distribution; and (ii) that a shift in the mean of the
distribution curve for "potential" disease development, if this
should occur in either of the two interacting populations would
result in a relatively smaller shift in the mean of the curve that
represents "realized" disease development. Since we think that
this reduction in effect will serve as a source of stability in
polygenically-determined systems we have given it the label
"phenotypic damping". We think it reasonable also to assume a
direct relationship between pathogenicity and resistance and the
reproductivities of pathogen and host, respectively. If this
assumption is valid, selection for pathogenicity and for resistance
will both be directional. But, as Figure 1 shows, the directions
taken by natural selection in the two interacting populations will
be diametrically opposed. This, we think, will also serve as a
source of variability in polygenically-determined systems.

STUDIES OF *Ustilago hordei*

It is our opinion that until more experimental data are available it will not be possible to determine whether in polygenically-based systems the genes interact specifically or nonspecifically. At the University of British Columbia during the past several years we have investigated this problem, and are now able to report a certain amount of progress. In our early work Emara (1972) and Emara and Sidhu (1974) showed that some of the variability of *Ustilago hordei* toward susceptible barley cultivars was polygenically determined. In continuing work, carried out co-operatively with Doctor Chris Caten at the University of Birmingham, the early results have been confirmed and extended (unpublished). In more recent work we have developed a new experimental approach that allows us to recognize the effects of both major genes and polygenes, and to study how they work in concert together in determining the extent of disease development.

The recent work takes advantage of the fact that *Ustilago hordei* forms an ordered tetrad. Three types of tetrad (ignoring sequence) are formed, two of which are "ditype"; the third is "tetratype". The four products of the tetrad can be used to establish four haploid cultures that can be mated in four different combinations to produce "selfed" progeny. As shown in Table 1 the ditype tetrads (parental and nonparental) give rise to progeny dikaryons all four of which are heterozygous. Ditype tetrads therefore give no segregation even though the parental teliospore was heterozygous. A tetratype tetrad, on the other hand, does give rise to a segregation, and where it is an ordered tetrad, as it is in *U. hordei*, the 3:1 ratio enables the assignment of genotypes to

Fig. 1. For these figures (1a and 1b), f(P) and f(S) represent the frequencies with which individuals with varying levels of pathogenicity (P) or susceptibility (S) occur in parasite and host populations. Because of their assumed polygenic determinations these are represented as normal frequency distributions. For the additive model (Fig. 1a), where it is assumed that the amount of disease is $D = P + S$, the matrix records the severity of disease for the various indicated classes of host-parasite interaction. The resulting frequency distribution for D, f(D), is normally distributed. For the multiplicative model (Fig. 1b) where it is assumed that $D = P \times S$, the shape of the resulting frequency distribution, f(D), depends on the parameters of the distributions for pathogenicity and susceptibility, and the distribution is one that would favour host resistance. (Fig. 1a was used, with our permission, by Robinson, 1976).

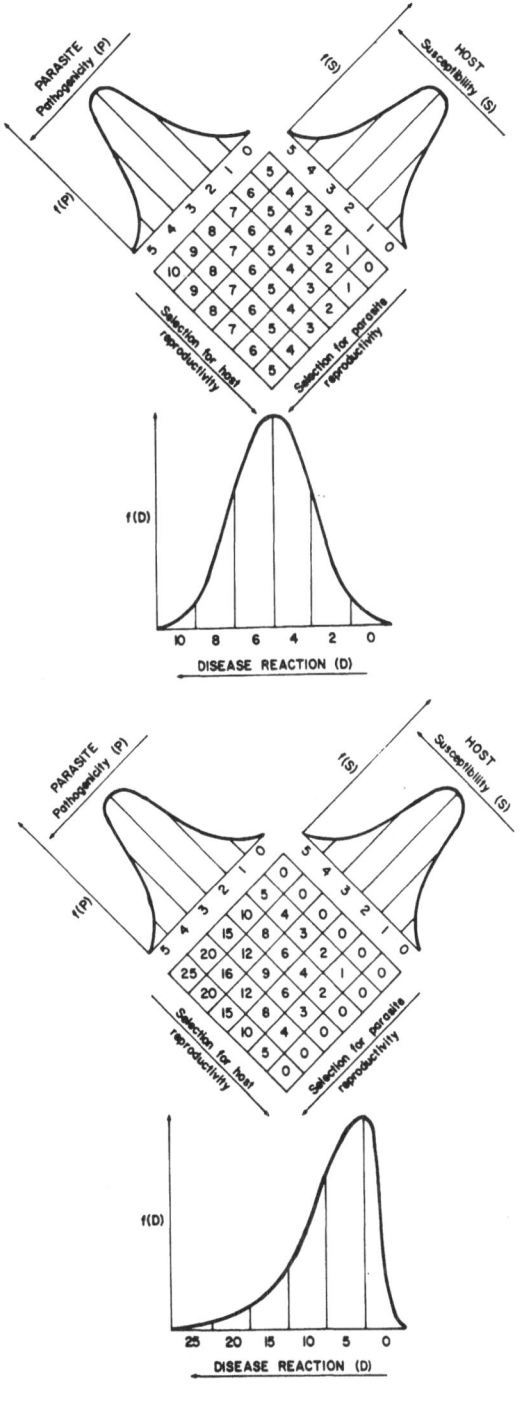

Table 1. The three kinds of tetrad, and the progeny obtained via
selfing from single teliospores, of a Vv dikaryon for which the
original mating was V+ X v-. Parental ditype (PD) and nonparental
ditype (NPD) tetrads do not segregate when "selfed". Tetratype (T)
tetrads yield 3:1 (V-:vv) segregations that lead to identification
of V and v sporidia.

Type of tetrad:			
Sporidium	PD	NPD	T
1	V+	v+	V+
2	V+	v+	v+
3	v-	V-	V-
4	v-	V-	v-
Selfing			
1 x 3	Vv	Vv	VV
1 x 4	Vv	Vv	Vv
2 x 3	Vv	Vv	Vv
2 x 4	Vv	Vv	vv*
Segregation (V_:vv)	4:0	4:0	3:1

dikaryon

*This culture identifies the two v-containing and, by extension, the
two V-containing sporidia of the tetrad.

Table 2. Expectation of "constant ranking" among dikaryons: when
plus and minus sporidia with varying numbers of polygenes (range
zero to five) are arranged in ascending series the resulting dikaryons
(having zero to 10 polygenes) are expected to show constant ranking
in respect of their pathogenicity.

Number of polygenes in "plus" sporidium

		0	1	2	3	4	5
	0	0	1	2	3	4	5
Number of polygenes in "minus" sporidium	1	1	2	3	4	5	6
	2	2	3	4	5	6	7
	3	3	4	5	6	7	8
	4	4	5	6	7	8	9
	5	5	6	7	8	9	10

each of the four products of the tetrad. For the example shown in
Table 1 the homozygous recessive dikaryon resulted from the mating
of cultures derived from sporidia 2 and 4. These sporidia therefore carried the recessive allele, and the other two (1 and 3) the
dominant allele. This assignment of genotypes to the four sporidia
of the tetrad is easily verified by combining them with other
cultures that are known to carry the recessive allele. Using this
procedure a large number of different tetrads, all produced by the
same dikaryon, can be analysed, and for each set-of-four sporidia
the genetic content (V and v) in relation to the major gene can be
determined. Where Pathogenicity is also determined in part by
polygenes it is also to be expected that these will be distri-
gate. But there is no reason to expect that these will be distributed equally to the four sporidia of a tetrad during meiosis, and
this in turn leads to the prediction that because of their variable
polygene content the V- and v- containing sporidia produced by a
single dikaryon will vary in their pathogenic behaviour. To test
this possibility the prediction was made that the V- and v-containing
sporidia would exhibit constant ranking, as shown in Table 2.

The results of a large experiment are shown in Table 3. The
expectation of constant ranking (which to us seemed evident on
visual inspection) was confirmed by mathematical analysis. For
each of the three major genotypes (VV, Vv and vv) there is a considerable range of variability that is attributable to polygenes.
The analysis of variance showed that this variation between gametes
in their contribution to the pathogenicity of dikaryons was significant. The analysis of variance also showed significant interactions, detected in the VV group and in one of the two Vv groups,
which indicate that the polygenes do not always act in an additive
manner as assumed by the general theory of polygenes.

Having established in our work with *U. hordei* a series of
cultures whose genetic content is fairly well known we are now
able to use these as test cultures in investigating the genetics
of the host. This work is now underway, and is being done with
the active participation of Doctor Eugene Sharp at Montana State
University. When our understanding of the genetics of the host is
sufficiently advanced we hope to be able to return to the problem
mentioned earlier, which was to determine how the polygenes of
host and pathogen interact, and how this interaction is integrated
with the interaction of the major genes.

As we see it the important problem is to identify the genetic
ingredients that lead to long-term stability in parasitic systems.
When these have been identified and their mechanisms of action are
understood, it should be possible to breed with greater efficiency
for durable resistance.

Table 3. Disease reactions (recorded as percent of infected plants) caused by *Ustilago hordei* on the barley cultivar Trebi. Data are for F₂ dikaryons all of which derive from a single parental F₁ dikaryon that was heterozygous for a major gene.

Sporidial Lines	Sporidia, mating type a(−) — Genotype: V						Genotype: v						Means			
	17 A 4⁻	21 C 4⁻	19 B 4⁻	20 A 4⁻	21 E 4⁻	20 C 1⁻	21 C 2⁻	21 B 4⁻	24 C 2⁻	24 A 4⁻	20 C 4⁻	20 A 3⁻	VV	Vv	vv	Over-all
Genotype: V																
18 D 1⁺	60	62	48	35	53	42	42	48	43	28	32	35	50	38		44
21 B 1⁺	42	42	50	41	34	27	35	41	34	26	28	26	39	32		36
21 B 3⁺	50	44	61	50	54	49	33	25	13	8	10	5	51	16		34
20 A 2⁺	53	34	34	47	31	25	34	43	29	24	14	20	37	27		32
23 C 3⁺	43	30	18	24	29	17	31	31	21	17	16	23	27	23		25
Genotype: v																
21 C 1⁺	50	48	50	56	45	38	16	16	11	4	10	3		48	10	29
19 A 3⁺	55	47	55	60	48	32	14	0	4	6	5	9		50	6	28
22 G 3⁺	47	35	27	32	27	23	12	17	18	14	8	8		32	13	22
20 A 1⁺	48	46	36	36	1	7	13	2	0	1	4	0		29	3	16
24 D 2⁺	45	31	33	25	13	19	5	2	3	1	2	2		28	3	15
24 A 3⁺	38	38	21	16	8	15	5	2	0	2	1	1		23	2	12
22 F 2⁺	24	16	24	8	10	2	8	4	4	1	3	1		14	4	9
Means																
VV	50	42	42	39	40	32							41			
Vv	44	37	35	33	22	19	35	38	28	21	20	22		30		
vv							10	6	6	4	5	3			6	
Overall	46	39	38	36	29	25	21	19	15	11	11	11				25

(Sporidia, mating type, A(+))

ACKNOWLEDGEMENTS

We wish to acknowledge the financial support received through
the Natural Sciences and Engineering Research Council of Canada.
We are also greateful to Rolando Robillo and to Ila Westergard-
Thorpe for their most valuable assistance.

REFERENCES

Driver, C.M., 1962, Breeding for disease resistance, *Scottish
    Plant Breeding Stat. Rep.*, 1-11.
Ellingboe, A.H., 1976, Horizontal resistance: an artifact of
    experimental procedure, *Aust. Pl. Path. Soc. Newsletter*, 4:
    44-46.
Emara, Y.A., 1972, Genetic control of aggressiveness in *Ustilago
    hordei* I. Natural variability among physiological races.
    Can. J. Genet. Cytol., 14:919-924.
Emara, Y.A., and Sidhu, G., 1974, Polygenic inheritance of
    aggressiveness in *Ustilago hordei*, *Heredity*, 32:219-224.
Fleming, R.A., and Person, C.O., 1982, The consequences of poly-
    genic determination of resistance and aggressiveness in non-
    specific host:parasite relationships, *Can. J. Plant Path.*
    (in press).
Johnson, R., and Law, C.N., 1975, Genetic control of durable
    resistance to yellow rust (*Puccinia striiformis*) in the wheat
    cultivar Hybride de Bersée, *Ann. appl. Biol.*, 81:385-391.
Parlevliet, J.E., and Zadoks, J.C., 1977, The integrated concept
    of disease resistance; a new view including horizontal and
    vertical resistance in plants, *Euphytica*, 26:5-21.
Robinson, R.A., 1976, "Plant Pathosystems", Springer-Verlag,
    Berlin, Heidelberg and New York.
Vanderplank, J.E., 1963, "Plant Diseases: Epidemics and Control",
    Academic Press, New York and London.
Vanderplank, J.E., 1968, "Disease Resistance in Plants", Academic
    Press, New York and London.

DISCUSSION

PARLEVLIET: Constant ranking as a measure of HR is obtained in
both models in principle. Can you agree with this?

PERSON: Yes.

PARLEVLIET: The Person/Fleming model and the addition model are
similarly constructed. Both lead in the long run to a "race for
weaponry" in the genetic sense, although phenotypically intermediate levels of disease are maintained. Phenotypically there
is a kind of stability but underlying it there is genetic build-

up in both host and parasite, so there is no stability. In the interaction model this race of weaponry is reduced. Can you agree with this?

PERSON: I agree that with the interaction model the race of weaponry is reduced. As for the Person/Fleming model the evolution of host and parasite can be visualized as a two dimensional system, one in which the normal curves representing the parasite and the host population can move along the x and y axes (respectively) as time progresses. But regardless of their rate of progress, at any point in time and two curves would interact as they do in our model. As for the weapons race I question whether genes for resistance and for pathogenicity would, or could accumulate without limit. The need to operate at an "optimal" rather than a "maximal" level of pathogenicity places a restraint on the parasite. While I do not see any similar constraint operating on the host I nevertheless incline to the view that parasitic systems evolve toward an equilibrium in which a "stable weaponry" is maintained via genetic feedback.

VAN DER GRAAFF: Regardless of whether the interactions are specific or nonspecific the system is not stable by itself. A race between genes for pathogenicity and genes for resistance is possible in both systems. Only if one assumes pleiotrophic "negative" effects of genes for increased pathogenicity, (and probably for increased resistance) will there be a maximum (= optimum) pathogenicity. The level of maximal pathogenicity may be partly determined by environment, and thus variable from one location to another, and from one agricultural system to another.

PARLEVLIET: Van der Graaff's remark suggests that stabilizing selection in both host and parasite could keep the pathosystem stabilized. In my opinion this type of selection, if it exists, may be effective in natural pathosystems but not in agricultural pathosystems.

ROBINSON: Are you suggesting that the microevolutionary mechanisms that take place in agricultural pathosystems and the macroevolutionary mechanisms that lead to the establishment of natural pathosystems, are different?

PARLEVLIET: In my opinion there are not two kinds of evolution. There is only one. I am suggesting the Person model is not really valid for a "man-guided" agricultural pathosystem in which man reproduces the host, and in which the pathogen is allowed to increase its virulence without "natural" constraints.

DINOOR: With natural selection for better performance of both host and parasite there are probably upper limits to both resistance and aggressiveness that are less than maximal. In our work

we have observed, both in the wild and under cultivation, that
maximal protection of the host is not the optimal situation. As
for the pathogen one may imagine that an overly aggressive pathogen
may colonize the host too rapidly and, in so doing, reduce its
capacity to produce resting spores for the coming year. Selection .
would not favour such a pathogen.

JOHNSON: Did you suggest in your paper that there was little
evidence for specificity involving minor genes?

PERSON: I agree that there is some evidence such as that provided
by Loegering and Burton using a computer for analysis of gene-for-
gene relationships of wheat with *Puccinia graminis*, where many
smaller but race-specific effects of resistance were observed.
However, it does not follow from the demonstration of a differential
interaction between host cultivars and pathogen isolates that all
genes in the system are interacting specifically. In any limited
test one is likely to find a number of host genes that do not interact specifically with pathogen isolates. And in further tests it
may be found that some, but not necessarily all of these do show
specificity. But having acknowledged these possibilities, we are
still a long way from being able to assert that in further investigation all interactions involving polygenes will turn out to be
specific. In my opinion there is at the present time no firm basis
for excluding the possibility of nonspecific interaction. The
problem at present is to design and carry out experiments that
will demonstrate whether polygenic interaction is specific or
nonspecific. I imagine that both kinds of interaction will be
found.

ESKES: You mentioned that it will not be easy to demonstrate
experimentally which one of the two models proposed by Parlevliet
and Zadoks is closest to being correct. I think this can best be
investigated by studying those host-pathogen combinations that
involve incomplete resistance. For these combinations the interaction model would predict that in the absence of any resistance
gene in the host there cannot be any pathogenic variability among
the pathotypes. The interaction model also predicts increasing
variability as the number of resistance genes is increased. These
results would not be predicted by the additive model. For facul-
tative pathogens there are, I think, examples in which great
variation in pathogenicity is observed even on highly susceptible
hosts. As an example, Nelson and McKenzie (1970) observed large
variability among the 161 isolates of *H. turcicum* when these were
made to interact with the highly-susceptible R_4 maize inbred.
This is in disagreement with Parlevliet and Zadok's interaction
model, and indicates that this model is not universally applicable
to all pathosystems. We will have to acknowledge, I think, that
at least for general pathogenicity the variability need not be
correlated with the presence of particular resistance genes in the

host.

PERSON: You have brought up an interesting point that deserves to
be investigated in further experiments.

DEPONTI: It is a general biological principle that intimate
interaction may exist and be determined at the minor-gene level
(e.g. congruity vs. incongruity; terminology of Hogenboom, 1973)
as well as at the major-gene level. Hogenboom studied crossing
barriers between tomato species and found that this was true of
interactions involving pollen and pistil. From the results of
other studies as well as his own he postulated that this would occur
as a general phenomenon.

DURABLE RESISTANCE IN CROPS : SHOULD THE CONCEPT OF PHYSIOLOGICAL RACES DIE?

J.E. Vanderplank

Plant Protection Research Institute
Department of Agriculture & Fisheries
Pretoria, R.S.A.

Developments especially in relation to *Puccinia graminis* and *P. recondita* force one to question the concept of physiological races, and to suggest an alternative. For simplicity the argument is stated around the gene-for-gene hypothesis, but the argument is quite general.

The first objection to the concept of physiologic races is that the number of potential races increases geometrically as the number of available resistance genes increases. If resistance gene R1 is available, it can be used to distinguish two races of the pathogen, namely, race (0) avirulent and race (1) virulent for gene R1. If genes R1 and R2 are available, they can distinguish $2^2 = 4$ races, namely, race (0) avirulent for both genes, race (1) virulent for R1 but avirulent for R2, race (2) virulent for R2 but avirulent for R1, and race (1,2) virulent for both genes. If genes R1, R2 and R3 are available, they can distinguish $2^3 = 8$ races, namely, races (0), (1), (2), (3), (1,2), (1,3), (2,3) and (1,2,3). And so on. Browder (1980) has listed 35 R genes (called Lr genes) in wheat for resistance to *Puccinia recondita*. There are therefore 2^{35}, i.e. approximately 34 billion races of *P. recondita* which could be distinguished if all the available Lr genes were used in the tests. Use of all the available genes would make race classification impossible. Worse, the discovery of a 36th Lr gene would potentially add another 34 billion races. A single gene pair in the fungus (if one assumes a gene-for-gene relation) would create 34 billion races. Can absurdity go further?

Neither geometric increase nor retrospective action is known in standard taxonomy, not even in races or their equivalents like varieties, cultivars or breeds. Discover a new race of mankind in

the Gobi desert, and you add ONE to the number of known races and
the previously known races are left intact without retrospective
division. Breed a new variety of rose and you add ONE to the number of known varieties, without retrospective division. In short,
a physiologic race of a pathogen is a taxonomic freak. Consider
general taxonomy. We can divide fungi into divisions, subdivisions,
classes, orders, families, genera, species, formae speciales and
physiologic races. In this sequence, from divisions to formae
speciales, taxonomic discoveries are dealt with additively, without necessary retrospective division. From formae specialies to
physiological races, taxonomic discoveries are dealt with geometrically (multiplicatively), without retrospective division. One
cannot bolster the concept of physiological races by appealing to
standard taxonomy.

The second objection to the race concept is that races are
fixed taxa, whereas the evidence demands gene flow not fixity.
Consider an example. Virulence and avirulence in *Puccinia graminis
tritici* exists on both sides of the United States–Canada border,
but with a difference. In the United States and Mexico, according
to surveys of thousands of isolates over many years published in
Plant Disease Reporter, virulence for resistance gene Sr6 is invariably associated with virulence for Sr9d. In Canada, according
to the Canadian Plant Disease Survey of thousands of isolates,
virulence for the gene Sr6 in most years is invariably dissociated
from virulence for Sr9d.

In the twelve years, 1967 through 1978, virulence for Sr6
associated with virulence for Sr9d was found only in 1975 and then
only in three samples. The Canadian data are analysed in Table 1;
Chi-square = 3263; n = 1; and P is vanishingly small. A clear con-
tinental picture emerges. In North America, from Mexico through
Canada, the greater part of the population of *P. graminis tritici*
is virulent for Sr9d. In this part of the population virulence for
Sr6 south of the United States–Canada border changes to avirulence
north of the border. Even in the relatively uniform area of the
Great Plains, the allele for virulence changes to the allele for
avirulence at the border. Clearly there is a swift gene (or allele)
flow between the times of the United States and Canadian surveys.
All this is incompatible with the concept of fixity embodied in
races. Indeed, to get at the facts it was first necessary to dis-
solve the United States and Canadian races published in the respective surveys and then crystallize the essentials.

Fluidity, the opposite of fixity, is reflected as gene flow.
The population of the pathogen is seen not in terms of fixed associations and dissociations of virulence and avirulence but in terms
of fluid associations and dissociations. This brings the topic
into the main stream of population genetics.

Table 1. The number of isolates of *Puccinia graminis tritici* in
Canada virulent and avirulent for the wheat stem rust
resistance genes Sr6 and Sr9d during the twelve years
1967 through 1978 (*).

	Virulent for Sr9d	Avirulent for Sr9d	TOTAL
Virulent for Sr6	3 (446.4)**	527 (83.6)	530
Avirulent for Sr6	3053 (2609.6)	45 (488.4)	3098
TOTAL	3056	572	3628

*   Data of Green published in Canadian Plant Disease Survey.

**   Figures in parenthesis are the numbers expected if the
distribution were random.

Chi-square (trivially corrected for continuity) = 3263

Consider for example, the classic example of gene arrangements
in the third chromosome of *Drosophila pseudoobscura*. Dobzhansky
(1943), in one of his studies of the genetics of natural populations,
determined the frequency of various gene arrangements, among them
"Chiricahua" and "Standard". The adaptive value of these arrangements varied from locality to locality and from month to month.
In a four year study on the San Jacinto Mountains the observed frequency of the Chiricahua arrangement rose month by month from 11.0
percent in January, when flies emerged after the winter to 34.6
percent in June, and then fell again to 10.1 percent in October.
Contrariwise the frequency of the Standard arrangement fell from
57.5 percent in January to 37.4 percent in June and then rose
again to 58.9 percent in October. These, and other changes in

frequency were constantly recurring and completely reversible. There were evidently spatial and temporal adaptations to the environment; and great changes in the frequency of gene arrangements could occur within two sexual generations. Dobzhansky calculated that the selection coefficient was as high as 0.4.

A system is needed in which complexity increases arithmetically, not geometrically, with increasing numbers of identified resistance genes and in which gene flow is recognized.

Record for each isolate of the pathogen the host resistance genes for which the isolate is virulent or avirulent, the data of isolation, the temperature at or preceeding isolation and any other information thought to be relevant. Continue the collection of isolates throughout the season in order to determine allele flow. Then with a computer estimate how the pathogen's virulence for selected host resistance genes associates and dissociates with time and how temperature or other factors affect the association. For example, how do date of isolation and temperature affect the association and dissociation of virulence genes Sr6 and Sr9d.

Our meeting is about durable resistance. Dissociation of virulence stabilizes appropriate resistance and is an essential topic for us. The dissociation of the virulence for genes Sr6 and Sr9d has been used for illustration, but is not unique. Like Sr6 are Sr9a, 9b and 15, which I call the ABC group of genes (Vanderplank, 1982). Like Sr9d are Sr7b, 9e, 10, 11, Ttl and Tmp, the XYZ group. Virulence for genes taken one from each group tends to dissociate and stabilize the resistance that the two genes give. There are possibly other combinations and relations. That is what a computer can probe; that is what the race concept obscures.

Throw out races. Bring in computers.

REFERENCES

Browder, L.E., 1980, A compendium of information about named genes for low reaction to *Puccinia recondita* in wheat, *Crop Sci.*, 20:775-779.
Dobzhansky, T., 1943, Genetics of natural populations, IX. Temporal changes in the populations of *Drosophila pseudoobscura*, *Genetics*, 28:162-186.
Vanderplank, J.E., 1982, *Host-Parasite Interactions in Plant Disease*. Academic Press. New York, London. 207pp.

THEORETICAL RESISTANCE MODELS

R.A. Robinson

Simon Fraser University
Burnaby, B.C.
Canada, V5A 1S6

INTRODUCTION

The pathosystem concept is essentially about new ideas and
any attempt to describe a new idea encounters special difficulties.
This is because the very newness of the idea means that words to
describe it do not exist. There are then three possibilities; we
can give new meanings to old words, but this is thoroughly
confusing; or we can coin new words, though this is also confusing;
or we can describe the new concept in terms of analogies. I am now
faced with this choice and I propose to use analogies.

ISLANDS

The first analogy is that of Janzen (1968) who thought of
host plants as islands. If each host individual is an island, then
each parasite individual can behave in one of two ways; it can
either migrate to an island, or it can colonize an island if it is a
descendent of a successful migrant. We can take this analogy even
further; migration necessitates flying, while colonisation is
possible with walking. In the case of aphids, this analogy becomes
reality because the winged and wingless forms serve this function
exactly. The distinction between flying and walking turns out to be
so important that it is one of those rare occasions when new terms
are justified. Migration (flying) is called allo-infection and
colonization (walking) is called auto-infection. This is a
clear-cut distinction; it is an "either-or" situation with no other

possibilities or intermediates. (Plant breeders may prefer an
alternative analogy in which a pathogen spore is compared with a
pollen grain; allo-infection is then the equivalent of
cross-pollination or allogamy, while auto-infection is the
equivalent of self-pollination or autogamy).

The second analogy introduces two kinds of island; there are
new islands, which are uninhabited, and old islands, which are
inhabited. A new island represents a young host seedling recently
emerged from the soil, or new leaf tissue on a deciduous tree.
Being new, it is free from parasites and it can become parasitized
(or inhabited) only by migration. Epidemiologically, the only
possible infection is allo-infection. An old island, on the other
hand, represents a perennial host individual; it has been in
existence (and inhabited) for a long time and parasites have been
colonizing it continuously for years. Allo-infection is of minor
significance to an old island, but auto-infection is of major
importance.

LOCKS AND KEYS

The third analogy concerns the the gene-for-gene
relationship (Flor, 1942; Person, 1959) which can be described in
terms of locks and keys. Each host has a lock and each resistance
gene is the equivalent of a tumbler in that lock; the lock may have
many tumblers. Each parasite individual has a key. And each
parasitism gene in the parasite is the equivalent of a tooth on that
key; the key may have many teeth. The key of any one parasite
individual either does or does not fit the lock of a host
individual. Once again, this is a clear-cut, "either-or"
distinction with no other possibilities.

Now let us bring the three analogies together. Consider
1,000 new islands (e.g. newly emerged host seedlings). The new
islands are parasite-free and can become parasitized only by
migration. Each island has a lock and we may assume that there are
1,000 different locks; each island thus has a lock which is
different from all the other locks. Let us also assume 1,000
parasite individuals, each one with a key which will open only one
lock; there are 1,000 different keys and each migrant thus has a key
which is different from all the other keys. Each island receives
one migrant. The chances of that migrant's key fitting the island
lock are only 1/1,000. On average, therefore, only one migrant will
open its island lock; and only one island will become inhabited;
only one allo-infection will be a matching infection.

That single "new" island becomes an "old" island; it is now parasitized and it is being colonized. Every component of the island (i.e. leaves, stems) has the same lock. And every descendent of the successful migrant has the same key, either because the progeny is a clone or because the sexually produced offspring quickly become homozygous. And these identical keys fit the lock. It follows that the lock cannot prevent colonization.

This brings us to our first conclusion which appears to be irrefutable. The gene-for-gene relationship cannot control colonization (auto-infection); and, because there are only two kinds of infection, it follows that the gene-for-gene relationship can only control migration (allo-infection). This is the most important conclusion to emerge from the pathosystem concept and all other conclusions stem from it.

If the lock and key analogy is accurate, we can also conclude that the gene-for-gene relationship must provide an effective control of migration (allo-infection) otherwise it would never have evolved in the first place.

Let us take the analogy further. Consider 1,000 motor cars, each with a different lock; and 1,000 car thieves, each with a different key. Each thief has only one attempt at car stealing; the probability of a key fitting a lock is only 1/1,000. On average, therefore, only one car would be stolen. Our third conclusion is obvious; the gene-for-gene relationship controls migration provided that there are many different locks and many different keys. In other words, there must be heterogeneity in both the host and the parasite populations.

Consider the alternative situation. There are 1,000 cars which all have identical locks; and there are 1,000 thieves who all have identical keys. There are now two possibilities. Either none of the keys fit, and no cars are stolen. Or every key fits, all the cars are stolen, and the whole system of car locking has been ruined by uniformity.

Now consider a modern crop. Each host individual is an island and has a lock. But the host population is either a clone (e.g. potatoes, sugarcane, citrus), a pure line (e.g. wheat, rice, beans) or a hybrid variety (e.g. maize, onions, cucumber). All the islands have identical locks. Following an initial period, when all the locks are effective, all parasite individuals soon possess identical keys which fit those identical locks. The whole system of "locking" has been ruined by uniformity.

Our fourth conclusion must be that we have have made two mistakes in our use of the gene-for-gene relationship in agriculture. Firstly, we have employed it to control both migration and colonization, even though it evolved to control migration only. Secondly, the gene-for-gene relationship can only control migration if it is employed on a basis of heterogeneity, but we have employed it on a basis of uniformity. We have depended on that period of adjustment when none of the keys fit. Occasionally, this period is quite long and the resistance appears durable. More often, however, the period is short and, once that period is over, all the keys fit and there is no control of either migration or colonization. The resistance is said to have "broken down" and the damage is so great that the cultivar is abandoned.

HARDWARE AND SOFTWARE

Now let us consider an entirely different set of analogies which involve computers. The pathosystem concept is based on the general systems theory which makes a clear distinction between structure and behavior. In computer terminology, structure is called hardware, and behavior is called software. A behavior pattern may be defined as a response to a stimulus. There are thus three components of a behavior pattern; the stimulus, the response and the control of that response. In computer terminology, software normally has three components; the command, the response and the control of that response by the program. Computer enthusiasts will know that programs are usually "strings" of instructions which say, in effect, if so-and-so, do such-and-such. A computer operates by being given a command (the stimulus) and responding to it in a controlled manner.

We must begin to think in terms of plant behavior, just as zoologists have only recently started thinking in terms of animal behavior (ethology). Behavior is controlled. The control of animal behavior may be either acquired or inherited; acquired behavior is controlled by learned experience and inherited behavior is controlled by the genetic code. In plant pathosystems, it is probably safe to assume that all behavior patterns are inherited in both the plant hosts and in all of their parasites. In other words, all behavior patterns in plant pathosystems are controlled by the genetic code.

Consider an obvious example of inherited plant behavior. In response to spring, a deciduous tree will produce new leaves; in response to autumn, it will shed those leaves. A parasite of the leaves will behave accordingly; in the absence of leaves it will either become dormant or it will find an alternate host.

A coherent system of behavior patterns is usually called a strategy. In the wide sense of the term, a computer program controls a strategy. So does the genetic code. The computer program was designed to impose that strategy. The genetic code evolved to impose its strategy. If it is to survive commercial competition, the computer program must have the best possible strategy; all earlier and inferior program designs are eliminated by commercial competition. Similarly, if it is to survive evolutionary competition, the genetic code must have the best possible strategy; all earlier and inferior strategies having been eliminated by evolutionary competition.

We can argue that a wild plant pathosystem has behavior patterns which are controlled by the genetic code; a coherent system of these behavior patterns (the strategy) is also controlled by the genetic code; that strategy is the best possible strategy because of the elimination of all earlier, inferior strategies by evolutionary competition. It follows that we can study wild plant pathosystems in terms of behavior strategies and that we can test our conclusions by searching for alternative and, possibly, superior strategies (Robinson, 1979, 1980, 1981).

SYSTEMS LEVELS

We must now consider a third series of analogies based on writing. An ordinary book is a perfect example of a system, even if it is only a relatively simple, static system. In particular, a book illustrates the concept of systems levels. The book itself is the system; it is a pattern of subsystems called chapters; each chapter is a pattern of subsystems called paragraphs; each paragraph is a pattern of subsystems called sentences; and so on down through words to individual letters. Equally, the book itself is part of a supersystem called a library.

At the lowest systems level, a word has the two basic properties of structure and behavior; structure is normally called spelling, and behavior is called meaning. However, these two properties can be recognised at all other systems levels also; the sentence, the paragraph, the chapter and so on.

The same is true of a plant pathosystem. At the highest systems level, a pathosystem is an entire species of a host and its interaction with an entire species of a parasite, on a global scale. Next down is the geographically defined pathosystem which Putter (1981) has called a pathotope by analogy with an ecotope (as in topology; *topos* = place). Then there are the subsystems of a pathotope, based on the interactions of populations which are defined in terms of parasitism; thus pathotypes of the parasite and

pathodemes of the host (i.e. populations in which all individuals
have a specified pathosystem character in common, Robinson, 1969).
Lower still are the interactions between individuals; at this level
it is possible to start referring to allo-infection or migration,
and auto-infection or colonization. At a still lower systems level
are the various mechanisms of resistance and parasitism. And then
the biochemical processes; and so on down to the units of genetic
code. And at each of these systems levels there is structure and
behavior.

The recognition of systems levels in plant pathosystems is
particularly important in terminology. A literal term can only be
precise at the systems level at which it was defined. Thus "field"
resistance may be precise at the systems level of the epidemic, but
it is meaningless at the level of the genetic code. And the
converse is true of, say, "monogenic" resistance. If we want terms
that are precise at all systems levels, we must employ abstract
words and find appropriate definitions for them at each systems
level. This is the great advantage of Vanderplank's (1963) terms
'vertical' and 'horizontal'.

We can now define the principle subsystems of a pathosystem
which can be called "pathosubsystems". Because an essential feature
of the pathosystem concept is that the behavior of the host
population and the behavior of the parasite population are both
treated as one system, it follows that each subsystem must be
defined in terms of both the host and the parasite. There are three
pathosystem characters which can only be defined in this way. The
first is infection, which is clearly impossible if either the host
or the parasite is absent. As we have already seen, infection is
either allo-infection or auto-infection. The second character is
the gene-for-gene relationship, which is either present or absent,
and which is also impossible if either the host or the parasite is
absent. The third is the differential interaction which means that
a series of parasite differentials is necessary to identify any one
resistance, and a series of host differentials is necessary to
identify any one parasitic ability. Once again, a differential
interaction is impossible if either the host or the parasite is
absent. In any pathosubsystem, there either is or is not a
differential interaction; if there is no differential interaction,
differences of resistance and parasitic ability exhibit a constant
ranking; there are no other possibilities.

The vertical subsystem is named after Vanderplank's (1963)
concept of vertical resistance; it has a gene-for-gene relationship,
it exhibits a differential interaction and it can only control
allo-infection. The horizontal subsystem is named after
Vanderplank's (1963) concept of horizontal resistance; it does not
have a gene-for-gene relationship, it does not exhibit a

differential interaction (i.e. there is a constant ranking), and it is the only subsystem which can control auto-infection. In addition, there are about a dozen other pathosubsystems (Robinson, 1979) which do exhibit a differential interaction but which do not have a gene-for-gene relationship; they are of minor importance and are not pertinent to the present discussion.

A THEORETICAL MODEL

It is now possible to bring all these ideas together and to synthesise models of wild plant pathosystems. We do this primarily with a view understanding how these various strategies and behavior patterns operate and why they evolved in the first place. Such knowledge is necessary if we are to improve our management of the crop pathosystem.

The model presented here is based on an idea of Cynthia D. Scott (in press) who applied various pathosystem principles to both wild and cultivated apples, and constructed pathosystem models using apple scab (*Venturia inaequalis*) as the parasite.

Let us consider the wild host population first. Each wild apple tree grew from a seedling which was heterozygous due to cross-pollination. The population is thus heterogeneous, particularly with respect to vertical resistance genes. We shall assume a total of twelve pairs of genes in the gene-for-gene relationship of this pathosystem, and there are sound mathematical reasons for thinking that each host individual will possess six of these genes. This means that 924 different vertical resistances (vertical pathodemes) are possible; for purposes of discussion, we can round this figure to 1,000. We may assume also that all of these vertical pathodemes occur with equal frequency and that they are distributed randomly in the host population.

When the epidemic begins in the spring, each individual in the parasite population is an ascospore. Each individual is thus the result of sexual recombination in a heterothallic fungus. For each vertical resistance gene in the host, there is a matching gene in the parasite and we may assume that each parasite individual possesses six of these genes. There are thus 924 different vertical pathotypes in the parasite population and, as with the host, we may assume that they occur randomly and with equal frequency throughout the parasite population; once again, we can round this figure to 1,000.

Now consider the wild pathosystem, which is the interaction of these two wild populations. The epidemic begins in the spring. Each tree produces new leaves which, being disease-free, can only be allo-infected; the only way in which a parasite individual can reach a host is by migration. Let us assume that there are enough ascospores for each host individual to be allo-infected only once. There are 1,000 different vertical pathodemes and 1,000 different vertical pathotypes, all randomly distributed. The probability of any allo-infection being a matching infection is thus 1/1,000. Let us asssume that the total wild apple population is 1,000,000 trees; there are thus 1,000 vertical pathodemes and each pathodeme consists of 1,000, randomly distributed individual trees. If the probability of matching is 1/1,000, each pathodeme will be matched only once; that is, 1,000 trees will be matched and, on average, each of them will belong to a different vertical pathodeme. Each vertical pathotype of the parasite will thus become established in only one tree. This migration is called the exodemic and, because there will be further exodemics, it is called the primary exodemic.

Colonisation begins in each matched tree; this colonisation is called the esodemic. Every part of the one tree has the same vertical resistance. Every descendent of the original migrant has the same vertical pathogenicity because the conidia are a vegetatively propagated clone. Every auto-infection is a matching infection; all the colonisation is matching infection. The vertical subsystem cannot control the esodemic which, consequently, can only be controlled by the hortizontal subsystem. Because every epidemic has esodemics, it follows that horizontal resistance is universal; it occurs in every plant host individual against every parasite species of that individual.

In addition to auto-infection, the descendents of the original migrant (conidia) can also allo-infect; they can become migrants themselves. This is the secondary exodemic. Once again, the probability of matching allo-infection is only 1/1,000. We can assume that the secondary exodemic differs from the primary exodemic only in that there are ten times as many pathogen individuals. Consequently, ten times as many trees are matched (i.e. 1% = 10,000). There are then more esodemics which lead to the tertiary exodemic in which 10% of trees are matched (i.e. 100,000). Finally, in the last exodemic, all the trees are matched.

The vertical subsystem has major effects on the epidemic as a whole. If a wild apple tree lives for 1,000 years, it will obviously have to endure 1,000 epidemics; the vertical subsystem will ensure that, on average, only one of these involves matching in the primary exodemic; only ten in the secondary exodemic; only ninety in the tertiary exodemic; and about nine hundred of the epidemics will involve matching only in the last exodemic which is the least damaging because the resulting esodemic is the shortest. Similarly, each tree suffers maximum damage only once in its lifetime; and minimum damage in about 90% of its lifetime. Because

of the sequence of exodemics, the whole epidemic is slowed down.
Only 0.1% of the population is matched at the begining of the
epidemic and it is not until the end of the epidemic that the last
90% of host individuals are matched. All vertical pathodemes suffer
equal damage and none has a survival advantage over the others. The
same is true of the vertical pathotypes. Neither the host nor the
parasite threatens the evolutionary survival of the other.

In the crop pathosystem, however, this delicate balance is
ruined. The apple orchard is a clone. Every tree has the same
lock; both the host and the parasite populations are homogeneous;
every allo-infection in every exodemic is a matching infection.
Coupled with a low level of horizontal resistance, this means that
the apple scab fungus is no longer a harmless parasite; it now
causes a very damaging disease.

GROUP SELECTION

The classic intepretation of Darwinian evolution involves
three systems levels; (i) genes which mutate, (ii) individuals which
are selected, and (iii) species which evolve. With this process,
selection can only occur at the systems level of the individual and
the evolution of the strategy controlled by a gene-for-gene
relationship is difficult to explain.

Consider what happens to the model of the wild apple scab
pathosystem when selection occurs at the systems level of the
individual. In the host, the individual with no vertical genes has
the minimum survival advantage because it can be matched by every
vertical pathotype; all allo-infections would match. Conversely,
the host individual with all the available vertical genes would have
the maximum survival advantage because it could be matched by only
one vertical pathotype.

Similarly in the parasite. The individual with no vertical
genes would have the minimum survival advantage because it could
match only one vertical pathodeme, while the individual with all the
vertical genes could match all pathodemes.

Such a system could not operate because it has only one
possible point of stability. The host population would change to
homogeneity of the one vertical pathodeme which possessed all the
vertical genes. The parasite population would become equally
homogeneous with the one vertical pathotype which also possessed all
the vertical genes. At this point, every allo-infection would be a
matching infection and the vertical subsystem would cease to
function. This is the "identical locks and identical keys"
sitation.

For purposes of definition, macro-evolution is taken to involve changes which are new, which are irreversible, and which normally occur over periods of geological time. Micro-evolution involves changes which are not new, which are reversible, and which normally occur in periods of historical time. The problem is to explain both the macro-evolution and the micro-evolution of the vertical subsystem in terms of selection at the level of the individual.

One explanation is that of "cost". If the optimum number of vertical genes is n/2, the possession of either fewer or more genes than this optimum might cost too much to permit survival. Thus, too few genes would lead to an unacceptable cost in susceptibility; and too many genes would lead to an unacceptable genetic cost.

An alternative explanation derives from a form of speculation which treats the genetic code as if it were computer software. If the gene-for-gene relationship and the genetic code were computer software, what would be the best design of program for it? This computer science approach to biology leads to an obvious possibility. The vertical subsystem might be programmed to prevent selection at the level of the individual.

Such programming of the genetic code would compel selection at the level of the population (group selection). It can even be suggested that any behavior strategy which controls a population and which is itself controlled by the genetic code, can only evolve by group selection and can only be maintained by group selection. That is, both the macro- and the micro-evolution of an inherited strategy depend on group selection; and the group selection depends on programming which prevents individual selection.

A possible mechanism of such programming is by a linkage of vertical genes and self-incompatibility genes in such a way that only the strategically advantageous combinations of vertical genes occur. Such incompatibility mechanisms are common in angiosperms and heterothallic fungi and they probably occur in other categories of parasite such as insects and nematodes.

ACKNOWLEDGEMENT

I am grateful to F.E. Williams for valuable and constructive comments.

REFERENCES

Flor, H.H., 1942, Inheritance of pathogenicity in *Melampsora lini*, *Phytopathology*, 32:653-669.
Janzen, 1968, Host plants as islands in evolutionary and contemporary time, *Amer. Natur.*, 102:592-595.
Person, C.O., 1959, Gene-for-gene relationships in host parasite systems, *Canadian Journal of Botany*, 37:1101-1130.
Putter, C.A.J., 1980, An epidemiological analysis of the *Phytophthora* and *Alternaria* blight pathosystem in the Natal Midlands. Doctoral Thesis, Univ. of Natal, Pietermaritzburg, 203pp.
Robinson, R.A., 1969, Disease resistance terminology, *Review of Applied Mycology*, 48:593-606.
Robinson, R.A., 1976, *Plant Pathosystems*. Springer-Verlag, Berlin. Heidelberg, New York, 104pp.
Robinson, R.A., 1979, Permanent and impermanent resistance to crop parasites; a re-examination of the pathosystem concept with special reference to rice blast, *Z. Pflanzenzuchtg.*, 83:1-39.
Robinson, R.A., 1980, New concepts in breeding for disease resistance, *Ann. Rev. Phytopath.*, 18:189-210.
Scott, C.D., 1981, in press, Simon Fraser University, British Columbia, Canada.
Vanderplank, J.E., 1963, *Plant Diseases: Epidemics and Control*, Academic Press, New York and London. 349pp.

MODELS EXPLAINING THE SPECIFICITY AND DURABILITY OF HOST RESISTANCE DERIVED FROM THE OBSERVATIONS ON THE BARLEY-*PUCCINIA HORDEI* SYSTEM

J.E. Parlevliet

Plant Breeding Department (I.v.P.)
Agricultural University
Wageningen, The Netherlands

SUMMARY

Hosts employ a wide range of defense mechanisms, which can be grouped into avoidance, resistance and tolerance. Especially with resistance problems of specificity and durability arise. Toward parasites in general hosts employ various mechanisms that are effective against whole groups of parasites. This broad resistance is probably very durable but can be race-specific. Specialized parasites, and breeding for resistance is often directed against such ones, have overcome these broad resistances. *Puccinia hordei* is such a specialized parasite. In barley two types of resistance to this rust can be discerned. The hypersensitive type of resistance (HyR) reduces or arrests the growth of the fungus accompanied by extensive host cell necrosis. Partial resistance (PR) reduces or arrests the growth of the pathogen not accompanied by host cell necrosis. The former is governed by major genes (Pa1-Pa9), is race-specific and not durable at all. PR, typically of a polygenic nature, appears near-race non-specific and durable. Both resistances are pathogen-specific; they operate only against *P. hordei*. Even against the related *P. striiformis* they are ineffective.

To explain the observations it is assumed that PR as well as HyR operate on a gene-for-gene basis with genes in the pathogen. In HyR the recognition is for incompatibility (hypersensitivity) when a resistance allele is confronted with its corresponding avirulence allele. In PR the recognition is for compatibility (basic susceptibility) when a susceptibility allele is confronted with its corresponding pathogenicity allele. In the former a loss mutation of the avirulence allele can lead to loss of recognition

and so to a restoration of the basic pathogenicity; virulence
therefore can be induced relatively easily. In the latter resistance can only be overcome by a gain mutation, neutralizing the
effect of the resistant gene. Such a mutation is not easy to
produce; the resistance tends to be durable. Race-specific effects
and lack of durability are disconnected here; they are considered
as independent traits of pathosystems.

Partial resistance is assumed to interfere with the basic
pathogenicity of the pathogen. The HyR system, superimposed upon
this basic pathogenicity, is used by the biotroph as a genetic feed
back mechanism to regulate its co-existence with the host at
intermediate levels of virulence.

INTRODUCTION

In natural ecosystems there are producers (plants) and
consumers (plants and animals), which satisfy their food requirements from the producers or other consumers. The consumers range
from herbivores (caterpillars, sheep) to highly specialized parasites (rusts, viruses). Hosts developed a variety of defense
mechanisms to limit biological damage caused by the consumers, and
the consumers in turn evolved adaptations to such defences. In this
process of coevolution some defence mechanisms are more readily
neutralized than others.

Defence mechanisms can be classified into avoidance, resistance and tolerance (Parlevliet, 1981a). Avoidance operates before
parasitic contact between host and parasite is established, and
decreases the frequency of parasitic contact. After parasitic contact has been established the host can resist the parasite by
decreasing its growth, or tolerate its presence by suffering
relatively little damage. Avoidance is especially employed against
animal parasites and includes such diverse mechanisms as volatile
repellents, mimicry, and morphological features like hairs, thorns
and resin ducts (Harper, 1977). Resistance is usually of a chemical
nature. Little is known about tolerance; it is very difficult to
measure without confounding it with partial resistance (Parlevliet,
1981b, 1981d).

To protect our crops from their parasites plant breeders make
ample use of the defense mechanisms present. Against animal
parasites both avoidance and resistance are widely employed, against
pathogens it is mainly resistance that is used. And it is with
resistance that the problems of specificity and lack of durability
arose. Van der Plank (1963, 1968) discussed this comprehensively.
He discerned vertical (VR) or race-specific resistance and
horizontal (HR) or race-non-specific resistance. The former is
assumed to be elusive and simply inherited, the latter durable and

polygenically inherited. Van der Plank too assumes that VR and
vertical pathogenicity or virulence operate on a gene-for-gene
basis, while HR and horizontal pathogenicity or agressiveness do
not do so. This classification of resistance into two types,
although quite convenient, does not satisfactorily explain the
great variety in observations reported anymore. A more comprehensive
view is needed.

EVOLUTION OF HOST AND PARASITE

For the parasite the host is an extremely important part of
its environment and this part is notoriously variable. There are
large numbers of host species to choose from and within each host
a variety of vegetative tissues exist, but the presence of these
tissues varies greatly over the seasons. In short the parasite has
to adapt to the seasonal growth patterns and the morphological,
anatomical and physiological features of the hosts in order to
survive. Such adaptations lead to specialization among parasites.
In this process of adaptation the parasite may choose between
lower or higher levels of specialization (generalists versus
specialists)

The hosts tend to evolve defence mechanisms to ward off,
resist or tolerate the parasites. Some of these mechanisms are
directed against whole groups of parasites others are more
specific, meant to check specialists. Parasites in turn may adapt
to such defenses. In this course of coevolution between host and
parasite remarkable levels of specialized interaction may develop
as shown by the *Passiflora-Heliconius* relationship. The *Heliconius*
butterflies are typical specialists; the larvae of most species
feed only on certain parts or tissues of a very limited number of
the 350 *Passiflora* species. The hosts employ a wide range of
defense mechanisms of chemical and morphological nature. Several
of these are specifically directed against *Heliconius* like the
glandular outgrowths from the leaves of some *Passiflora* species
that mimic *Heliconius* eggs at the point of hatch. *Heliconius*
detects and rejects shoots of *Passiflora* that carry eggs of young
larvae (Gilbert, 1975).

The rate at which the co-evolution takes place varies. Some
defences are more likely to be overcome by the parasites than
others, but probably no one is sacrosanct. Parasites appear to
adapt more readily to chemical defenses than to morphological
ones. And it is chemical defense (mostly resistance) that is used
so often in breeding for disease and pest resistance. This paper
therefore restricts its discussion to the specificity and
durability of chemical resistance.

BROAD RESISTANCE

    Hosts employ a wide range of chemicals to control all sorts
of parasites. Many of these chemicals have a general or broad
effect. They operate against groups of parasites. The terpenes
produced by pines are effective against most insects, but some
like the pine saw fly (*Neodiprion sertifer*) have learned to deal
with them. Phyto-alexins, produced by many plants following cell
damage, are effective against numerous fungi. However, some
pathogens are able to prevent or suppress the production of phyto-
alexins (biotrophs), or to tolerate or to neutralize the phyto-
alexins produced by the hosts on which they have specialized.

    Broad or general resistances normally represent an acquired
trait, a positive function. The parasite in order to overcome such
a resistance must develop a mechanism, a positive function, to deal
with this resistance. The avenacins present in oat roots represent
such a broad resistance mechanism. They have fungicidal properties
protecting them from most but not all soil fungi. Despite their
high avenacin contents oat roots are colonized by fungi such as
Gaeumannomyces graminis var. *avenae* and *Fusarium avenaceum*
(Lüning et al., 1978). Both fungi convert the highly active
avenacins into inactive compounds by means of enzymes absent in
fungi not able to colonize oat roots (Turner, 1961; Lüning et al.,
1978). Especially the case of *G. graminis* var. *avenae* (take-all)
is interesting as the var. *tritici* is completely inhibited by the
oat avenacins (Table 1). Both the oat resistance to the *tritici*
form and the pathogenicity of the *avenae* form are positive
functions. This represents the basic model for resistance/patho-
genicity (Model I).

    Table 1. Resistance (-) or susceptibility (+) of wheat
             and oats to *Gaeumannomyces graminis* var. *tritici*
             and var. *avenae* (take-all).

Host	G. graminis var.	
	tritici	*avenae* (avenacinase)
wheat	+	+
oats (avenacin)	-	+

Although the avenacins represent broad resistance the take-all fungus appears to vary in its pathogenicity to this mechanism. The *tritici* and *avenae* forms can be considered as races in the pathogenic sense. Considering the durability of resistance from a human time scale point of view resistance is probably of a very durable nature, despite the incidental occurrence of races. The occurrence of such a new race like the *avenae* form of take-all is most likely an extremely rare event, because the acquisition of a new positive function must be very difficult.

Highly specialized parasites normally have overcome the broad resistance mechanisms of their hosts by suppressing them or by tolerating or neutralizing their effects like the pine saw fly in pines and most biotrophs, such as the cereal rusts and powdery mildews and it is especially when resistance breeding against such specialized parasites is carried out that the problems of race-specificity and lack of durability of resistance come forward. Leaf rust of barley, caused by *Puccinia hordei*, is such a specialized pathogen, and the results obtained by studying this pathosystem are used to discuss the resistances employed by the hosts when dealing with such specialists.

THE BARLEY-LEAF RUST SYSTEM

The leaf rust occurring on barley seems restricted to this host and to the wild barley, *Hordeum spontaneum*. The alternate host is formed by *Ornithogalum* species, but is not functioning in most barley growing areas outside the Eastern Mediterranean area.

In barley two types of resistance can be discerned. The cultivars Quinn, Sudan and EP 75 in Table 2 exemplify one type. The resistance is governed by major genes, it is race-specific and it is of a hypersensitive nature. This hypersensitivity reaction is characterized by host cell collaps round the point of entry accompanied by reduced or arrested growth of the fungal mycelium. The manifestation of this reaction is a necrotic flecking of the leaves. Some cultivars, although fairly resistant in the field, do not show any signs of this hypersensitive reaction, they give a susceptible infection type (cultivars Berac, Julia and Vada in Table 2). This partial resistance (Table 3) results from a reduced infection frequency (IF), a longer latent period (LP) and a reduced spore production (SP) per urediosorus (Parlevliet and Van Ommeren, 1975; Neervoort and Parlevliet, 1978). LP, the most important component of the partial resistance, is governed by polygenes (Parlevliet, 1978b).

Table 2. Infection types on seedlings of six barley cultivars
inoculated with five races of *Puccinia hordei*.
S denotes a susceptible IT, MR (some sporulation)
and R (no sporulation) denote resistant IT's (of a
hypersensitive nature).

Cultivar	Race				
	11-1	1-2	18	22	24
Quinn	MR	S	R	MR	R
Sudan	S	S	S	R	S
EP 75	MR	MR	R	MR	S
Berac	S	S	S	S	S
Julia	S	S	S	S	S
Vada	S	S	S	S	S

Table 3. Partial resistance of five barley cultivars to
Puccinia hordei, race 1-2, expressed as number of
urediosori/tiller, infection frequency (IF), latent
period (LP) and spore production (SP) relative to
the most susceptible cultivar, and the estimated
number of polygenes governing LP (all measurements
relate to adult plants).

Cultivar	Number of urediosori per tiller	IF	LP	SP	No. of polygenes for LP
L94	2800	100	100	100	0
Sultan	750	65	130	80	3
Volla	115	70	130	110	3
Julia	17	65	160	50	5
Vada	1	40	190	50	6

When the race-specificity of this resistance was tested by exposing three partially resistant cultivars to five different leaf rust races (Table 4) the low to moderate disease severity was by and large consistent with race-non-specific resistance. Berac was the least resistant and Vada the most resistant cultivar against the five races. However, there are two small exceptions. Julia always had less leaf rust than Berac except when exposed to race 18 against which Julia was the more susceptible one. This differential interaction, was due to a difference in LP. Berac and Julia gave a similar LP for the other four races while race 18 had a LP 1 to 1½ days shorter on Julia than on Berac. This corresponds   well with the estimated effect of one polygene. The genetic and epidemiological data suggest that one of the polygenes for LP in Julia has been "broken" by race 18 (Parlevliet, 1978a), although the size of this race-specific effect is too small to be used to readily identify races. The second even smaller interaction is shown by Berac when exposed to isolate 22. This gave a slightly too high disease severity in this experiment (Table 4) as well as in two consecutive experiments, in 1979 and in 1980 (Parlevliet, unpublished). Similar small, race-specific differences in partial resistance have been reported for the potato-*Phytophthora infestans*, barley-*Rhynchosporium secalis*, wheat-*Septoria tritici*, wheat-*Puccinia recondita* and other systems (reviewed by Parlevliet, 1979).

Table 4. Disease severity (% sporulating leaf area) on three
         partially resistant barley cultivars caused by five
         races of *Puccinia hordei* just prior to maturation.

Cultivar	Race				
	11-1	18	1-2	22	24
Berac	8.1	6.7	3.1	5.0[b]	0.9
Julia	4.5	12.1[a]	1.8	1.1	0.6
Vada	0.8	0.5	0.6	0.2	0.1

[a] In absence of interaction appr. 3.0%

[b] In absence of interaction appr. 1.8%

    The genes governing the two systems, hypersensitivity and
partial resistance appear to be expressed independently of each
other, the genes for partial resistance acting before those
controlling hypersensitivity (Clifford, 1974; Parlevliet, 1980).
Niks (1982a and b) and Niks and Kuiper (1982) compared the partial
and hypersensitive resistance and the non-host reaction
histologically. With partial resistance part of the fungal colonies
aborted very early, the others developed at a reduced rate and many
of these sporulated eventually. There was no accompanying host cell
collaps. The hypersensitivity genes caused either early aborted
colonies or slowly developing colonies, both accompanied by cell
collaps. In non-hosts most colonies aborted very early without cell
collaps, but the few colonies that passed this stage grew very
slowly, the slow growth often but not always being accompanied by
cell collaps. With early abortion, caused by partial resistance,
and non-host reactions the mycelia were arrested before haustoria
were formed, in the case of hypersensitivity this was after the
formation of haustoria. Niks concluded that partial resistance and
the non-host reaction had many features in common. Carver and Carr
(1980) reported a similar observation for the oat-powdery mildew,
Erysiphe graminis f.sp. *avenae* pathosystem. Here too the main
difference between race-specific and race-non-specific resistance
is the frequency of host cell collaps accompanying the infections,
very high in the former and low in the latter type of resistance.

    Both the hypersensitivity and the partial resistance genes
appear pathogen-specific (Parlevliet, 1981a). The Pa-genes are
effective only to certain races of *P. hordei* not to any other
pathogen. Equally so is partial resistance to *P. hordei* not
effective to other pathogens not even to the related *P. striiformis*.
Vada with a high level of partial resistance to *P. hordei* is highly
susceptible for *P. striiformis*. Table 5 shows the independence of
the partial resistance to the two rusts.

    The durability of the resistance differs markedly between the
two types. Hypersensitivity genes when present in commercial
cultivars are not effective against the prevalent races (Rintelen,
1979). Of the nine identified Pa-genes only three are effective
in Europe, Pa-3, Pa-7 and Pa-9 (Parlevliet, 1981c). Only Pa-9 has
been used in commercial cultivars, Trumpf and Nadja. It losts its
effectiveness within four years (Walther, 1979; Walther and Thiele,
1979). With partial resistance the situation is different.
Parlevliet (1978b) showed, that some of the polygenes for a longer
LP occur with a high frequency in Western Europe and these genes
must have been present in European barley already for a long time
because they have not been consciously introduced or selected for
recently. As they are still effective they must be very durable.

Table 5. Level of resistance of six spring barley cultivars to *Puccinia striiformis* and *P. hordei* according to the list of recommended cultivars for England and Wales, 1979. (1 = extremely susceptible; 9 = extremely resistant).

Cultivar	*Puccinia*	
	striiformis	*hordei*
Sundance	4	6
Armelle	4	5
Mazurka	4	3
Midas	4	1
Wing	2	3
Lofa Abed	5	5

RACE-SPECIFICITY

Van der Plank (1963, 1968) assumes that VR is based on gene-for-gene systems, HR not. Parlevliet and Zadoks (1977) showed, using a simple model, that, if polygenes for resistance and polygenes for pathogenicity operate on a gene-for-gene basis, small differential interactions occur. If the trial error is óf a size similar or even larger than those of the differential interactions it is not possible to discern these interactions, and one speaks of HR and horizontal pathogenicity.

Table 6 shows this. The differential interaction (DI) characterizing VR is of the same size as the average gene effect. In case of major genes these interactions are large enabling us to recognize races easily. When the effects of the genes become smaller it becomes progressively more difficult to discern races unambiguously from one another because the DI's become smaller. When the gene effects have become so small that we cannot recognize them individually anymore (polygenes) it has also become impossible to recognize DI; they are covered up by the trial error. There are no races to be discerned and one speaks of HR. So VR and HR are in principle the extremes of a continuum, VR representing large gene effects, HR small gene effects.

Although this explains in a satisfactory way the presence of large (hypersensitivity) and small (partial resistance) DI in the barley-leaf rust and other pathosystems, it does not explain the

Table 6. Percentage of barley plants affected by loose smut
(*Ustilago nuda*) when two host loci each with two
alleles operate on a gene-for-gene basis with two
pathogen loci each with two alleles. The gene effects
are assumed to be large, each resistance gene reducing
the percentage of affected plants with 40%, or small,
each resistance gene reducing the percentage of
affected plants with 5%.

Host	pathogen							
	AB	aB	Ab	ab	AB	aB	Ab	ab
rr ss	80	80	80	80	40	40	40	40
RR ss	40	80	40	80	35	40	35	40
rr SS	40	40	80	80	35	35	40	40
RR SS	0	40	40	80	30	30	30	40

gene effects	large ⟷	small
[a]differential interactions	large ⟷	small
recognition of races	easy ⟷	not really possible
resistance	VR ⟷	HR

apparent difference in durability, major gene resistances being
often but not always elusive, polygenic resistance being durable.
It should be mentioned that race-specificity does not necessarily
need to be based on a gene-for-gene system. The production of the
avenacins in oat roots and the production of avenacinase by the *avenae*
form of take all (Table 1) might in both cases be based on several
genes. Studies carried out uptill now, however, indicate that a
simple inheritance underlies most race-specificity.

DURABILITY

Race-specificity is generally considered as evidence for the probable breakdown or erosion of resistance but this is not necessarily true. Race-specific resistance against biotrophic pathogens, generally of a major-genic and hypersensitive type, is often highly unstable, while the major-gene resistance against non-biotrophs, although race-specific too, often lasts considerably longer. Partial resistance seems even more durable like barley-leaf rust, (section 4); several other cases of slow rusting in cereals (Wilcoxson, 1981) and potato-*Phytophthora infestans* (Estrada and Turkesteen, 1979).

Typical examples of elusive major genic resistance are formed by the Pa-genes in barley to *P. hordei*, the Pm-genes in barley and wheat to *E. graminis*, the Sr, Lr and Yr-genes in wheat to *P. graminis*, *P. recondita* and *P. striiformis* respectively, the Pc genes in oats to *P. coronata*, the R genes in potato to *Phytophthora infestans*, the Pi and Xa genes in rice to *Pyricularia oryzae* and *Xanthomonas oryzae* respectively and the H genes in wheat to the Hessian fly, *Mayetiola destructor*. In all these cases large numbers of resistance genes appear to be present. Here race-specificity and absence of durability go clearly together.

Other monogenic resistances have been used in commercial cultivars for a considerable period of time without losing their effectiveness, although the corresponding races of the parasite are known. The Ht1 resistance gene in maize to northern leaf blight, *Helminthosporium turcicum*, the Nx resistance gene of potato to virus A and the Er resistance gene of apple to the woolly aphid, *Eriosoma lanigerum* are some examples. In some cases of monogenic resistance no corresponding races have yet been observed, despite a wide use of these genes (monogenic resistances in cabbage to *Fusarium oxysporum* f.sp. *conglutinans*, in cucumber to *Cladosporium cucumerinum*, and in maize to *Helminthosporium carbonum*).

The durability of this latter group of resistances is considerably greater than that of the former although at least in several cases the corresponding races have been found. With polygenic, partial resistance too small race-specific effects have been reported although the resistance appears durable as in the cases of partial resistance in barley to *P. hordei* and in potato to *Phytophthora infestans*.

Race-specificity of resistance apparently does not necessarily mean lack of durability of that resistance. Specificity and durability should be considered as two different, sometimes independent, characteristics.

PATHOGEN-SPECIFIC RESISTANCE

As mentioned before broad resistance rarely functions against
specialized parasites. The resistances against such specialists
that appear to exist (and used by the plant breeder) are
predominantly of a pathogen-specific nature, i.e. effective to one
parasite species only, like the hypersensitivity and partial
resistances in barley to *P. hordei*. The products of the host genes
apparently recognize those of the pathogen in a very specific way
resulting in the host species x pathogen species interaction. When
the genes of the host and the genes of the parasite are assumed to
operate on a gene-for-gene basis, as shown in Table 6, the race-
specific effects and the pathogen-specific nature of both types of
resistances can be explained. However, the other observations, the
remarkable difference in durability and the apparent different
modes of action as indicated by the histological data, are very
hard to explain on the basis of one gene-for-gene model where the
only difference between the two resistances is found in the size
of the gene effects. Nevertheless race-specific effects of resistance to specialized parasites are best explained with gene-for-
gene systems. This leads to the hypothesis that the two types of
resistance genes, operate on gene-for-gene systems different from
each other. That two different systems may exist is indicated by
the following:

Susceptibility and resistance to specialized parasites can be
considered to represent compatibility and incompatibility between
dissimilar, living tissues respectively. Esser and Blaich (1973)
reviewed the (in)compatibility between living tissues within
biological species. It regulates the co-existence of different
populations, different individuals and different tissues that are
in close proximity. They discerned two types of incompatibility,
homogenic and heterogenic incompatibility. With the former incompatibility occurs when identical alleles or their products meet.
All other combinations lead to incompatibility. With the latter
compatibility occurs at the confrontation of identical alleles;
incompatibility results from all other combinations. Within fungal
species both systems have been observed to exist regulating sexual
mating and vegetative hybridization respectively.

If fungi exploit two such systems to regulate the cooperation
of individuals within the biological species it is likely that they
exploit similar systems to regulate the co-existence with their
hosts. It is therefore assumed, that there are two gene-for-gene
systems, one corresponding with homogenic incompatibility, the
other with heterogenic incompatibility. In the former
incompatibility (or hypersensitivity) results from the meeting
between the product of the pathogen allele (the avirulence allele)
and the product of the host allele (the resistance allele). In the
latter *compatibility* results from the confrontation of the product

of the pathogen allele (allele for pathogenicity) and the product
of the host allele (allele for susceptibility). In both models a
gene-for-gene recognition takes place, the subsequent result of
the recognition being compatibility in the one (Model II) and
incompatibility (Model III) in the other. Table 7 shows this. In
the normal host-pathogen system where several genes may operate
and where not all gene combinations are represented it will be
very difficult to distinguish between the two models. In both
models differential interactions may occur.

The hypersensitive reaction in crops to biotrophic fungi are
considered to follow model III. Ellingboe (1979) concluded that in
the interaction between host and biotrophic parasite resistance
and avirulence appear to be positive functions i.e. the products
of functional alleles (R and A, S and B in model III of Table 7).
Compatibility arises (or is restored) when either the host or the
pathogen produce the wrong product or nothing at all (Samborski,
1978). Absence of the locus for susceptibility/resistance
(Loegering, 1978; Loegering and Sears, 1981) or of the locus for
virulence/avirulence (Flor, 1960) result in compatibility.

When the host or the pathogen produce the wrong product no
recognition for incompatibility occurs, i.e. absence of incompatibility in model III means compatibility. Incompatibility is
superimposed upon basic compatibility. And this basic compatibility
may follow model II (Ellingboe and Gabriel, 1977). The specialized
biotrophic pathogens undoubtedly employ mechanisms by which they
recognize their host. These recognition mechanisms presumably form
the key to the compatible reaction establishing pathogenicity. The
work of Ouchi et al (1974) can exemplify this. *Sphaerotheca
fuliginea*, the powdery mildew of melon, cannot infect barley,
while *Erysiphe graminis* f.sp. *hordei*, the powdery mildew of barley,
cannot infect melon. If the barley and the melon are infected by
their own, compatible, powdery mildew and the pathogen is wiped
off with moist cotton two to three days later, these infected
spots on barley and melon have become susceptible to the melon and
the barley powdery mildew respectively. The barley powdery mildew
and the melon powdery mildew are apparently capable to induce
compatibility with their own host but not with the other one. Once
compatibility is induced colonization, albeit restricted, appears
possible in the non-host.

That recognition for compatibility and for incompatibility
both occur can be concluded from the interaction between the
bacteria *Rhizobium*, *Agrobacterium* and *Pseudomonas solanacearum*
and their hosts. Compatibility (mutualism for *Rhizobium*,
pathogenicity for *Agrobacterium*) or incompatibility (avirulence
through hypersensitivity for *P. solanacearum*) results from the
interaction between structural components of bacterial and host
cell wall material (Sequeira, 1979). An example of recognition for

Table 7. Compatibility or susceptibility (+) and
         incompatibility or resistance (-) of 16 combinations
         between two alleles at each of two host loci and two
         alleles at each of two pathogen loci when the
         specific interaction is for compatibility (model II)
         or for incompatibility (Model III).

	Pathogen							
Host	Model II[y]				Model III[z]			
	AB	aB	Ab	ab	AB	aB	Ab	ab
rr ss	−	+	+	+	+	+	+	+
RR ss	−	−	+	+	−	+	−	+
rr SS	−	+	−	+	−	−	+	+
RR SS	−	−	−	−	−	−	−	+

[y]Compatibility (+) results when a and r and/or b and s meet.

[z]Incompatibility (−) results when A and R and/or B and S meet.

pathogenicity is given by Ellingboe (1981). In the oats-
Helminthosporium victoriae system the pattern of compatibility and
incompatibility appears to be the reverse of the normally observed
pattern, Model III. Here the alleles for susceptibility and
virulence seem to represent the active functions (Model II) leading
to pathogenicity. All other combinations representing absence of
recognition for pathogenicity, result in resistance. The resistance
of oats to *H. victoriae* and of virulence (toxin formation) of
H. victoriae to oats are both monogenically controlled indicating
a gene-for-gene system. The resistance has been shown to be durable.

Change from avirulence to virulence is assumed to result
from a loss of function in model III, and from a gain of function in
model II. Loss of function in model III means loss of recognition
for incompatibility, restoring basic compatibility. This is easy
to accomplish. In model II the resistance allele prevents or
hinders the recognition for compatibility. The pathogen can only
restore full pathogenicity if it can restore this recognition
reaction. This asks for a very specific, positive change not easy
to accomplish. Resistances operating according to model III are

therefore expected to be rather elusive, while resistances following model II tend to be durable.

In both models VR and HR may occur, VR representing large gene effects, HR small gene effects. Durability of resistance is a different characteristic not primarily depending on the race-specificity but on the type (model) of resistance, as shown in Table 8.

If the major resistance genes in the barley-leaf rust system (and other cereal-rust systems) follow model III, and the partial resistance genes model II the observations, summarized in the section "the barley-leaf rust system" can be explained.

Most probably there is another, third source of variation in the growth of the pathogen on or in the host. Like in any organism pathogen genotypes may vary slightly in inherent growth rates; host genotypes may vary somewhat in their suitability to sustain a good growth of the pathogen. The pathogen and the host genes involved in influencing the inherent growth rate of the pathogen in a given host are most likely non-specific in nature. There are some indications for it. The barley cultivar L94 does not seem to carry any genes for an increased latent period (LP) to *P. hordei*. If all partial resistance and partial pathogenicity follows the model II gene-for-gene system L94 would have an LP that is identical for all leaf rust isolates. This is not the case. With different isolates small differences in LP were observed on L94 (Parlevliet, 1976).

Some of the differences in the observed partial resistance could be due to non-specific effects of this nature. Most of the partial resistance, though, cannot be explained with such non-specific effects because partial resistance is predominantly pathogen-specific as discussed before. Host cultivars that are partially resistant to one *Puccinia* species can be highly susceptible to a closely related *Puccinia* species. A cultivar that would be a rather poor substrate for one *Puccinia* species, because of non-specific factors, would be a rather poor substrate for a related species as well.

EVOLUTIONARY SIGNIFICANCE OF SPECIALIZATION

Specialization is not only a process forced upon the parasite by the host evolving defense mechanisms, it is also a process pursued by the parasite to avoid competition from similar parasites on the same hosts. The partially sympatric seed head weevils, *Larinus sturnus* and *L. jacea* exemplify this. Their hosts belong to the genera *Arcticum*, *Carduus*, *Centaurea* and *Circium*. In the areas of overlapp they have specialized on species in different genera (Zwölfer, 1970). Also the often remarkable specialization of the passion vine butterflies (*Heliconiini*) in terms of *Passifloraceae*

Table 8. Type and durability of resistance in relation to
the size of the gene effects when the host and
pathogen operate on a model II or model III
gene-for-gene basis.

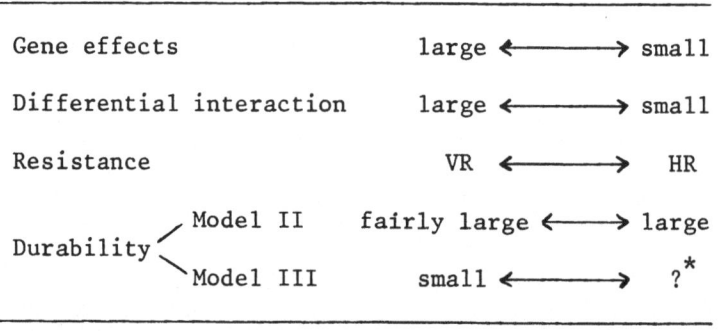

*Does this exist?

host species and host tissues strongly suggests mutual avoidance
of competition (Benson, 1978). Specialization, especially when
evolved to avoid competition, means the development of mechanisms
by which the parasite can find and recognize the right host tissues
to be parasitized. Animal parasites use their sensorial capacities,
pathogens have to employ biochemical means in order to recognize
their host tissues. The induction of basic pathogenicity as
exemplified by the powdery mildews of barley and melon
may represent this recognition system. By interfering with it
(partial resistance) the host can resist the specialized parasite.

Specialization creates another problem; dependence on the host.
Biotrophs like the cereal rusts and powdery mildew have specialized
to such an extent that they became totally host-dependent.
Endangering the host endangers the pathogen. Because the host plant
competes with other plants, a slight reduction in fitness and thus
in competitive ability could result in a serious decline of the
host population. A biotroph therefore can affect the fitness of
its host considerably and so its own fitness. The fitness of the
pathogen population is greatest at some intermediate level of
virulence. More virulence could result in less host and so in a
smaller pathogen population. Less virulence leaves some of the
niche unexploited. It is assumed that the elusive hypersensitive
type of host-pathogen interaction is exploited by the pathogen as
a feed-back mechanism to regulate the mean virulence level of the
population. Genetically the pathogen is adapted to shift easily
from compatibility to incompatibility and vice versa. And the
plant breeder has selected especially these incompatibility genes
in the host to protect his crops; without lasting success of course.

Not only may too high levels of virulence endanger the host population, large fluctuations in virulence can equally do so. A system that regulates and stabilizes the level of pathogenicity at an intermediate level would be selected, because populations carrying such a system would have a better chance to survive in the long run. The gene-for-gene relationship for incompatibility (Model III) provides such a system. With a substantial number of different avirulence alleles at varying frequencies, the pathogen population has a feed-back system that keeps the "mean population virulence" equilibrated around an intermediate level. Within the host population the frequencies of R-genes tend to increase because they confer an advantage to the plants carrying them, but when the frequencies of R-genes become too high, i.e. the pathogen population too small, the frequencies of the corresponding virulence alleles increase to restore the balance.

There are some problems attached to this feed-back hypothesis.
i) Does virulence of a pathogen tend to equilibrate towards moderate levels?
ii) If so, how does selection for lower virulence operate because the avirulence allele lowers the fitness of its individual carrier, even if it enhances the fitness of the population as a whole.
iii) In Model III the change from avirulence to virulence is assumed to be easy because a loss of function is involved. The reverse would be a gain of function and therefore not easy to perform. This is not what one expects from a feed-back mechanism. On the contrary the change should be relatively easy both ways. The scarce information available indeed suggests that the change to avirulence is not more difficult to accomplish for biotrophic fungi than the change to virulence.

i) Selection for moderate levels of virulence does occur. The balance between the Myxoma virus with its rabbit host is a good example (Fenner, 1965). The Brazilian Myxoma virus, which is highly pathogenic, was introduced into the Australian rabbit population in 1950, and it killed 99.5 percent of the rabbits. By 1958, this strain was completely replaced by less pathogenic ones, and the rabbits had become moderately resistant. This co-evolution for a decreased pathogenicity and an increased resistance, which resulted in a survival rate of over 75 percent in the rabbit population, occurred in only eight years. The Phleum mottle virus in Wales (Catherall and Chamberlain, 1977) provides another example. Many grasses were symptomless carriers for the mild common strain of the virus. The strain that caused severe symptoms was quite rare. Pimentel (1968), allowed the house fly, *Musa domestica*, and the parasitic wasp, *Nasonia vitripennis*, to interact over a large number of generations. The host population became more resistant, the parasite population less virulent while the fluctuations in population numbers decreased significantly for both the host (at increased numbers) and the parasite (at decreased numbers).

ii) Although moderate levels of virulence do indeed enhance the co-existence of pathogens with the hosts they depend on, it is not easy to explain the evolutionary processes leading to such a situation. The allele for avirulence may enhance the fitness of the population as a whole, it certainly lowers the fitness of the individual that carries it. And because evolution is not an anticipatory process the Darwinian selection, favouring the fittest individuals, cannot explain the presence of such altruistic traits. Many authors have dealt with the problem of how altruistic traits, which occur quite commonly, can be selected for. Kinship and group selection are generally considered to be the forces behind the selection of such traits. Kin selection is not expected to be of any importance in pathogens, because they do not normally occur in kinship groups nor is there any kinship recognition known to operate. Group selection on the contrary might play a significant role. It is considered as an extension on the group (deme) level of the three Darwinion principles of natural selection that operate among individuals within demes (Uyenoyama and Feldman, 1980). These principles (for demes) are: i) Phenotypic variation among demes. ii) Continuity in phenotypes of demes between generations. iii) Different demes have different reproductive success. Group selection therefore is the process by which certain demes make a greater genetic contribution to the next generation than other demes do. The ideal conditions for group selection to occur consist when the biological species occupies a variable and flunctuating environment by means of many smaller and larger demes between which a restricted gene flow exists (Cohen and Eshel, 1976; Ayenoyama, 1979). These conditions probably exist with many specialized pathogens. They ensure genetic variation among demes for altruistic traits such as avirulence through genetic drift (occurring in small demes, in larger demes that go through bottle necks between seasons, and when demes are re-established from a few founders) and differential and fluctuating selection pressure.

The rabbit-Myxoma virus system in Australia is a good example. For the virus the rabbits form hospitable islands that emerge from and sink into a hostile ocean. The islands are generally reached by a few virus particles only, the founders. These founders may vary in virulence. When an island is colonized by highly virulent founders it will be over-exploited and destroyed rapidly. Less virulent founders exploit the islands better and over a longer period, allowing for more emigrants to leave such islands. New islands therefore are more often reached and colonized by the less virulent types than by the more virulent ones. When founders of varying virulence level reach a new island, the most virulent one tends to become the dominating type. The selection between founders within an island favouring the more virulent types is of the individual or Darwinian type. The selection between groups or demes (the virus population on one island or rabbit is a group or deme) favouring the less virulent groups represents group selection. Group selection is enhanced by the phenomenon of cross protection that prevents the establishment of viruses once a related one is already present. This reduces the chance of mixed populations on one island and so the individual selection for virulence.

The two opposing forces, group selection favouring demes with relatively high frequencies of avirulence alleles and Darwinian selection favouring within a deme individuals carrying the virulence alleles form the genetic feed back mechanism. They tend to equilibrate, in dependence on the frequencies of the host incompatibility alleles, around a virulence level that allows a vigorous host population and so an equally vigorous parasite population over the years.

iii) In Model III, where the interaction occurs between the products of the avirulence and of the resistance allele, the change from virulence to avirulence would be difficult because it represents a change toward a positive function. In a feed back system as assumed above the change from virulence to avirulence should be about as easy as its reverse. It is therefore assumed that the difference between the products of the avirulence and of the virulence allele consists of a certain small change, easily reversible. In the one condition the product is recognized by the product of the host allele (incompatibility), in the other condition it is not (compatibility).

CONSEQUENCES FOR MODERN AGRICULTURE

The co-existence of host and specialized pathogen, regulated by the gene-for-gene system for incompatibility, operates satisfactorily in nature, where the genetically heterogeneous host and pathogen populations are exposed to a highly variable environment. In modern agriculture, a totally different situation, the genotypic variation of the host population has been reduced enormously. Plant breeders, when breeding for resistance to biotrophic pathogens, have selected and used the incompatibility genes rather than the genes that confer resistance to basic pathogenicity. These genes have been exposed to the pathogen one by one. Because they were not meant to protect the host but to prevent the pathogen from becoming too aggressive, it is not surprising that these genes, when not embedded in the natural system, capitulated to the pathogen population quickly. The manifestation has been the frequent loss of resistance and the rapid evolution of pathogen races. The finely tuned balance existing in natural host-pathogen populations got lost: The pathogen population has been allowed to specialize to a high degree on a host made uniform and abundant by man. The pathogen has become more virulent because this no longer endangers the host because man multiplies it.

To increase the durability of resistance to such specialized
parasites man must choose between the more durable partial resistance and the elusive hypersensitivity. The latter genes should
only be used embedded in a system that retards or prevents the
racial development in the parasite population.

REFERENCES

Benson, W., 1978, Resource partioning in passion vine butterflies,
     Evolution, 32:493-518.
Carver, T.L.W. & Carr, A.J.H., 1980, Some effects of host resistance
     on the development of oat mildew. *Ann.Appl.Biol.*, 92:290-293.
Catherall, P.L. & Chamberlain, J.A., 1977, Relationships, host
     ranges and symptoms of some isolates of Phleum mottle virus,
     Ann.Appl.Biol., 87:145-157.
Clifford, B.C., 1974, Relation between compatible and incompatible
     infection sites of *Puccinia hordei* on barley, *Trans.Brit.Myc.
     Soc.*, 63:215-220.
Cohen, D. & Eshel, J., 1976, On the founder effect and the
     evolution of altruistic traits, *Theor.Pop.Bio.*, 10:276-302.
Ellingboe, A.H., 1979, Inheritance of specificity: the gene-for-
     gene hypothesis. In:"Recognition and Specificity in plant
     host-parasite interactions", eds. J.M. Daly and I. Uritani,
     Univ. Tokyo Press, 3-17.
Ellingboe, A.H., 1981, Changing concepts in host-pathogen genetics,
     Ann.Rev.Phytopathol., 21:125-143.
Ellingboe, A.H. & Gabriel, D.W., 1977, Induced conditional mutants
     for studying host/pathogen interactions. In: Induced mutations
     against plant disease, IAEA (Vienna),:35-46.
Esser, K. & Blaich, R., 1973, Heterogenic incompatibility in plants
     and animals, *Adv. in Genetics*, 17:107-152.
Estrada, N. & Turkesteen, L., 1979, Breeding for resistance to
     late blight at CIP. Reports of the planning conference on the
     control of important fungal diseases of potatoes, CIP, Lima,
     Peru, 1978:57-63.
Fenner, F., 1965, Myxoma virus and *Oryctolagus cuniculus*; two
     colonizing species. In:"Genetics of colonizing species", eds.
     Baker, H.G. and Stebbins, G.L. Academic Press, New York and
     London, 485-499.
Flor, H.H., 1960, The inheritance of X-ray induced mutations to
     virulence in a urediospore culture of race 1 of *Melampsora
     lini*, *Phytopathology*, 50:603-605.
Gilbert, L.E., 1975, Ecological consequences of a coevolved
     mutualism between butterflies and plants, *Proc. First Int.
     Congr.Syst. & Evoln.Bio.*, Boulder, Colorado, 5:210-240.
Harper, J.L., 1977, Population biology of plants. Academic Press,
     London, New York, San Francisco.
Loegering, W.Q., 1978, Current concepts in interorganismal
     genetics, *Ann.Rev.Phytopath.*, 16:309-320.

Loegering, W.Q & Sears, E.R., 1981, Genetic control of disease
    expression in stem rust of wheat, *Phytopathology,*71:425-428.
Lüning, H.U., Waiyaki, B.G. and Schlösser, E., 1978, Role of
    saponins in antifungal resistance. VIII. Interactions
    Avenae sativa - Fusarium avenaceum, Phytopath.Z., 92:338-345.
Neervoort, W.J. & Parlevliet, J.E., 1978, Partial resistance of
    barley to leaf rust, *Puccinia hordei.* V. Analysis of the
    components of partial resistance in eight barley cultivars,
    Euphytica, 27:33-39.
Niks, R.E., 1982a, Haustorium formation of *Puccinia hordei* in
    hypersensitive, partially resistant and non-host genotypes,
    Phytopathology, submitted.
Niks, R.E., 1982b, Comparative histology of partial resistance and
    non-host reaction in barley seedlings to leaf rusts,
    Phytopathology, submitted.
Niks, R.E. & Kuiper, H.J., 1982, Histology of the relation between
    minor and major genes for resistance of barley to leaf rust,
    Phytopathology, submitted.
Ouchi, S., Oku, H., Hibino, C. & Akiyama, I, 1974, Induction of
    accessibility to a non pathogen by preliminary inoculation
    with a pathogen, *Phytopathologisch Zeitschrift,* 79, 142-154.
Parlevliet, J.E., 1976. Evaluation of the concept of horizontal
    resistance in the barley/*Puccinia hordei* host-pathogen
    relationship, *Phytopathology,* 66:494-497.
Parlevliet, J.E., 1978a, Race-specific aspects of polygenic
    resistance of barley to leaf rust, *Puccinia hordei, Neth.J.Pl.
    Path.,* 84:121-126.
Parlevliet, J.E., 1978b, Further evidence of polygenic inheritance
    of partial resistance in barley to leaf rust, *Puccinia hordei,
    Euphytica,* 27:369-379.
Parlevliet, J.E., 1979, Components of resistance that reduce the
    rate of epidemic development, *Ann.Rev.Phytopath.,* 17:203-222.
Parlevliet, J.E., 1980, Minor genes for partial resistance epis-
    tatic to the Pa7 gene for hypersensitivity in the barley-
    Puccinia hordei relationship, *Proc.5th Eur. & Medit. Cereal
    Rusts Conf.,* Bari,:53-57.
Parlevliet, J.E., 1981a, Race-non-specific disease resistance, in:
    "Strategies for the control of cereal disease", J.F. Jenkyn &
    R.T. Plumb, eds., Blackwell Scient.Publ., Oxford, 47-54.
Parlevliet, J.E., 1981b, Crop loss assessment as an aid in the
    screening for resistance and tolerance, Manual on crop loss
    assessment methods, Suppl. 3, FAO/Commonwealth Agr. Bureaux,
    Rome/London, 111-114.
Parlevliet, J.E., 1981c, Stabilizing selection in crop pathosystems:
    an empty concept or a reality?, *Euphytica,* 30:259-269.
Parlevliet, J.E., 1981d, Disease resistance in plants and its
    consequences for breeding, In:"Plant Breeding II", K.J. Frey,
    ed., The IOWA-State Univ. Press, Ames, Iowa, 309-364.

Parlevliet, J.E. & Ommeren, A. van, 1975, Partial resistance of barley to leaf rust, *Puccinia hordei*. II. Relationship between field trials, micro-plot tests and latent period, *Euphytica*, 24:293-303.

Parlevliet, J.E. & Zadoks, J.C., 1977, The integrated concept of disease resistance; a new view including horizontal and vertical resistance in plants, *Euphytica*, 26:5-21.

Pimentel, D., 1968, Population regulation and genetic feedback, *Science*, 159:1432-1437.

Rintelen, J., 1979, Verfügen unsere Gerstensorten über spezifischen Resistenzen gegen physiologische Rassen der Zwergrostes?, *Bayer. landwirtsch Jahrbuch*, 56:391-397.

Samborski, D.J., 1978, Concepts dealing with specificity in host-parasite systems, *Proc. 3rd Int.Congr.Plant Path.*, München, 1978, Abstract:220.

Sequeira, L., 1979, Recognition between plant hosts and parasites, In "Host-parasite interfaces", B.B. Nichol, ed.,Acad. Press, New York, San Francisco, London, 71-84.

Turner, E.M.C., 1961, An enzymic basis for pathogenic specificity in *Ophiobolus graminis*, *J.Exp.Bot.*, 12:169-175.

Uyenoyama, M., 1979, Evolution of altruism under groupselection in large and small populations in fluctuating environments, *Theor.Pop.Biol.*, 15:58-85.

Uyenoyama, M. & Feldman, M.W., 1980, Theories of kin and group selection: a population genetics perspective, *Theor.Pop.Biol.*, 17:380-414.

Van der Plank, J.E., 1963, Plant diseases: Epidemics and control, Academic Press, New York & London.

Van der Plank, J.E., 1968, Disease resistance in plants, Academic Press, New York & London.

Walther, U., 1979, Die Virulenz- und Resistenzgensituation bei *Puccinia hordei* West, *Archiv fur Zuchtungsforschung* 9:49-54.

Walther, U. & Thiele, M., 1979, Zur Rassen situation beim Zwergrost, *Puccinia hordei* Otth., in der DDR. *Tagungsber. Akad.Landwirtsch.Wiss.DDR*, Berlin, 175:67-71.

Wilcoxson, R.D., 1981, Genetics of slow rusting in cereals, *Phytopathology*, 71:989-993.

Zwölfer, H., 1970, Der regionale Futterpflanzwechsel bei phytophagen Insecten als evolutionares Problem, *Z.Ang.Ent.*, 65:233-239.

DISCUSSION

DINOOR: Concerning complete resistance, you assumed that hypersensitivity is the cause of resistance, the death of the host cells preventing further growth of, for example, the rust colony. We have shown that crown rust haustoria in a resistant oat plant die first and only those cells in which the haustoria had died were necrotized. This indicates that hypersensitivity is a result rather than a cause of resistance. How will this influence your distinction between models II and III ?

PARLEVLIET: In my models, it is not assumed that hypersensitivity is the cause of resistance. In my models there are two independent types of resistance: partial resistance and hypersensitivity. The latter is characterized by cell collapse. I also assumed that the hypersensitivity is not a defense mechanism operated by the host but a feed back mechanism operated by the pathogen to regulate its virulence around some intermediate value. Your observation is certainly not in contradiction with this feed-back hypothesis.

PERSON: Do you know how the reduced virulence in the rabbit myxoma virus in Australia evolved?

PARLEVLIET: Yes. Let us consider rabbits to be individual islands to be colonized by the virus and let us assume that the colonization has to be carried out by founders (carried by the mosquito vector) from already colonized islands. If one island is colonized by a highly aggressive strain and another by a mild strain, then the former has over-exploited its island far more quickly than the latter, resulting in a reduced chance to colonize new islands. This will favour milder strains against more aggressive ones.

SHARP: Do the yield and yield components of a partially resistant cultivar differ from those in a cultivar without partial resistance? Have you obtained quantitative information on your second model? The quadratic check may be used for model III but what about model II ?

PARLEVLIET: There is no obvious association between yield or its components and partial resistance. High yielding cultivars exist with little and with a lot of partial resistance. Results of detailed research are unknown to me. Concerning the second part of your question, one could make the quadratic check:

	Pathogen	
Host	A	a
r r	−	+
R R	−	− ,

recognition being between r and a. A gives non-pathogenicity and will not occur, the quadratic check thus being hypothetical. If a mutates to a' making recognition of R possible, it would produce:

	Pathogen				Pathogen	
Host	a	a'		Host	a	a'
r r	+	+	or	r r	+	−
R R	−	+		R R	−	+

In the second case a' does not recognize r. More likely, a second locus becomes involved, thus not a⟶a' but aB⟶ab.

	Pathogen	
Host	aB	ab
r r	+	+
R R	−	+

The difference of the quadratic check of model III resides in the genotypes of the pathogen, which is difficult to trace in the experiment.

KRANZ: Tolerance may mean reduced symptoms and is then identical to resistance. If it is used to describe reduced damage, then only leaves are diseased that no longer contribute to yield production; this may provide high levels of inoculum for neighbouring, less tolerant, varieties or lead to a pathogen build-up, which may backfire in the future.

PARLEVLIET: I agree with your first comment; in virology tolerance is often used to mean resistance. Concerning tolerance as reduced damage, if it exists, it is too difficult to select for in a breeding programme.

PATHOGEN FITNESS IN CEREAL MILDEWS

M.S. Wolfe,* J.A. Barrett† and S.E. Slater*

*Plant Breeding Institute
 Trumpington, Cambridge, CB2 2LQ, England
†Present address:
 Genetics Department, University of Liverpool
 Liverpool, L69 3BX, England

SUMMARY

The opposed processes of selection for adaptation towards a particular host, and of selection for adaptability towards the range of
hosts, affects the structure of populations of the powdery mildew
pathogen developing on barley varieties with different resistance
genes grown in a region. In the UK, from 1975 to 1981, on a relatively large range of host varieties, it was found that the distribution
of non-matching pathogenicity genes was similar on all hosts. The
values obtained were correlated with the frequencies of the pathogenicity genes found in the air spora, and with the contribution of
each host variety to the overall infection of the barley crop. On
some recently introduced varieties, however, certain combinations of
pathogenicity genes occurred less commonly than expected, revealing
interactions between some host resistance genes and particular non-
matching pathogenicity genes.

Despite these deviations, selection against non-matching pathogenicity genes generally appeared to be slight and of similar magnitude on all hosts. Nevertheless, complex races did not appear to
increase in frequency during the period. Reasons for this apparent
lack of response to selection for adaptability probably included the
large size of the pathogen population on individual varieties, and
the greater efficiency of selection for adaptation.

If a single variety is grown on a large proportion of the crop
area there may be a rapid response in the pathogen to selection for
adaptation to it. The response will be limited if the variety has
inherent durable resistance. The probability of a pathogen response

will be further reduced, however, if the variety is used in a way
that provides system durability, for example, by exposing different
varieties in sets of mixtures whose composition is regularly changed.
Durability of such a system will be improved by using varieties that
have inherently durable resistance.

INTRODUCTION

Cereal powdery mildews (*Erysiphe graminis* f.spp.) consist of
large populations of haploid individuals that can exhibit a wide
range of variation in host pathogenicity. Each population, whether
in the natural or agricultural ecosystem, is subject to two major,
but opposed, forms of selection, for adaptation and for adaptability.
For *adaptation* to a single host, particularly in a single environment, there is presumably only one ideal pathogen phenotype, although
this may be provided by a limited number of different genotypes. It
is unlikely that this range of genotypes will provide perfect adaptation to other hosts. For *adaptability* to a fixed range of hosts in a
single environment, the required pathogen phenotype may again be
derived from a number of genotypes, but it is likely that a considerably different range of genes will be required compared with those
providing specific adaptation.

Fitness in the pathogen, which is the ability to leave survivors, will therefore be determined largely by the structure of the
host population, and the limitations on pathogen reproduction caused
by the nature of the organism and by selection for adaptation and
adaptability.

Wolfe and Schwarzbach (1978) described this evolutionary dilemma
for cereal mildews, pointing out that selection for flexibility or
adaptability occurred for one or two cycles only at the beginning of
the epidemic, acting on the airborne spore population drifting and
settling over emerging crops. At this early stage, the migrant spore
population in the atmosphere is small, but its influence in determining the structure of the initial pathogen population is relatively
large, because infections within crops have not yet developed.
Subsequently, as the infections within the crop sporulate and spread,
their numbers quickly exceed those of immigrant spores whose influence diminishes and can be disregarded within a short period.

During selection for adaptability, the relative fitness of different pathogen genotypes is a simple function of the pathogenicity
genes carried and the numbers and relative areas of hosts that these
genes allow them to establish upon, e.g.

$$wi = g \frac{m}{n} (1-s)^{m-1}$$

where wi is the fitness of the ith pathogen genotype, and m is the number of pathogenicity genes it possesses among the required set of n, where s, the selection coefficient, is the same for all non-matching loci. g is the proportion of the crop area occupied by matching hosts (assuming all hosts to have only single gene resistances).

Following establishment, selection for adaptation now operates within the crop and this may proceed for as many as 10 to 15 cycles. Considering an isolated host j, and using the model provided by Barrett (1980), the numbers of a pathogen genotype i under exponential increase for n generations would be

$$N_{i_n} = N_{i_o} \cdot wi \, (1 + \alpha)^n$$

where N_{i_o} is the initial number of pathogen genotype i, wi is its

relative fitness on host j (=(1-s)), and (1 + α) represents the effective reproduction rate.

The value wi confounds relative establishment and relative growth rate of different genotypes; these were separated in an essentially similar model by Østergaard (1982).

For simulations using his model, Barrett (1980) assumed that fitness was multiplicative, and that non-matching at any locus reduced the fitness of the pathogen by equal amounts. The most fit pathogen genotype that becomes prevalent at the end of the epidemic will thus be determined by

a) the range of pathogen genotypes infecting the host at the
   beginning of the epidemic
b) their relative numbers at the beginning of the epidemic
c) their relative rates of reproduction
d) the number of generations that occur

More recently, Skylakakis (1980) pointed out that the relative rate of replacement of one pathogen genotype by another within the crop will be different from that suggested by the model because pathogen increase is not exponential, except in the very early stages of epidemic development (Vanderplank, 1963; see also Leonard & Czochor, 1980). Pathogen development is subject to logistic regulation and there may also be competition for available infection sites. Skylakakis (1980) and the second author (JAB) showed that under these more realistic conditions, replacement of one genotype by another would be slower. The relative fitness of pathogen genotypes may also change during the course of the epidemic, for example, if their apparent infection rates change differentially, or if the carrying capacity of the host for different pathogen genotypes varies, or if the genotypes are affected by competition.

In a constant environment with an unchanging host distribution from season to season, an unstable equilibrium may develop between pathogen genotypes highly adapted to individuial hosts (and which therefore have the highest individual growth rates), those with extreme flexibility, and forms intermediate between these extremes. The equilibrium can only be produced under these extremely limited conditions (Barrett, 1980): any perturbation, for example, of host distribution (which could occur simply by variation in host genotype-environment interactions) will change the relative fitness of the pathogen genotypes and therefore their relative frequencies. If very few hosts predominate, the disease will become most severe because of the selective advantage for highly adapted pathogen genotypes, i.e. those with the fastest growth rates. If there are many different hosts, although all may become infected by prevalent widely-adapted genotypes of the pathogen, the infection level may be less because of the slower growth rates of such pathogen genotypes individually on each cultivar.

A SIMPLE MODEL OF PATHOGEN POPULATION STRUCTURES

Given that there are several different resistant varieties in use simultaneously, let us assume that the pathogenicity genes are independent, that each is subject to a similar selective disadvantage on all non-matching hosts, and that the sizes of the selective disad-vantage associated with each gene is similar. Let us further assume that there are equal areas of cvs. R1, R2, R3, R4, all equally infec-ted. Cv. R1 will thus receive spores from R2, R3, R4... in equal proportions. Amongst these spore populations there will be equal proportions of pathogen genotypes with combined pathogenicity for R1R2, R1R3, R1R4... These will increase to a particular frequency on cv. R1 dependent on the selection coefficient. A similar process will occur on cvs. R2, R3, R4... and this will generate a symmetrical pattern as shown in Table 1a.

From field data, we might expect to observe deviations from this pattern which can be classified into three groups. The first, (Table 1b) occurs where a particular pathogenicity gene has a general cost to the pathogen, which is greater than for other genes, e.g. p1 in this example. The second group (Table 1c) occurs where a particular host imposes a more severe constraint on the pathogen than do other hosts, as with R1. The third group (Table 1d) is the result of specific interactions between host and pathogen genes, as for example with p2/R1 and p1/R2; specific but non-reciprocal interactions might also occur.

The implication of the occurrence of these kinds of deviation is that correlated responses can occur between host resistance genes and non-matching pathogenicity genes, which might be general or specific. This aspect has often been overlooked in consideration of the gene-

Table 1, a-d.   Models of pathogen population structures on a range of
                host varieties with different resistance genes. The
                numbers represent relative frequencies of matching and
                non-matching pathogenicity genes on the host range

a Equal selection against all non-   b Greater selection against
   matching pathogenicity genes   one non-matching pathogeni-
                                           city gene on all hosts

Resistance genes	Pathogenicity genes					Resistance genes	Pathogenicity genes			
	p1	p2	p3	p4...			p1	p2	p3	p4..
R1	100	30	30	30...		R1	100	30	30	30..
R2	30	100	30	30...		R2	10	100	30	30..
R3	30	30	100	30...		R3	10	30	100	30..
R4	30	30	30	100...		R4	10	30	30	100..
.	.	.	.	.		.	.	.	.	.
.	.	.	.	.		.	.	.	.	.
.	.	.	.	.		.	.	.	.	.

c All non-matching pathogenicity   d Specific interaction involv-
   genes at a greater disadvantage   ing two non-matching patho-
   on one host than on others   genicity genes

Resistance genes	Pathogenicity genes					Resistance genes	Pathogenicity genes			
	p1	p2	p3	p4...			p1	p2	p3	p4..
R1	100	10	10	10...		R1	100	10	30	30..
R2	30	100	30	30...		R2	10	100	30	30..
R3	30	30	100	30...		R3	30	30	100	30..
R4	30	30	30	100...		R4	30	30	30	100..
.	.	.	.	.		.	.	.	.	.
.	.	.	.	.		.	.	.	.	.
.	.	.	.	.		.	.	.	.	.

for-gene system, although it could be important in considering strategies for the use of host resistances and fungicides.

    The values given in Table 1 are relative and it is unlikely that
the pathogen will have progressed to equal pathogenicity simultaneously on a series of hosts, or, indeed that it is capable of doing
so. Consequently we may also expect differences to occur in the
values for matching pathogenicity along the diagonal in Table 1. If
a pathogenicity gene becomes fixed in the whole population, we may
expect it to be equally pathogenic on all varieties.

OBSERVED POPULATION STRUCTURES IN BARLEY POWDERY MILDEW IN THE UK

During the period 1978-80, Wolfe and Slater (1981) analysed populations of mildew obtained from nine different groups of barley cultivars distinguished by differences in their resistance genes (Table 2).

Table 2. BMR (Barley Mildew Resistance) groups of varieties used in the surveys, and the resistance genes that identify each group

BMR Group	Common resistance gene	Test varieties
0	–	Golden Promise
1	Mlh	37/136, 41/145, Astrix
2	Mlg	Goldfoil, Julia, Union, Zephyr
3	Mla6	Maris Concord, Midas
4	'Mlv'	Lofa Abed, Vada
5	Mla12	Hassan, Sultan
6	Mla4/7	H.1063, Ark Royal*, Tern†, Wing
7	Mla	Algerian, Tyra
8	Mla4/9	Akka

* used in 1980 only
† used in 1979 and previously

Isolates of the pathogen were obtained from field plots and crops of the test cultivars and others containing recognised resistance genes. They were inoculated in a settling tower on to the set of test cultivars, using the susceptible cv. Golden Promise as a control. After eight days incubation under controlled conditions (12 h day, 15°C day and night temp., approx. $50\mu E/m^{-2}/sec^{-1}$ light intensity: all cultures maintained and tested on detached seedling leaf segments on 0.5% agar containing 150 ppm benzimidazole), the colony numbers were counted by eye. The numbers were expressed relative to those on the appropriate Golden Promise control (= 100) to eliminate differences in inoculum density and condition between tests. Values for different cultivars within each test group were averaged. Test data obtained from different cultivars within each source group were also averaged. The final table (Table 3) presents the average values for the three years since there was little variation between years.

The values for pathogenicity confound different pathogenicity values for different clones in each population with their frequencies: they indicate the relative amounts of disease that might be caused by each population. The matching pathogenicity values, however, can be considered as maximum values caused by individuals that have the same pathogenicity as each other. Dividing each non-matching pathogenicity value by the appropriate matching pathogenicity value thus gives a closer indication of the frequency of the pathogenicity genes on non-matching hosts.

Table 3. Averaged values (1978-80) for matching (underlined) and non-matching pathogenicity of pathogen populations obtained from each of the major BMR resistance groups used in the UK

Population source group	BMR test group							
	1	2	3	4	5	6	7	8
0	47	67	41	19	28	33	2	1
1	42	67	24	13	21	42	3	0
2	40	76	45	21	27	19	4	2
3	45	62	73	19	33	32	1	0
4	29	61	22	66	23	3	2	0
5	40	75	15	13	67	9	0	0
6	47	74	20	5	15	87	0	4
7	42	71	19	6	41	22	74	2
8	56	64	3	3	16	82	0	85
Means	43	69	29	18	30	31+	10	10
non-matching / matching	1.02	0.89	0.33	0.20	0.42	0.26+	0.01	0.01

+ excludes value for pathogenicity for BMR 6 in populations from BMR 8

The diagonal values for matching pathogenicity, relative to pathogenicity on Golden Promise, varied from the lowest, 42, for BMV 1** on BMR 1 , to 87 for BMV 6 on BMR 6. Thus the best adapted genotypes of the pathogen on BMR 1 were able to produce only half as many colonies on their 'own' host as those of BMR 6 on their 'own' host, under the experimental conditions. However, this test only measures the ability to produce colonies in a single cycle; spore production and other features of infection, dispersal and survival may vary in other directions between the host cultivars and their matching pathogen populations.

**Barley Mildew Virulence factor 1; Barley Mildew Resistance factor 1

Variation between columns in Table 3 (corresponding to Table 1b) was considerable, but probably due to differences in the relative use of the host cultivars rather than to differences in inherent select- ive disadvantage between pathogenicity genes. BMR 0 cultivars have been used for many years, and BMR 1 and 2 were introduced into European barley cultivars in the 1930's; BMR 7 and 8, on the other hand, were introduced only recently. Consequently, because of the longer period and larger scale of usage of the earlier introduced resistance genes, there has been greater selection on the pathogen for adaptation to them.

The approximate correlation between each row and the row of column means suggests that selection against non-matching pathogeni- city gene is similar on each host resistance group. Values for BMR 4 tend to be lower than average both in rows and in columns, even though this resistance group has been in use for a long time and on a large scale.

To explore the occurrence of specific variation within Table 3, an analysis was made of the variation in occurrence separately of each pathogenicity gene on the range of hosts sampled, i.e. the within-column variation. Compared with the column mean, the single large value for matching pathogenicity should deviate positively. Where the pathogenicity gene is non-matching it will deviate nega- tively. If the sizes of the negative deviations are similar, then selection against the pathogenicity gene is similar on all non- matching hosts. Differences in size of negative deviations will indicate interactions between resistant varieties and non-matching pathogen races, an indication of linkage disequilibrium (Wolfe & Knott, 1982). The deviations were consistent between years, irrespective of the range of varieties within each BMR group from which isolates were collected. This suggests that the variations in negative deviations that did occur, were due to interactions between the resistance genes and non-matching pathogenicity genes.

The nature of the data makes it difficult to devise a rigorous statistical test for the significance of the deviations. Assessment of significance was therefore based on their reproducibility. The data for each of the years 1978-80 were analysed separately, and it was found that there was little variation between years. Further, an analysis of earlier data for 1975-77, obtained in a slightly differ- ent way, revealed a pattern closely similar to the more recent data. For simplicity, the accumulated deviations for 1978-80 are given in Table 4.

Deviation values greater than 32 are almost certainly signifi- cant, while those between 27 and 32 may be somewhat less reliable. Deviations less than 16 are almost certainly non-significant. Within the general pattern, positive deviations for non-matching genes are uncommon, and two of the five involve the longer established resist-

ance genes. Of the 16 negative deviations, the majority involve
pathogenicity for at least one of the more recently introduced
resistance genes. This may indicate that the pathogen has not yet
undergone sufficient selection to overcome conflicting requirements
on particular combinations of those hosts. In populations from BMR
4, all non-matching values were negative indicating that this group
was particularly restrictive on the pathogen population, as in the
example of R1 in Table 1c.

Table 4. Deviations from column means in Table 3 summed for the
          three years 1978-80. Matching pathogenicity deviations
          are boxed. Test groups 7 and 8 omitted because of the
          small size of the original values. Values considered
          significant are underlined.

Population source group	BMR test group					
	1	2	3	4	5	6
0	+13	-4	+35	+3	-6	+6
1	-14	+12	-18	-17	-28	+32
2	-8	+15	+47	+8	-9	-36
3	+5	-15	+137	+5	+10	+2
4	-42	-22	-23	+142	-21	-84
5	-10	+19	-43	-16	+110	-66
6	+11	+18	-29	-40	-46	+171
7	-4	+7	+7	-30	-38	-27
8	+38	-13	-80	-46	-41	-*

* omitted from calculations

A more detailed survey of the deviations is as follows:

a) Positive deviations

1. Pathogenicity for BMR 1 was similar for isolates from most varieties; only populations from BMR 8 were more pathogenic than others.
This may be related to the higher than average pathogenicity of BMR 6

on BMR 1, since BMR 6 and BMR 8, which have one gene in common, share some common characteristics.

2.   Pathogenicity for BMR 3 was higher than average among populations from BMR 0 and BMR 2.   The first deviation probably occurred because most of the BMR 0 isolates were from Golden Promise grown in the north of England and Scotland.   Next to Golden Promise itself, Midas (BMR 3) has been the most popular variety in that area so that selection for combined pathogenicity on these two groups has been considerable.

The high value for pathogenicity for BMR 3 among isolates from BMR 2 may relate to the widespread cultivation in the late 1960's of varieties that combined BMR 2 and 3 resistance which caused the matching pathogenicity combination to become prevalent at the time (Wolfe and Barrett, 1976).   Varieties with this resistance combination have not been grown since about 1970.

3.   Pathogenicity for BMR 5 was common among isolates from BMR 7. The reason is not known, although it may be related to the fact that these resistances are allelic, or closely linked at the Mla locus on chromosome 5.

b)   <u>Negative deviations</u>

1.   Pathogenicity for BMR 1 was similar on most varieties: only populations from BMR 4 were less pathogenic than others.

2.   Pathogenicity for BMR 3 was low among populations from BMR 5 and 7, which are related, and among those from BMR 6 and BMR 8, which have a resistance gene in common.   There were no reciprocal effects.

3.   Pathogenicity for BMR 4 was low among populations from BMR 6 and 8; there was a reciprocal relationship in the low pathogenicity for BMR 6 in populations from BMR 4.   Pathogenicity for BMR 4 was also low among populations from BMR 7, and, to a lesser extent, from BMR 5.

4.   Pathogenicity for BMR 5 was low in populations from BMR 1.   It was also low in populations from BMR 6 and 8, but this deviation was repeated reciprocally in the low pathogenicity for BMR 6 in populations from BMR 5 and 7.   Low pathogenicity for BMR 6 also occurred in populations from BMR 2.

Large negative deviations in pathogenicity for BMR 1-6 on non-matching hosts occurred despite the widespread and long-term use of these resistance groups in the UK.   Indeed, the deviations involving BMR 5 and 6 had been noted at the time of introduction of the two resistance groups, several years previously (Wolfe et al., 1975).

For the examples given, including those that involve reciprocal deviations, it is not possible to determine whether selection was operating against the combination of pathogenicity genes, or against the non-matching pathogenicity gene alone.

THE RELATIONSHIP BETWEEN FIELD INFECTION, SPORES LIBERATED INTO THE ATMOSPHERE AND PATHOGEN POPULATION STRUCTURE

Surveys of cereal foliar diseases made by the Harpenden Laboratory (Agricultural Development and Advisory Service) provides information on the distribution of cultivars of spring barley and their infection levels during the period under discussion, for England and Wales (King, pers. comm.). Unfortunately, the data was sufficiently extensive only for comparison of BMR groups, 2, 4, 5 and 6. Table 5 was obtained by calculation of the survey data for the appropriate cultivars to obtain the product, (relative area x relative infection) for each BMR group, representing the relative contribution of each group to the overall amount of infection. There was relatively little variation between the years: the decreasing value for BMR 6 may have been due to the increasing area of cv. Triumph during the period, which remained highly resistant to powdery mildew.

Table 5. Products of relative area and infection on four BMR groups calculated from survey data from the Harpenden Laboratory (King, pers. comm.) for the years 1978-80, compared with non-matching pathogenicity values from the population surveys

| Year | Source | BMR group | | | | Correlation coefficient |
		2	4	5	6	
1978	Area x inf^n	56	23	29	28	1.00
	non-matching pathogenicity	72	9	22	22	
1979	Area x inf^n	52	35	29	23	0.80
	non-matching pathogenicity	60	12	28	21	
1980	Area x inf^n	58	32	25	21	0.91
	non-matching pathogenicity	71	17	27	25	
Means	Area x inf^n	55	30	28	24	0.93
	non-matching pathogenicity	68	13	26	23	

For the four BMR groups the mean values for (area x infection) correlated closely with those for non-matching pathogenicity. This suggests that selection against pathogenicity for BMR 2, 5 and 6 is similar on all non-matching hosts, because the relation between the non-matching pathogenicity values is similar to that for the relative amounts of infection generated by the three cultivar groups.

Pathogenicity for BMR 4 was considerably less common on non-matching hosts than the data for (area x infection) in Table 5 suggest. This could have occurred either because the quantity of spores disseminated from infected BMR 4 cultivars was less than from BMR 2, 5 or 6 cultivars so that non-matching hosts received correspondingly fewer spores with pathogenicity for BMR 4, or because there was greater selection against pathogenicity for BMR 4 on non-matching hosts, than against pathogenicity for BMR 2, 5 or 6 on their non-matching hosts.

This question was resolved by the use of a spore trap fitted to a car roof for sampling the air spora (Wolfe, et al., 1981; see also Schwarzbach, 1980). The trap directs moving air on to whole seedlings which were exposed in batches over distances of 50-100 km throughout the major barley-growing area of England and southern Scotland. The bulk populations obtained were tested on the standard differential cultivars. Different exposures gave similar results: the mean values for the whole area are therefore presented in Table 6, together with the mean contributions to overall infection from Table 5, and the mean values for non-matching pathogenicity from Table 4.

Table 6. Data for 1980 for the products of relative area and relative infection, pathogenicity in airborne spore samples, and non-matching pathogenicty from population samples

| | | | BMR group | | | |
Factor	1	2	3	4	5	6
(Rel. area x rel. infection)	-	58	-	32	25	21
Airborne spora	40	76	21	41	25	33
Non-matching pathogenicity	37	71	26	17	27	25

From Table 6, excepting the data for BMR 4, the three sets of values are closely correlated which again suggests that selection

against pathogenicity for BMR 1, 2, 3 5 and 6 is probably similar on
all non-matching hosts, so that the pattern of distribution of non-
matching pathogenicity genes is closely correlated with the area and
infection of each BMR group. The relatively high incidence of pathogenicity for BMR 4 in the air spora indicates that the major reason
for the unexpectedly low frequency of this gene on non-matching hosts
is probably that there is greater selection against it than against
other non-matching genes (see Jørgensen, this volume).

Sampling the air spora provides a rapid and cheap technique for
obtaining an estimate of the overall frequency of individual pathogenicity genes. However, it cannot provide an accurate estimate of
specific interactions between host and pathogen genes because it
confounds differences between the individual populations (Wolfe &
Knott, 1982).

VARIATION IN MATCHING POPULATIONS

For convenience, the populations of host and pathogen in the
barley mildew surveys were analysed in terms of interacting resistance and pathogenicity genes. However, each population exhibits
variation in resistance betweeen individual varieties, and in pathogenicity towards these varieties. Examples of such interactions were
provided by Wolfe and Slater (1980) and they are reproduced in a
simpler form in Table 7a-c.

In each example, control populations from non-matching hosts
were tested on different varieties from a particular BMR group. This
indicated the general level of pathogenicity towards the varieties in
question. Isolates from matching varieties clearly showed greater
adaptation to the test varieties, but in each case there was an
indication of specific adaptation of an isolate to the test variety
from which the isolate had originally been collected.

Chin (1979) collected isolates of the pathogen from Hassan (BMR
5) and Wing (BMR 6) which had combined pathogenicity for the two
varieties. Those collected from Hassan were more pathogenic on
Hassan than on Wing, while those collected from Wing were more
pathogenic on Wing. Varietal adaptation in the field following
selection has been recorded in other host-parasite systems (Habgood,
1976; Caten, 1974).

A more detailed analysis was made on isolates from a mildew
infection of the barley variety Keg (BMR 6). The isolates were collected as two groups, one from the oldest available colonies on a
random selection of 50 plants, and the others from the youngest
available colonies on the same plants; the two populations were probably separated by about four pathogen generations. Both populations
exhibited variation in pathogenicity towards test seedlings of Keg,

Table 7a-c. Examples of specific adaptation of *E. graminis* f.sp.
hordei to the variety from which population samples were
isolated. Data are colony numbers per unit leaf area,
relative to those on the susceptible control variety

a)

Source	Test variety (BMR 4)	
	Vada	Lofa Abed
Non-matching varieties	64	32
BMR 4	174	90
Lofa Abed	110	79

b)

Source	Test variety (BMR 5)	
	Sultan	Hassan
Non-matching varieties	43	47
Hassan	109	159

c)

Source	Test variety (BMR 2+5)	
	Aramir	Porthos
Non-matching varieties	30	20
Aramir	70	59
Porthos	50	55

but in the 'old' population, the variation was greater, and the distribution about the mean was symmetrical and normal. The isolates
from the 'young' population, which had undergone further selection,
had the same mean pathogenicity, but the distribution was more peaked
around the mean, and considerably more skewed. This indicated that
the less fit isolates, with low pathogenicity, had been eliminated
(Wolfe et al., 1981).

The data for barley mildew identify two levels of host adaptation in the pathogen. The first is towards the major effects of
commonly-used resistance genes (the BMR system), and the second
towards individual varieties within each group, distinguished by
minor differences in resistance. However, this must be regarded as a
classification of practical convenience. As more host-isolate interactions are recognised, and as techniques of recognition become more
refined, it becomes increasingly difficult to determine the numbers
of genes involved in each interaction. It also seems increasingly
less likely that all interactions are strictly gene-for-gene (see

also Schwarzbach, 1981). Nevertheless, there was no doubt that the identified resistance genes of major effect had a dominant influence on the pathogen populations.

DISCUSSION

The data on barley mildew populations provide some indication of the size and variation in the selection coefficients, applicable to pathogenicity for a range of widely used resistance genes. First, since none of the non-matching pathogenicity genes are completely eliminated in a single season on any host, we assume that the selection coefficients must be relatively small (probably less than 10%). Second, since there was a correlation between the values for non-matching pathogenicity, the relative pathogenicity in the air spora, and the area of use and infection of the resistant hosts, for most of the pathogenicity genes, it seem that the selection coefficients for each pathogenicity gene are of similar magnitude.

If the selection coefficients are small and similar, we would expect to see a response to selection for adaptability in the form of an increase in the relative frequency of complex races. This should be reinforced by the system of sequential introduction of resistance genes. Selection for pathogenicity on each subsequently introduced resistance gene must occur in pathogen populations already selected for adaptation to previously used hosts. This appears to be so for BMR 1 and 2, the longest-used resistance genes, for which pathogenicity has become virtually fixed in the pathogen population. Thus the majority of isolates selected for pathogenicity on other hosts are also pathogenic on BMR 1 and 2.

In the field, however, progress towards adaptability appears to be slow. Despite the continued large-scale use of the same major resistance groups, there has been no detectable evidence of a shift towards more complex races during the last six years. Several factors may be involved in this apparent paradox:

a) the size of the pathogen population selected on any one component may be large enough for independent existence. Sufficient spores are available at the beginning of each new season to start a large-scale epidemic; there is no likelihood of a population crash between years due to lack of an appropriate host or unfavourable environmental conditions. The extra advantage of flexibility may therefore be small.

b) the number of cycles of selection for adaptation within a crop is about ten times as many as for adaptability between crops.

c) selection for adaptation is more efficient than for adaptability. Within a homogeneous crop, most of the spores that are generated are subject to selection for adaptation to that crop (autoinfection phase). During selection for adaptability (alloinfection phase), however, the spores in the atmosphere are not randomly distributed over all hosts; local sources of infection are likely to be more important in initiating epidemics, and these will tend to be less variable than the range from all sources.

d) because of the range of resistance groups, and of the different varieties in each, the numbers of genes required for a high level of adaptability may be very large and represent too great a physiological burden for the pathogen, or the combinations may not occur.

e) the occurrence, and persistence, of linkage disequilibrium in the levels of pathogenicity for certain combinations of hosts must limit the response to selection for adaptability across some combinations of pathogen alleles.

The relative importance of the sexual stage in the selection for adaptation and adaptability is difficult to determine. Segregation occurs during the autumn. This has the advantage of generating a maximum amount of variation when the winter crop is emerging since the latter may provide a less favourable environment to the selected pathogen population of the previous season. This advantage of adaptability may be greater than the disadvantage caused by the continual break-up of adapted gene complexes. However, if recombination occurs within a population selected on a particular host, and that host is maintained in the population, then there may also be more rapid progress towards selection of pathogen genotypes better adapted to that host.

The balance of the argument suggests that the current structure of the pathogen population is determined largely by the usage of particular resistance genes and by specific adaptation to them, and that adaptability is relatively less important. However, a definitive answer cannot be found and the pathogen population could still be evolving in one of three directions:

a) the complex races that exist at the moment are declining in frequency. Eventually, larger populations of specifically adapted pathogen genotypes will cause more disease than at present, or

b) the complex races are slowly increasing in frequency and, when they become prevalent, they will cause more disease than occurs at present, or

c) a 'compromise' has been reached between the advantages of adaptation and adaptability so that the pathogen population is already causing the maximum amount of disease of which it is capable. If the population was less adaptable, then the starting frequencies for initiation of epidemics would be smaller and maximum levels of epidemic development would be less. On the other hand, if the population was more adaptable, starting frequencies would be higher, but growth rates less, again resulting in lower maximum levels of disease.

Because the barley mildew population continues to adapt quickly to new varieties and, indeed, to fungicides (Fletcher & Wolfe, 1981), but does not show rapid shifts to greater complexity, even on variety mixtures, it seems that a) or c) above may represent the real situation, rather than b). It has therefore been possible to establish a recommendation for variety diversification in the UK (Priestley & Wolfe, 1979). Direct control of disease cannot be large because the period for interaction between field epidemics is small (see also Østergaard, 1982). However, there is an indirect benefit from spreading the risk of severe disease.

Multilines and Variety Mixtures

Disease control in multilines and variety mixtures depends on slowing down the development of the best adapted race on each component. This causes increased selection for adaptability relative to that in homogeneous crops because the number of cycles of selection for adaptation and adaptability are more or less equal. However, the efficiency of selection for adaptation remains greater than that for adaptability, because the spores produced in each pathogen generation, and the parent colonies, are not randomly redistributed over all components: they tend to remain on the plant on which they were produced. In other words, at the level of individual plants, autoinfection may be more important than alloinfection, just as it is at the field level.

The importance of spore distribution on the relative fitness of pathogen genotypes in heterogeneous crops was explored by Barrett (1980) and Østergaard (1982). Indeed, where we have followed population dynamics of the pathogen in mixtures in the field (Barrett & Wolfe, 1980), there has been no evidence of rapid shifts towards the development of super-races. The risk remains, however, particularly if certain multilines or mixtures are exploited on a large scale. For this reason, several sub-strategies need to be considered and used where practicable:

a) the numbers of components within a multiline or mixture can be increased. This has the practical disadvantage of increasing the difficulty of matching the components.

b) increasing the complexity of resistance within components: this is difficult to achieve in classic multilines since the components are selected for genetic similarity.

c) by changing the composition of the crop at frequent intervals.

Again this may be difficult for classic multilines because few suitable components may be easily available, but can be achieved easily
in a simple mixture system (Wolfe & Barrett, 1980).

Inherent, or Varietal Durability and System Durability

Using mixtures of varieties and changing the composition of the
mixtures at intervals, is a means of reducing the response of the
pathogen to selection for adaptation and adaptability. Put another
way, it is a means of limiting the fitness of the pathogen both on
individual varieties and on the whole population of varieties: it can
therefore be described as a means of obtaining system durability.

Johnson (this volume) has described the characteristics of durable resistance which is inherited, or, in other words, the ability of
a variety itself to limit pathogen fitness. However, such inherent
or varietal durability may be affected by the size and nature of the
pathogen population to which it is exposed. The probability of a
variety maintaining its inherent durability will be increased if the
area and frequency of exposure is limited, and if the variety is
challenged by only a small pathogen population coming from diverse
resistant hosts. The probability will be decreased if the remaining
host population is highly susceptible. Thus, inherent durability of
a variety may be extended if it is used in a way that provides system
durability.

The converse is also true. The probability of maintaining system durability, such as for example, by changing mixtures of varieties, will be increased by using components that have inherent durability.

ACKNOWLEDGEMENTS

JAB wishes to thank the Agricultural Research Council for financial support; the three authors thank Sandie Clements for technical
help.

REFERENCES

Barrett, J.A., 1980, Pathogen evolution in multilines and variety
    mixtures. *Z. Pflanzenk. und Pflanzen.*, 87: 383-396.
Barrett, J.A. and Wolfe, M.S., 1980, Pathogen response to host
    resistance and its implication in breeding programmes. *EPPO
    Bull.* 10: 341-337.
Caten, C.E., 1974, Intra-racial variation in *Phytophthora infestans*
    and adaptation to field resistance for potato blight. *Ann.
    Appl. Bio.* 77: 259-270.

Chin, K.M., 1979, Aspects of the epidemiology and genetics of the foliar pathogen, *Erysiphe graminis* f.sp. *hordei* in relation to infection of homogeneous and heterogeneous populations of the barley host, *Hordeum vulgare*. PhD Thesis. U. of Cambridge.

Fletcher, J.T. and Wolfe, M.S., 1981, Insensitivity of *Erysiphe graminis* f.sp. *hordei* to triadimefon, triadimenol and other fungicides. *Proc. 1981 Br. Crop Prot. Conf.*, 2: 633-640.

Habgood, R.M., 1976, Differential aggressiveness of *Rhynchosporium secalis* isolates towards specified barley genotypes. *Trans. Br. mycol. Soc.* 66: 201-204.

Leonard, K.J. and Czochor, R.J., 1980, Theory of genetic interactions among populations of plants and their pathogens. *Ann. Rev. of Phytopath.* 18: 237-258.

Østergaard, H. 1982, Predicting the development of an epidemic on a variety mixture. *Phytopath.* (in press).

Priestley, R.H. and Wolfe, M.S., 1977, Crop protection by cultivar diversification. *Proc. of the 1977 Br. Crop Prot. Conf.*, 135-140.

Schwarzbach, E., 1979, A high throughput jet trap for collecting mildew spores on living leaves. *Phytopath. Z.*, 94: 165-17.

Schwarzbach, E., 1981, Progress and problems with breeding for resistance. *Proc. of the 4th Barley Genetics Symp.*, (in press).

Skylakakis, G., 1980, Estimating parasitic fitness of plant pathogenic fungi: a theoretical contribution. *Phytopath.*, 70: 696-698.

Vanderplank, J.E., 1963, Plant Diseases: Epidemics and Control. Academic Press, New York.

Wolfe, M.S., 1968, Physiologic race changes in barley mildew, 1964-67. *Plant Path.*, 17: 82-87.

Wolfe, M.S. and Barrett, J.A., 1976, The influence and management of host resistance on control of powdery mildew on barley. *Proc. of the 3rd Barley Genetics Symp.* 433-439.

Wolfe, M.S. and Barrett, J.A., 1980, Can we lead the pathogen astray? *Plant Dis.* 64: 148-155.

Wolfe, M.S. and Knott, D.R., 1982, Some considerations of the collection and analysis of data on variation for pathogenicity in populations of plant pathogens. *Plant Path.*, (in press)

Wolfe, M.S., Minchin, P.N. and Wright, S.E., 1975, Powdery mildew of barley. *Ann. Rep. of the Plant Breeding Inst.*, 1974.

Wolfe, M.S. and Schwarzbach, E., 1978, Patterns of race changes in powdery mildews. *Ann. Rev. of Phytopath.*, 16: 159-180.

Wolfe, M.S., Slater, S.E. and Minchin, P.N., 1981, *Ann. Rep. of the Cereal Pathogen Virulence Survey* 42-56.

DISCUSSION

PARLEVLIET: Considering the dilemma between adaptation and adaptability facing this pathogen, I wonder which is the best solution from an agricultural point of view to control the disease, the variety mixture or the multiline, if the resistance genes in the components are the same.

WOLFE: The variety mixture is the best solution because the additional background resistance genes make the response to both adaptation and adaptability more difficult for the pathogen. In addition, the variation in genetic background of the variety mixture can provide control of non-target pathogens; a number of multilines have proved to be uniformly susceptible to non-target pathogens.

KRANZ: Do you use the term virulence and fitness as synonyms?

WOLFE: No. Virulence is only a component of fitness. In different situations, the same virulence character may be of crucial importance, or quite unimportant, in its contribution to fitness.

SHARP: Travelling at 110 km/hr, what percentage of spores are trapped on barley leaves in the WIST trap?

WOLFE: We do not know. It is likely that many are lost, but the important point is that we trap sufficient to develop many colonies on the seedlings in the WIST.

JØRGENSEN: The variety mixtures have two 'components' of heterogeneity. The specific resistance genes, and the genetic background of the variation. What are the relative importance of the two?

WOLFE: The genes of major effect are of greater importance. However, we did have one trial in which three mixtures were grown, each comprising three varieties with the same major gene resistance, but in different genetic backgrounds. The difference in epidemic development between mixtures and pure stands took longer to become evident than in conventional mixtures, but by the end of the season, the final amount of disease was about half of the expected value.

JØRGENSEN: This means that it may be profitable to grow mixtures even of susceptible varieties, i.e. varieties without any known resistance genes.

WOLFE: Mixtures of varieties without identified resistance genes can certainly have a useful effect, but it may not be of a profitable size. If they are normally very susceptible, then a 50 percent reduction in disease is large, but not large enough to control all of the yield loss.

PROBLEMS IN ESTIMATING PARASITIC FITNESS

Hanne Østergaard*

Agricultural Research Department
Risø National Laboratory
DK-4000 Roskilde, Denmark

INTRODUCTION

Durable resistance is a characteristic of a pathosystem and not only of the host population as the expression implies. It is measured in terms of disease quantity, e.g. diseased leaf area, and this entity is influenced by the resistance of the cultivar as well as by the mean fitness of the parasite population. Therefore, parasitic fitness is an important component in studies of durable resistance. One group of problems in estimating parasitic fitness comes from the possible effects of density dependent growth of the parasite. The analysis of a mathematical model including density dependence in spore production and in survival of infections has shown that parasitic density has to be taken into account in studies on characteristics of pathosystems.

MODEL FOR A HOMOGENEOUS PARASITE POPULATION

The life cycle of a fungus is divided into four processes, spore production, dispersal, infection, and survival. The total number of spores produced from all sporulating infections on one plant during the infectious period is described by two parameters: the potential spore production per sporulating infection (B), and the density dependence coefficient of the spore production process (I). The model assumes that for a low density of infections there is an approximately linear relationship between the number of sporulating infections and the total number of spores produced, but when

---
* This work was supported by a grant from the Danish Agricultural and Veterinary Research Council.

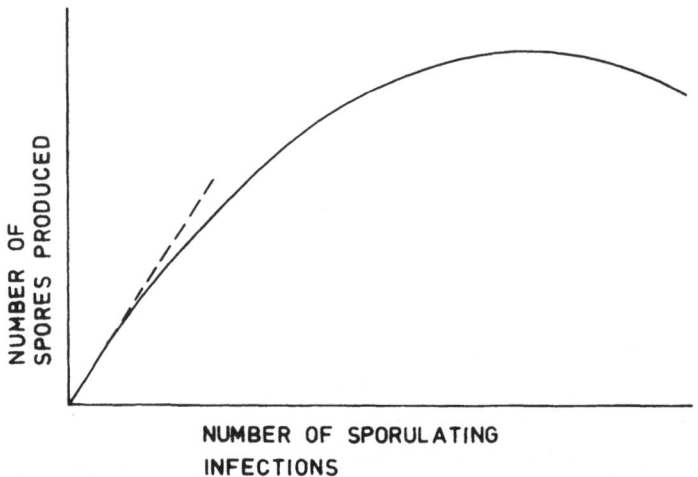

Fig. 1. The spore production processes (numbers in arbitrary units).
         The linear relation is y = xB (broken line) and the density
         dependent relation is y = x(B-Ix) (full line).

the density becomes higher the number of spores produced per infection decreases. This relationship is expressed mathematically
by a parabola as shown in Fig. 1 (full line). It is seen from the
density dependent relation that a large value of the density dependence coefficient (I) implies that the spore production per infection
is much restricted.

     The dispersal process is simplified by assuming deposition of
spores on new plants, and removal of old plants at a fixed time.
The number of spores deposited is assumed to be small compared to
the leaf area so the infection process is density independent. Thus
a fixed proportion of spores, given by the infection efficiency (P),
germinate and establish infections.

     The survival of infections during the latent period is density
dependent because the infections are competing for the resources in
the host. The relationship between the number of spores deposited
and successfully infecting, and the number of sporulating infections
produced is expressed by three parameters: the potential survival
efficiency (D), the density dependence coefficient of the survival
process (j), and the latent period (T). Since the number of infections decreases continuously during the latent period, the mathematical relationship, as shown in Fig. 2 (full line), is different
from the one used for the spore production process, even if the
biological assumptions are similar, e.g. the relationship is approximately linear when the number of spores infecting is small.

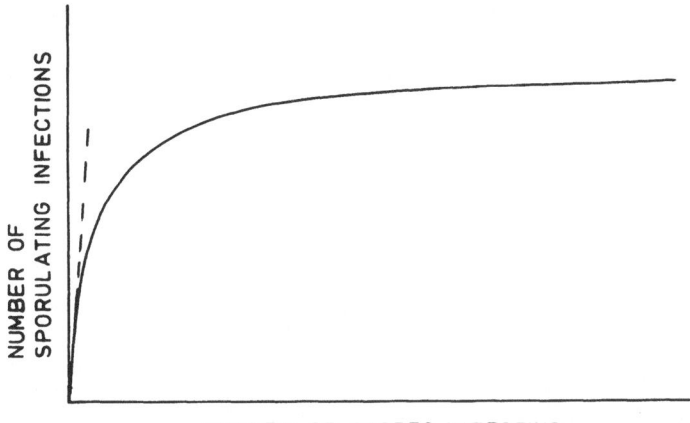

Fig. 2.   The survival process (numbers in arbitrary units). The
         linear relation is y = xD (broken line) and the density
         dependent relation is y = xD/(I-x(I-D)jT/log D) (full line).

        By combining the four processes successively, the expression
for the potential reproduction value (fitness) is found as the product of the potential spore production per sporulating infection,
the infection efficiency, and the potential survival efficiency (BPD).

PREDICTING FOR A HETEROGENEOUS PARASITE POPULATION

        The extension of the model to describe the growth of a popu-
lation consisting of different genotypes with synchronous development is done by specifying the parameters for each genotype and
defining the interactions among genotypes. Most previous studies
of heterogeneous populations have dealt with mixtures of only two
genotypes. The usual way to present data from such experiments is
to plot the logit of the proportion of genotype 1 against the generation number. When omitting density dependence (I=0 and j=0) this
would theoretically result in a straight line with a slope measuring
the logarithm of the relative fitness of genotype 1, i.e. $\log (B_1P_1D_1/B_2P_2D_2)$ (cf. Leonard, 1969). Including density dependence, as
done in the present model, a straight line is in general not ex-
pected. The reason is that even if each generation is initiated
with the same amount of spores, the total number of sporulating
infections changes from generation to generation as long as the
composition of the spore population changes. Furthermore, the relative fitness changes from generation to generation and its average over several generations might differ considerably from the
relative fitness defined as above.

In a few cases different inoculum densities have been included in the experiments (Katsuya & Green, 1967; Rastegar, 1976) and it was observed that different isolates of cereal rust dominated at light and heavy infection. A similar result is expected in the model if the density dependence is determined by the total number of sporulating infections and by the total number of spores infecting, respectively, i.e. even if there are no antagonistic effects.

REFERENCES

Katsuya, K., and Green, G.J., 1967, Reproductive potentials of races
    15B and 56 of wheat stem rust, *Can. J. Bot.,* 45:1077-1091.
Leonard, K.J., 1969, Selection in heterogeneous populations of
    Puccinia graminis f.sp. *avenae. Phytopathology,* 59:1851-1857.
Rastegar, M.F., 1976, Competitive ability of an induced mutant race
    of *Puccinia hordei* otth in mixtures, *Proc. 4th Eur., Meditt.
    Cereal Rust Conf.,* 58-59.

RESPONSE OF A POWDERY MILDEW POPULATION TO A BARLEY VARIETY MIXTURE

Lisa Munk

The Royal Veterinary and Agricultural University
Department of Plant Pathology
Copenhagen, Denmark

INTRODUCTION

It has been suggested that durable resistance in self-pollinated crops can be achieved by using the limited number of known specific resistance genes in multilines and variety mixtures (Browning and Frey, 1969; Wolfe and Barrett, 1977). To evaluate this strategy it is necessary to investigate the effects of mixtures of lines or varieties with different resistance genes on the composition of the pathogen population. Such effects are being studied in a project at The Royal Veterinary and Agricultural University, Copenhagen, where a model experiment on the flexibility of a barley powdery mildew population is in progress.

Under field conditions the pathogen population is forced to produce many conidial generations. During the summer it is grown on seedlings of a mixture composed of seven varieties (with resistance from Algerian, *Hordeum spontaneum*, Lyallpur, Monte Cristo, Arabische, Weihenstephan, and *Hordeum laevigatum*) and during winters it survives on a susceptible winter barley. The initial powdery mildew population consisted of the natural population at the locality of the trial. The composition of the pathogen population is determined every spring and autumn by testing randomly selected isolates on appropriate differential varieties.

RESULTS

Data on frequencies of six necessary (in relation to the variety mixture) and ten unnecessary virulences are available for the period from spring of 1978 to the autumn of 1980. During the three summer seasons the frequencies of the necessary virulences generally increased, and during the winter 1978-79 they decreased, while during the second winter 1979-80 only minor changes were observed. The frequency of the necessary virulence corresponding to Monte Cristo resistance showed the opposite trend. During both summer and winter, six out of ten unnecessary virulences did not change from spring 1978 to autumn 1980, whereas a small increase was found for the remaining four.

The frequencies of occurrence with combinations of different numbers of virulences have been calculated and those with a given number of necessary virulences (except Ml-(La)) are shown in the accompanying table. The number of isolates tested is given in parentheses.

Number of virulences	Time of sampling (S= spring; A= autumn)					
	S.78 (243)	A.78 (60)	S.79 (90)	A.79 (98)	S.80 (100)	A.80 (99)
0	2.1		2.2	1.0		
1	8.6	1.7	5.6			1.0
2	19.3	5.0	15.6	4.1	11.0	
3	42.4	36.7	34.4	12.2	11.0	2.0
4	25.1	51.7	34.4	52.1	29.0	15.2
5	2.5	5.0	7.8	28.6	39.0	77.8
6				2.0	10.0	4.0

It is seen that frequencies of simple genotypes with nil, one and two virulences decreased during summer season, and increased during winter seasons, while the opposite occurred for complex combinations with four virulences. Further, the frequency of genotypes with three virulences decreased and frequency of genotypes with five virulences increased during the three years, the largest changes occurring in the summer seasons. From the autumn of 1979 the first genotype possessing all six necessary virulences was observed. Frequencies of combinations based on the ten unnecessary virulences showed a decrease of simple and an increase of complex combinations.

A comparison of the influence of the environment on the development of a powdery mildew population was made by growing a sample of the spring 1980 population from the field on the same variety mixture under isolated conditions in the greenhouse. After 16 discrete generations, which is approximately the number of generations produced during a summer season in the field, the greenhouse population showed a frequency distribution of number of virulences very similar to the autumn population in the field. One particular genotype possessing five necessary virulences predominated both in the final greenhouse population and in the autumn population in the field.

CONCLUSION

The variety mixture used in the field experiment imposed a selection favouring the necessary virulences. During the winter, where all the virulences were unnecessary, the trend seen for virulences corresponding to the varieties in the mixture indicates a selection force in the opposite direction. This force can hardly be a simple effect of unnecessary virulences and the resultant lack of selection pressure, as virulences unnecessary all year round did not decrease. Therefore, the decrease of combinations possessing many unnecessary virulences in winter most likely was due to them having a lower degree of general fitness. Such genotypes, however, increased in summer due to the selection for virulences which were then necessary.

An important question is how simple and complex genotypes will compete over the years. If simple combinations remain frequent, a multiline or a variety mixture will probably provide durable disease control. However, if the simple genotypes disappear, then the durability of such variety mixtures will be relatively short. Finally, I would like to stress that the experimental system described, which favours allo-infections and which excludes ascospore-infections, probably gives a pattern different from that which can be expected in a system where the plants grow until maturity.

REFERENCES

Browning, J.A., and Frey, K.J., 1969, Multiline cultivars as a means of disease control, *Ann. Rev. Phytopath.* 7:355-382.
Wolfe, M.S., and Barrett, J.A., 1977, Population genetics of powdery mildew epidemics, *Ann. N.Y. Acad. Sci.*, 287:151-163.

CHANGING GENE FREQUENCIES IN POPULATIONS

E. L. Sharp

Department of Plant Pathology
Montana State University
Bozeman, Montana USA 59717

SUMMARY

Recurrent selection populations (RSP's) have been effectively
utilized for changing gene frequencies in populations of various
crop plants and for pyramiding genes for resistance to specific
plant diseases. The method has been particularly adapted for cross-
pollinated crops and has been especially successful in developing
disease resistance in maize and alfalfa. In regard to these two
crops broad-based resistance to both fungus and bacterial diseases
has been obtained. The method has also been used for developing
resistance to various insect problems in maize and alfalfa.
Among the self-pollinated crops barley is particularly suitable
for improvement by RSP's due to numerous genetic male sterility
genes. In a program at Montana State University to improve the
control of barley diseases for the semi-arid areas of the world,
the method is being used with the major diseases of barley. The
objective is to pyramid genes for resistance into suitable
agronomic backgrounds for both 2 row and 6 row barley. Separate
RSP's are being developed with major and minor gene resistance.
The base cultivars containing one male sterile entry were selected
for their genetic diversity and general adaptability. Each cycle
of recurrent selection in each year consisted of two generations:
a selection for resistance and a recombination generation. In
the latter, seeds were harvested only from male sterile plants to
assure recombinations. The populations for resistance to specific
diseases were grown each year in several locations and resistant
plants were selected for further cycling. At the same time new
resistance sources, as identified, were added into the population.
The RSP's for scald and net blotch have shown a high percentage of

resistant plants wherever grown and have been combined into one
population. Significant progress has also been obtained on some of
the RSP's containing only minor effect additive genes for specific
barley diseases.

INTRODUCTION

        Pathogen populations have great diversity and ability to change.
Thus it is desirable to combat this by developing broad-based
genetic diversity in their hosts. A number of methods may be
used for combatting pathogen populations but the method discussed
here involves changing gene frequencies in populations by recurrent
selection.

        Jenkins et al., (1954) used recurrent selection for concentra-
ting genes for resistance to *Helminthosporium turcicum* Pass. (leaf
blight of maize). Earlier studies indicated that resistance to
this disease was influenced by numerous genes. They selected the
most disease resistant plants at pollination time and pollen from
these plants was used in bulk and placed on the silks of the same
plants. Each group of plants they worked with consisted of 7
progenies -- a susceptible corn belt inbred, a resistant inbred,
the cross between these lines, backcrosses to the susceptible
corn belt inbred and 3 successive generations of recurrent selection. Plants were scored for disease into 11 classes from very
resistant to complete susceptibility of 0, 0.5, 1.0, 1.5, 2.0, 2.5,
3.0, 3.5, 4.0, 4.5, and 5.0. Each generation of recurrent
selection reduced susceptibility with the greatest advances noted
in the first and second cycles. At the end of the third cycle
32% of the plants rated 1.0 or less. The susceptible inbred
lines generally showed ratings between 3.0 and 5.0. Three generations of recurrent selection were performed for 9 groups of progenies. Effective build-up of disease resistance occurred for 24 of
a total of 27 comparisons. Recurrent selection was used for
obtaining resistance to stalk rot (*Diplodia zeae* Schw.) of maize
(Jinahyon and Russell, 1969). Resistance to this disease is known
to be conditioned by genes at several loci on a total of at least
9 chromosome arms. Materials were developed from the open pollinated maize cultivar Lancaster. Most lines from this cultivar originally had good combining ability but poor stalk quality. The
plants were crossed in diallel in each cycle. Inoculations were
made when 50% of the silks had emerged. Significant shifts to
more resistance occurred with each cycle of recurrent selection.
After 3 cycles the population was as resistant as the selected
resistant single cross check while the original population was
more susceptible than the susceptible check. Additive gene
action was most important in conferring resistance and heritabilities were often higher than 60%. At the end of the third cycle of
recurrent selection an average of 26-50% of the inoculated inter-
node was rotted as compared to 76-100% for the unselected lines.

Recurrent selection populations have also been utilized for obtaining resistance to alfalfa rust (*Uromyces striatus* Schroet. var. *medicaginis* Pass. Arth.). Eight cycles of recurrent resistant phenotype selection in the field effectively shifted the field reaction of 2 pools of alfalfa germ plasm from predominately susceptible to predominately resistant (Hill et al., 1963).

Resistance to bacterial diseases was improved by use of recurrent selection populations. Barnes et al. (1971) started with 2 pools of alfalfa germ plasm representing the 14th generation of mass selection. Resistance to bacterial wilt (*Corynebacterium insidiosum* McCull Jens.) was greatly improved following 3 cycles of recurrent selection. The plants were artificially inoculated and otherwise treated to obtain maximum expression of disease. Sixty days after inoculation plants were rated for disease on the basis of discoloration of the tap roots. Classes 0 and 1 were considered highly resistant, 2 = moderately resistant, 3 = moderately susceptible and classes 4 and 5 = highly susceptible. Germ plasm pool A at the onset showed 9% of the plants in a 0, 1 very resistant category. After the second cycle of recurrent selection, 66% of the plants were in a 0, 1 disease category. Germ plasm pool B, on the other hand, contained no resistance prior to recurrent selection. At the end of 4 and 5 cycles this population showed an increased number of plants in the moderately resistant category and a decreased number of plants showing susceptibility. It was speculated that resistance in germ plasm pool A was controlled by a few major genes and that resistance in germ plasm pool B was controlled by many genes with minor effects.

Recurrent selection populations have also been used to obtain resistance to insects. One example is resistance in maize to the European corn border, (*Ostrinia nubialis* Hubner) (Penny et al., 1967). Seven chromosomes are known to have loci influencing borer resistance and additive gene action has a major role in differences in resistance. Dominance and epistasis had a minor but detectable effect. The starting materials were 5 synthetic cultivars of maize. Two cycles of recurrent selection were sufficient to shift the frequencies of resistance genes to a high level of resistance in these cultivars and 3 cycles of selection appeared sufficient to convert all to essentially borer resistant cultivars.

All previously discussed applications of recurrent selection for improving disease resistance were conducted without the benefits of male sterility. Singh (1977) used male sterility and recurrent selection populations for simultaneous conversion and improvement of alien sorghums. Recessive height and maturity genes were accumulated from a source population along with cold tolerance genes from highland introductions of sorghums. The potential use of the method for other traits such as disease resistance was noted. The advantages of using male sterility genes were considered as the following: (a) facilitates crossing (b) allows simultaneous integration of all sources of cold tolerance into an intermating pool (c) allows easy incorporation of any other desirable genes lacking in alien parents and/or intermating pool at any stage of the conversion program.

BARLEY

Barley is uniquely suited for improvement by recurrent selection populations via manipulation and pyramiding of genes resistant to specific diseases. This is largely because of numerous genetic male sterility genes which can be used to facilitate recombinations. In early work with barley improvement, composite crosses along with natural selection were used to develop superior barley (Harlan and Martini, 1929). Later bridge crosses were used where F_1's from various lines were crossed to combine desirable types. Without male sterility in the population, this was very time consuming and recombinations were limited. Suneson (1956) first described male sterility in barley and subsequently 28 loci for male sterility were described by Hockett et al., (1968). Many of these loci have been assigned to specific chromosomes.

Montana State University is currently involved in an extensive program to improve the control of barley diseases in the semi-arid areas of the world. Male sterility with recurrent selection populations is being utilized to pyramid resistance genes and develop a broad-based resistance to specific diseases of barley. Attempts are being made to incorporate major and minor gene resistance separately into both 6 row and 2 row barley types. The diseases being emphasized are scald [Rhyncosporium secalis (Oud.) J. J. Davis], net blotch [Pyrenophora teres (Died.) Drechsl.], leaf rust (Puccinia hordei Otth.), Xanthomonas streak [Xanthomonas translucens (L. R. Jones, A. G. Johnson and Reddy), Dowson] and barley yellow dwarf virus (Sharp, 1980).

The objective is to develop germ plasm with multigenic resistance in a background of good agronomic types. The base cultivars used in the 2 and 6 row barley populations are listed in Tables 1 and 2. Compana msg 10 and Manchuria msg 10 are used as male sterile sources in the 2 row and 6 row barley, respectively. Both of these cultivars have good combining ability. The other cultivars in the base were chosen for various specific traits. For example, in the 2 row population, Herta, Ingrid, Bruens Wisa and Zephyr have a history of good adaptation and yield, Erbet is early maturing, Dekap from Turkey does well under very dry conditions, Summit has good malting quality, Vireo has good stature and yield with late maturity and both Nude Compana and Toonucier are hulless types.

The 6 row base barley cultivars were also selected for various attributes. Several have been high yielders under dry conditions, CM 67 does well in dry areas and is also salt tolerant, Minn 21 has short stature, Athenais is a very early maturing cultivar from Cyprus Arimont combines a large number of seeds per spike with large seed and Nude Glacier and Nude Vantage are both hulless types.

Table 1. List of base germplasm in RSP-4 -- 2 row barley.

Herta	Union
Ingrid	Virio
Bruens Wisa	Summit
Zephyr	Maris Mink
Erbet	Nude Compana
Dekap	Cumhuriyet 50
Hector	Toonucier
Waxy Compana	Compana msg 10

Table 2. List of base germplasm in RSP-5 -- 6 row barley.

Gem	Steptoe
Unitan	Nude Vantage
Arimont	Nude Glacier
Waxy Titan	Minn 21
Nordic	Athenais
Atlas 68	Atsel
Beecher	Calif Mariout 67
Galt	Manchuria (2330) msg 10

In initiating the populations, each cultivar was crossed to the male sterile component and equal amounts of seed were saved from each cross. A separate population has been established for each important disease and as new sources of resistance were determined, these have been crossed into the populations.

The recurrent selection cycle consists of two generations: selection for resistance and recombination. Selection for resistance has been conducted each year in disease nurseries in many barley growing regions of the world. Seed from about 200 fertile heads of resistant plants was harvested at each location. Seed returned from these disease nurseries was then combined and planted in a recombination nursery in the desert southwest of the United States during the winter. Only seeds set on the male sterile plants were harvested to continue the next cycle and to assure out-crossing of germplasm. Using this method, it is possible to obtain 50,000 or more F_1 seeds from each population. One cycle of recurrent selection is completed each year. After several cycles

of selection and recombination there should be a pyramiding and
accumulation of genes for resistance.

The virulence pool for specific diseases is sampled by growing
nurseries in several locations throughout the world and selection
of resistant plants assures that resistance to these virulence
types is incorporated (Reinhold and Sharp, 1980). At the same time
a number of diseased barley samples have been collected at the
different sites and seedling tests were conducted in controlled
environment chambers to detect new virulence types. Barley
cultivars and lines showing resistance to these virulence types
were also crossed into the recurrent selection populations. Some
virulence types of *P. teres* affecting different barley cultivars
are shown in Tables 3 and 4. Four main virulence types were
determined for Montana isolates and 5 virulence types were determined isolates from the Middle East (Bjarko, 1979). These 9
virulence types were then used for evaluating resistance of a
number of barley cultivars reported at some time or place to be
resistant to *P. teres*. Of more than 160 barley cultivars and lines
evaluated to the 9 virulence types, 9 lines of barley were
resistant to all virulence types and 15 barley lines were resistant
to eight of the nine virulence types. A number of sources of
resistance were also determined to be resistant to scald *(Rhynchosporium secalis)*. These barley cultivars and lines are listed
in Table 5 and have been crossed into the 6 row population for
scald resistance. Since both the 6 row barley recurrent selection
populations for net blotch and scald show a high percentage of
resistant plants wherever grown, resistance to the two diseases
has been combined into one population.

Table 3. Virulence groups of Montana isolates of *Pyrenophora teres*
in relation to 15 different barley cultivars.

Virulence Groups	1	2	3	4	5	6	7	8	9	10	11	12	13	14	15
Group A	R	R													
Group B	R	R									R	R	R	R	
Group C	R	R		R							R	R	R	R	R
Group D	R	R									R				

1. Unitan	6. Ingrid	11. CI 7584
2. Steptoe	7. Betzes	12. CI 9776
3. Shabet	8. Firlbecks III	13. CI 9819
4. Hypana	9. Mona-Arivat	14. CI 5791
5. Dekap	10. Arimont	15. Tifang

Table 4. Virulence groups of Middle East isolates of *Pyrenophora teres* in relation to 15 different barley cultivars.

Virulence Groups	1	2	3	4	5	6	7	8	9	10	11	12	13	14	15
Group A	R	R									R	R	R	R	R
Group B	R											R	R	R	
Group C	R	R	R			R							R	R	
Group D	R	R	R	R		R	R	R			R		R	R	R
Group E	R		R			R	R				R	R	R	R	R

1. Unitan	6. Ingrid	11. CI 7584
2. Steptoe	7. Betzes	12. CI 9776
3. Shabet	8. Firlbecks III	13. CI 9819
4. Hypana	9. Mona-Arivat	14. CI 5791
5. Dekap	10. Arimont	15. Tifang

Evidence has been obtained that minor effect additive genes also occur for some diseases of barley (Bjarko, 1979). Several crosses of barley resulted in transgressive segregation for resistance to net blotch (Table 6). Recurrent selection populations containing only base cultivars exhibiting susceptibility to net blotch have been established and plants showing some evidence of resistance have been saved in each generation to accumulate minor genes. Following two cycles of recurrent selection more than 50% of the plants in the population were resistant or intermediate in reaction when evaluated with a mixture of virulence types of *P. teres*. Similar evaluations are underway with the scald disease of barley.

Table 5. List of resistant germplasm in RSP-5 Rrs. (*Rhyncosporium secalis*).

Trebi	CI 1218	La Mesita
Jet	CI 1257	CI 8158
Kitchin	CI 1622	CI 8159
Steudelli	CI 2107	CI 9255
Atlas 46	CI 3086	CI 9556
Modoc	CI 3940	CI 11577
Nigrinudum	CI 3944	CI 11579
Turk	Atlas	CI 11628
CI 668	CI 4354	CI 11839
CI 691	CI 4686	CI 13644
CI 935	CI 5581	CI 13682
CI 1021	CI 5823	CI 10078

Table 6. Evidence for additive resistance to *Pyrenophora t res.*

Cross	F$_1$ Reaction	F$_2$ Reaction Type				Isolate
		1	2	3	4	
Tifang (4) [a] x Georgie (4)	4	1[b]	6	18	158	Mt 77-1
Betzes (4) x Georgie (3,4)	3,2	0	7	40	254	Mt 77-6
Georgie (3,4) x Firlbecks III (4,3)	3,2,4	25	65	126	155	Mt 77-6
Betzes (4) x Firlbecks III (4,3,2)	4,3	0	13	28	328	Mt 77-7
Tifang (4,3) x Firlbecks III (3,4)	3,2	4	69	136	182	Mt 77-51

[a] Parental reaction type in parentheses.

[b] Indicates number of F$_2$ plants in each reaction type.

CONCLUSION

Male sterile facilitated recurrent selection populations should be considered as germ plasm banks that can be continually exploited and improved (Bockelman *et al.*, 1980a, 1980b, 1980c, 1980d). Local plant breeders are encouraged to grow the populations and improve them agronomically for local environments. Additional locally adapted cultivars may be added to the population as the cycles of selection and recombination continue. Following addition of cultivars to the population, disease resistance may temporarily decrease but will be later be recovered. Individual plant selections may be made from the population at any time and evaluated in the same manner as segregating material in a pedigree system. Ramage (1980) has used seed size, determined by use of screens or gravity separators, as the primary criteria for selecting improved barley from recurrent selection populations. The proportion of plant selections that are male sterile can be easily removed from the segregating lines by progeny testing. Direct crossing of selections from the population with external susceptible breeding material is not recommended as this will result in a break-up and loss of some of the resistance.

REFERENCES

Barnes, D. K., Hanson, C.H., Frosheiser, F. I., and L. J. Elling, 1971, Recurrent selection for bacterial wilt resistance in alfalfa, *Crop Sci.*, 11:545

Bjarko, M. E., 1979, Sources of and genetic action of resistance in barley to different virulence types of *Pyrenophora teres*, the causal agent of net blotch, M.S. thesis, Montana State University.

Bockelman, H. E., Eslick, R. F., and Sharp, E. L., 1980a, Description of RSP-5A Rpg., *Barley Newsletter*, 23:35.

Bockelman, H. E., Eslick, R. F., and Sharp, E. L., 1980b, Description of RSP-5A Rpt., *Barley Newsletter*, 23:36.

Bockelman, H. E., Eslick, R. F., and Sharp, E. L., 1980c, Description of composite cross XXXVI, *Barley Newsletter*, 23:38.

Bockelman, H. E., Eslick, R. F., and Sharp, E. L., 1980d, Registration of composite cross XXXVI, *Crop Sci*, 20:675.

Harlan, H. V., and Martini, M. L., 1929, A composite hybrid mixture, *Jour. Amer. Soc. Agron.*, 21:487.

Hill, R. R., Sherwood, R. T., and Dudley, J. W., 1963, Effect of recurrent selection on resistance of alfalfa to two physiologic races of *Uromyces striatus medicaginis*, *Phytopath.*, 53:432

Hockett, E. A., Eslick, R. F., Reid, D. A., and Wiebe, G. A., 1968, Genetic male sterility in barley, II Available spring and winter stocks, *Crop Sci.*, 8:754.

Jenkins, M. T., Robert, A. L., and Findley, W. R., Jr., 1954, Recurrent selection as a method for concentrating genes for resistance to *Helminthosporium turcicum* leaf blight in corn, *Agron. Jour.*, 46:89.

Jihahyon, S., and Russell, W. A., 1969, Evaluation of recurrent selection for stalk rot resistance in an open-pollinated variety of maize, *Iowa State Jour. Sci.*, 43:229

Penny, L. H., Scott, G. E. and Guthrie, W. D., 1967, Recurrent selection for European corn borer resistance in maize *Crop Sci.*, 7:407.

Ramage, T., 1981, Personal communication, University of Arizona, Tucson, Arizona 85717 USA.

Reinhold, M., and Sharp, E. L., 1980, Some virulence types of leaf rust of barley (*Puccinia hordei*) in semi-arid areas Proceedings of the 5th European and Mediterranean Cereal Rusts Conference, Bari, Italy, May 28 - June 5.

Sharp, E. L., 1980, Annual report on control of barley diseases for lesser developed countries of the world, Agency for International Development, Washington, D.C. 46 pp.

Singh, S. P., 1977, Use of male sterility for simultaneous conversion and improvement of alien sorghums, *Crop Sci.*, 17:482.

Suneson, C. A., 1956, An evolutionary plant breeding method, *Agron. Jour.*, 48:188

DISCUSSION

DE PONTI: Recurrent selection is a good selection procedure to accommodate vertical genes in a breeding population as there is a high frequency of recombination. I believe this is right if disease or pest attack is very large. If not, there is a large chance of escape and alternate line selection with recombination to select out 'escape progenies' may be better.

SHARP: Yes, in cases of low disease prevalence escapes may occur and mistakenly be considered as resistant. In these cases, it may be possible to select for resistance to an alternate disease or plants may be selected for favorable agronomic traits.

DE MILIANO: Gravity selection was used in Zambia on a segregating population (F3 and F4) of wheat to select for stem rust resistance. The parents were susceptible cultivars but selection of large seeds by this method resulted in a greater level of susceptibility. Possibly there is an association of grain size and susceptibility as plants which became diseased early may have produced smaller heads with larger seeds.

SHARP: Screening or gravity separation to obtain large seeds is applicable to some diseases but one must know the disease concerned. In some cases large seeds may be an indication of little effect while in other cases the main effects of the disease may be on kernel number. Large seeds may be the result of small heads with fewer but larger seeds.

DURABLE RESISTANCE AND DISEASE COMPLEXES

G.S. Sidhu and J.M. Webster[*]

Department of Plant Pathology
University of Nebraska-Lincoln
Lincoln, Nebraska, USA
[*]Department of Biological Sciences
Simon Fraser University
Burnaby, Canada

The present paper is an attempt to determine whether the concept of durable resistance (Johnson and Law, 1975) is applicable to disease complexes which are of common occurrence in agro- and eco-parasitic systems.

Meloidogyne-Fusarium complex. The root-knot nematode (*Meloidogyne*) species and *Fusarium* species are common parasites of most plants, and usually exist simultaneously in the soil. Together they may form a disease complex in a common host by inducing modifications of the host response through intra- and inter-specific interactions. Crop cultivars genetically resistant to the *Fusarium* spp. often become susceptible to these fungi when challenged by *Meloidognye* spp. (Sidhu and Webster, 1977a). The breakdown of fungal resistance is usually related to the intensity of nematode population present in the rhizosphere. However, in some cases host resistance is not modified any further after a certain level of nematode population is reached (Sidhu and Webster, 1981), and this level varies between different cultivars. Some cultivars, therefore, exhibit an acceptable level of resistance even after their resistance response has been modified by the nematode. Surprisingly, such levels of resistance remain relatively constant under various environmental conditions over a long period

of time – an observation which seems to be analogous to the characteristics of durable resistance. Therefore, it can be safely concluded that 'disease complexes' may also exhibit 'durable resistance'.
The following example of a specific disease complex will illustrate
our point.

    Root-knot nematode–*Fusarium* wilt complex. Twenty-five tomato
cultivars were tested routinely for resistance to *Meloidogyne-
Fusarium* complex for seven years under greenhouse conditions.
Cultivars which were genetically resistant to the wilt fungus but
susceptible to the root-knot nematode were evaluated for wilt response after inoculations by both of the pathogens sequentially
and simultaneously using 5,000–6,000 nematodes and 6×10^6/ml
Fusarium propagules per plant. Based on the wilt disease index
(Sidhu and Webster, 1977b), the tomato cultivars fell into three
arbitrarily designated reaction classes. Disease indices below
two, two to three, and three to five were equated with resistance,
moderate resistance and susceptible reaction classes, respectively.
Cultivars representing the three reaction classes, along with their
responses to single and concomitant inoculations, are shown in
Table 1.

Table 1. Response of tomato cultivars to inoculation with
        Meloidogyne incognita (N) and *Fusarium oxysporum* f.sp.
        lycopersici (F) separately and together.

Cultivar	General reaction to*		Disease index when inoculated with:	
	N	F	F	N + F
Ohio W.R. 25	S	R	0.03	2.40
Walter	S	R	0.04	2.05
Chico	S	R	0.05	2.90

* S = susceptible; R = resistant.

The data indicate that cultivars 'Ohio W.R. 25' (hereafter referred to as Ohio) and 'Walter' exhibited only a moderate resistance whereas 'Chico' fell into a susceptible class. To determine
whether these disease responses would be influenced by a further
increase in nematode population, the three tomato cultivars were
inoculated with different population levels ranging from 0–10,000
per plant (Table 2). The disease index based on leaf chlorosis
increased significantly with the corresponding increase in the
nematode population up to 5,500 and 6,500 nematodes per plant on
Ohio and Walter, respectively (Sidhu and Webster, in press).

Table 2. Modifications in *Fusarium* wilt disease index on tomato cultivars Ohio W.R. 25, Walter and Chico after inoculations with different levels of root-knot nematode populations.

| Nematode | Disease Index | | |
number	Ohio W.R. 25	Walter	Chico
0	0.03^{a*}	0.01^{a*}	0.05^a
1000	0.18^a	0.13^a	0.25^a
2000	0.99^b	1.21^b	1.15^b
4000	1.44^c	1.41^{bc}	1.75^c
4500	1.76^{cd}	1.54^{cd}	2.00^c
5000	2.09^{de}	1.72^{cd}	2.50^d
5500	2.56^f	1.70^{cd}	2.70^d
6000	2.56^f	1.85^d	2.90^e
6500	2.31^{ef}	2.31^e	3.50^f
7000	2.35^{ef}	2.35^e	3.95^g
7500	2.31^{ef}	2.31^e	4.35^h
8000	2.41^e	2.41^e	4.75^i
10000	2.65^f	2.35^e	5.00^j
S.E.	0.87	0.73	0.85

* Values within a column followed by the same letter are not significantly different (0.01) by Duncan's Multiple Range Test.

Thereafter, an increase in the nematode number had an insignificant influence on the disease index of these cultivars. By contrast, 'Chico' showed a constant increase in disease index with the corresponding increase in the number of infective nematode larvae. Thus the resistance of the cultivars varies in its response to increasing nematode populations as shown in Fig. 1. Data from 1972-1978 indicated that this type of modification in *Fusarium* response remains fairly stable over a long period of time (Fig. 2). The resistance shown by 'Ohio' and 'Walter' to *Fusarium* wilt in the presence of virulent nematode populations is fairly stable and, therefore, can be regarded as 'durable' in nature. Furthermore, the population levels of both the pathogens used in the greenhouse tests are normally found under field conditions (see Baker and Nusbaum, 1971). Kappelman (1975) working with the *Meloidogyne-Fusarium* complex on cotton showed that greenhouse and field results on breakdown of *Fusarium* resistance were positively

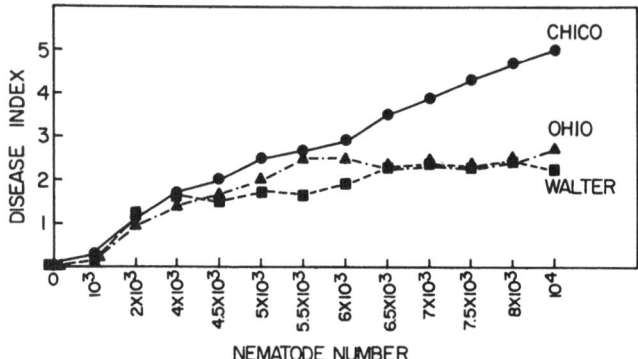

Fig. 1.   Influence of various levels of root-knot nematode
          populations on wilt severity of three tomato cultivars.

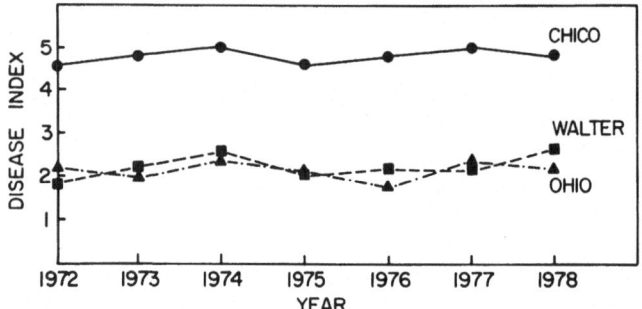

Fig. 2.   Maximum modification in wilt severity of three tomato
          cultivars induced by root-knot nematode as observed
          for seven years.

correlated. These results support the relevance of our data to
the field situation. Because the type of physiologic specialization found in rusts and mildews is almost lacking in *Fusarium*
wilt and root-knot nematode pathogens of tomato, rapid pathogen
virulence changes are unlikely.

        Under field conditions in Florida, many tomato cultivars
known to be resistant to race 1 and 2 of *Fusarium* wilt exhibited
tolerance in the field (Jones and Crill, 1974) even though the
tomato fields are usually found to be infested with root-knot
nematode - a resident parasite of most soils.

        Recently Kappelman (1980) analysed field data from 1969-
1978 on mean wilting of several cotton cultivars grown under uniform infestation of soil by *Fusarium* and root-knot nematode pathogens. The results clearly demonstrated that a uniform level of
Fusarium response persisted from 1968-1975, and a further improvement in cotton resistance by selection also remained stable

since 1975. Such a response is typical of durable resistance because resistance is incomplete and is stable under field conditions
for many years. The modifications in host response by nematodes
generally indicate a polygenic type of resistance (Sidhu and
Webster, unpublished). Smith and Dick (1960) demonstrated that the
genetic basis of resistance to *Fusarium* wilt in cotton under field
conditions was controlled by polygenes and a simple Mendelian segregation was found only after the soil was fumigated with ethylene
dibromide to kill the root-knot nematode. Thus monogenic resistance
to a single pathogen can be transformed into a polygenic resistance
in a disease complex situation, which indicates that durable
resistance to diseases involving pathogen interactions is determined by polygenic action. This is readily observed under field
conditions were environmental factors are variable and indeed, many
examples of polygenic resistance are based on field data (Sidhu,
1975).

"Disease complexes, durable resistance and polygenes. The
interactions in a disease complex involve epistatic, hypostatic
and neutral parasites. An epistatic parasite is one which influences
the expression of another parasite, and a hypostatic parasite is one
whose expression is influenced, whereas in a neutral interaction the
parasites do not interact with each other. Epistatic and hypostatic
are relative terms in the sense that epistatic and hypostatic parasites may change positions depending upon the components of a given
disease complex. Nevertheless, the phenomena of parasitic epistasis seems to operate through cyclic interactions leading to progressive modification of the host response. Three types of modifications in a disease-phenotype can be observed: (i) induced resistance, (ii) induced susceptibility and (iii) neutral interaction.
In a resistance response a genetically susceptible host to a hypostatic parasite is made relatively resistant to the same parasite,
whereas in an induced susceptible response the reverse happens
(Sidhu and Webster, 1977a). These modified responses through
parasite epistasis often lead to intermediate levels of disease
reactions, which because they are also influenced by other abiotic
and biotic factors, show further variability in their levels of
expressions. However, the fact remains that the induced modification either towards resistance or susceptibility is never
qualitative; the range of variability is similar to that expressed
by polygenic resistance, and is stable under field conditions over
a long period of time. This suggests that disease complexes
can modify oligogenic resistance into a relatively stable polygenic type of resistance with inherent features of durable resistance. Therefore, the possibility of durable resistance operating
through disease complexes should be kept in view.

REFERENCES

Baker, K.R., and Nusbaum, C.J., 1971, Diagnostic and advisory

programs, In: Plant Parasitic Nematodes, Vol. 1. (B.M.
    Zuckerman, W.F. Mai, and R.A. Rohde, eds.), 281-301, Academic
    Press, New York and London.
Johnson, R., and Law, C.N., 1975, Genetic control of durable
    resistance to yellow rust (*Puccinia striiformis*) in the wheat
    cultivar Hybride de Bersee, *Ann. appl. Biol.*, 81:385-391.
Jones, J.P., and Crill, P., 1974, Susceptibility of "resistant"
    tomato cultivars to *Fusarium* wilt, *Phytopathology*, 64:1507-
    1510.
Kappelman, A.J., Jr., 1975, Correlation of *Fusarium* wilt of cotton
    in the field and greenhouse, *Crop Sci.*, 15:270-277.
Kappelman, A.J., Jr., 1980, Long-term progress made by cotton
    breeders in developing *Fusarium* wilt resistant germplasm,
    Crop Sci., 20:613-615.
Sidhu, G.S., 1975, Gene-for-gene relationships in plant parasitic
    systems, *Sci. Prog.* (Oxford), 62:467-485.
Sidhu, G.S., and Webster, J.M., 1977a, Genetics of simple and
    complex host-parasite interactions, *Proc. Intl Atomic Energy
    Agency, Vienna 1977*, 59-79.
Sidhu, G.S., and Webster, J.M., 1977b, Predisposition of tomato
    to the wilt fungus (*Fusarium oxysporum lycopersici*) by the
    root-knot nematode (*Meloidogyne incognita*), *Nematologica*,
    23:436-442.
Sidhu, G.S., and Webster, J.M., 1982, Influence of population levels
    of root-knot nematode on *Fusarium* wilt, *Phytoprotection* (in
    press).
Smith, A.L., and Dick, J.B., 1960, Inheritance of resistance to
    Fusarium wilt in Upland and Sea Island cotton as complicated
    by nematodes under field conditions, *Phytopathology*, 50:
    44-48.

DURABLE RESISTANCE AND HOST-PATHOGEN-ENVIRONMENT INTERACTION

J.C. Zadoks and J.A.G. van Leur

Laboratory of Phytopathology
Binnenhaven 9
6709 PD WAGENINGEN
The Netherlands

SUMMARY

   Pathosystems may be horizontal, vertical, or mixed. They can be
characterized epidemiologically by the degree of statistical interaction between host and pathogen phenotypes. Three levels of interaction are distinguished. (i) The 'unnamed interaction phenomenon',
(ii) the 'small interaction phenomenon', and (iii) the 'common
cultivar-race interaction'. Future development may lead to reduced
verticality and increased horizontality of agricultural pathosystems.
If so, the three-way interaction cultivar-isolate-environment will
become more important, with possibly more races in a better stabilized
pathosystem.

INTRODUCTION

   Durable resistance in a cultivar is expected to hold at least
during the economic life of that cultivar. Durability of resistance
does not necessarily mean that the cultivar is and remains free of
a pathogen; it may also mean that the cultivar puts a strong brake
on the development of the pathogen. By pathogen we mean all its
phenotypes, those presently existing and those to be found later.

   The concept of durable resistance is not limited to partial or
general resistance as a mechanism. Nevertheless, a convention seems
to be spreading: 'The real trade mark of durable resistance is the
presence of just a few well-developed lesions'. This convention,
imposed by the technicalities of selection, is difficult to accept
for conservative plant breeders and disease managers, and even more
so for seed merchants and farmers. The technicalities are simply
these: To avoid hypersensitivity, the breeder selects those lines

125

that carry at least one well developed lesion; under conditions of
high disease pressure, more lesions are acceptable but they should
be few and scattered. The lesion becomes the pictogram for guaranteed
quality not unlike the one for '100% virgin wool'. Who believes?

Who of the prospective customers will buy a cultivar with so-
called durable resistance when the first lesions are already visible?
These customers reason that, if a few pustules are apparent now and
here, how can they be sure that the cultivar will stay undamaged
during the seasons to follow? This question, already asked so many
times, has remained unanswered. The following is an attempt to scan
existing knowledge on forms of partial resistance for a possible
answer, in which we will concentrate on intermediate resistance.

THE INFECTION CYCLE

A pathogen, to be successful, has to go through a reproductive
cycle usually called the infection cycle. The infection cycle consists of a number of consecutive phases, which can be recognized
morphologically. A quantitative breakdown of the infection cycle is
possible by counting the numbers of infection units that have developed up to a certain phase. The counts have to be repeated peri-
odically from the day of inoculation until the day on which the last
lesion dies.

Following the mathematical techniques of the life insurance com-
panies and ecologists the data can be transformed into a components
analysis, a life table and a survivorship table (Zadoks and Schein,
1979). The components analysis (Table 1) immediately shows at what
phase the brakes are put on: the running product makes a large drop.
The components analysis is a flexible tool. One can include as many
phases and sub-phases of the infection cycle as desired under only
one condition: the phases should be consecutive, that is, the following phase begins where the preceding phase ends. In other words: it

Table 1. Components analysis of the uredial cycle of brown
          leaf rust (*Puccinia recondita*, isolate 1037) on
          wheat. (From Zadoks and Schein, 1979).

		Observed Value	Running Product
GTR	(Germinated spores/Total spores Ratio)	0.93	0.93
AGR	(Appressoria/Germinated spores Ratio)	0.89	0.83
VAR	(Vesicles/Appressoria Ratio)	0.45	0.37
HVR	(Hyphae/Vesicles Ratio)	0.84	0.31
CHR	(Colonies/Hyphae Ratio)	0.67	0.21
PCR	(Pustules/Colonies Ratio)	1.53	0.32

is irrelevant whether the distance is covered in a few long strides
or in many short steps, but no jumps are allowed. The components
analysis is highly informative but laborious and it has a scientific
disadvantage: it lacks the time perspective.

The deficiency is remedied by constructing a life table. In the
cohort life table (Table 2) one recognizes the running product of
the components analysis in column 7. The reproductivity table, which
we will not discuss, summarizes reproduction of the successful infection unit. Reproduction may be low or high, early or late. Reproduction is characterized by three cardinal figures (Table 3),

R_o = the net reproduction rate per individual,
T_g = the mean generation time, and
r_{max} = the maximum relative growth rate.

A quantitative epidemiological definition of durable resistance could
be that durable resistance is characterized by a positive but small
R_o, a positive and large T_g, and a positive but small r_{max}, which,
in addition, are approximately equal with all phenotypes of the
pathogen.

Table 2. Cohort life table of brown leaf rust (*Puccinia
recondita*, isolate 1037) on primary leaves of
wheat. (From Zadoks and Schein, 1979).

(1) t	(2) State*	(3) Infection Units (per 1000 stomata)	(4) q_t	(5) $1-q_t$	(6) Ratio	(7) l_t	(8) e_t
0	SPD	294	0.07	0.93	GTR	1.00	6.3
1	GED	273	0.11	0.89	AGR	0.93	5.7
1	APD	243	0.55	0.45	VAR	0.83	6.4
3	VED	110	0.16	0.84	HVR	0.37	11.1
3	HYD	ˋ92	0.33	0.67	CHR	0.31	13.2
9	COD	61	-0.53	1.53	PCR	0.21	12.2
9	PUD	94	1.00	0.00	SPR	0.32	8.0
25	SUD	0					

*States of the infection process in cereal rusts. The
numbers of infection units are expressed as densities
(per 1000 stomata) in column 3. SPD, total spore density;
GED, germinated spore density; APD, appressorium density;
VED, substomatal vesicle density; HYD, hyphal density;
COD, colony density; PUD, pustule density; SUD, surviving
pustule density.

Table 3.   Reproductivity table of brown leaf rust (*Puccinia recondita*, isolate 1037) on primary leaves of wheat (cv Rubis).   (From Zadoks and Schein, 1979).

t	m_t	l_t	$l_m m_t$	$t l_t m_t$
9	1158	.319	369	3325
11	2000	.279	558	6138
13	1922	.239	459	5972
15	1026	.199	204	3063
17	1118	.159	178	3022
19	1080	.119	129	2442
21	526	.079	42	873
23	0	.039	0	0
25	0	.000	0	0
Σ			1939	24835

$$R_o = \Sigma l_t m_t = 1939$$

$$T_g = \frac{24835}{1939} = 12.8$$

$$r_{max} = \frac{\log_e 1939}{12.8} = \frac{7.57}{12.8} = .59$$

In principle, r_{max} is a predictor of the logistic infection rate r. The value r cannot exceed the value r_{max}. A cultivar with a low r_{max} in the monocyclic experiment indoors also shows a low r in the polycyclic experiment outside. Nevertheless, there is a weakness: the theory assumes that the infectivity of the infectious units is constant through time. When the pathogen has been influenced by a fungicide during the infection cycle, the resulting infectious units (spores) may have a lower infectivity than the original spores (Drandarewski and Schicke, 1975). The same might happen after passage through a partially resistant host.

Two supplementary remarks must be made. One: most quantitative studies of the infection cycle cover only part of the cycle. They concentrate on those components of resistance that seem most promising for analysis. With some effort, nearly all of these studies could be fitted in with the theory presented here. Two: with appropriate modifications the theory is applicable also to bacteria, viruses, nematodes, and insects as harmful agents. For insects, either as harmful agents or as vectors thereof, additional components are behavioral: searching, probing, feeding and ovipositing. Here, the borderline between partial resistance and exclusion resistance becomes vague.

PATHOSYSTEM CHARACTERISTICS

Epidemiologic reasoning may lead to conclusions on the epidemiologic behavior of a pathosystem: horizontal, vertical or two-dimensional. The epidemiologist should not, however, jump to conclusions about the genetic nature of his pathosystem. Genetic conclusions must be based on genetic experiments. Errors of judgement are easily made for a variety of reasons.

An example is given from the wild pathosystem of sea lavender rust: *Limonium vulgare* and *Uromyces limonii*. The results pointed towards a horizontal pathosystem with constant ranking, but when rust isolates had been hybridized, two rust isolates in the offspring behaved like typical races in the classic sense. Possibly, an error was made by relying on too small a sample of host genotypes or pathogen isolates. It may be, however, that the natural pathosystem is horizontal indeed. In a wild pathosystem with a population of over one million different host genotypes a new race will have difficulties to find its compatible host plant. As in this particular pathosystem no uredo-race can live longer than one summer, there is little scope for verticality. Horizontality may be the major characteristic of the pathosystem and verticality a marginal trait, if it occurs at all in nature, but verticality does exist in potential. The *Limonium-case* is as yet unsolved, but far more difficult cases exist. The problem is that the available descriptions are inadequate for our purpose.

Nilsson (1969, 1972) made an extensive study on *Gaeumannomyces* (*Ophiobolus*) *graminis* var. *graminis* of wheat. His results, not written for the present purpose, indicate a typical constant ranking pathosystem, constant ranking being obvious for both host cultivars and pathogen isolates. As highly virulent isolates form a minority, a balancing mechanism must operate in this pathosystem, in which grass hosts play an important role. The pathosystem is only partially domesticated.

THE UNNAMED INTERACTION PHENOMENON

Another partially domesticated pathosystem is that of rice and *Helminthoporium* (*Drechslera*) *oryzae* in Surinam (South America), where cultivated rice is a relatively recent introduction and the brown spot fungus is common on native grasses. The impression is that newly bred rice cultivars used to be clean at the time of introduction. After some two years of cultivation or four crop generations they were more or less spotted by Helminthosporium, and, according to weather conditions, losses occurred. Klomp (1977) established four facts: (i) grasses were the primary source of inoculum, (ii) the disease occurred mainly on weakened plants, (iii) no evidence for physiologic specialization was found, and (iiii) isolates from rice showed morphological characteristics which were

(a) definitely related to their respective host cultivars and (b) were temporary as they disappeared after a few transfers on agar. The available facts suggest the following interpretation. A new and clean rice cultivar confronted with an infinitely large pathogen population exerts a selection pressure so that a transient sub-population of the pathogen arises which is slightly more pathogenic to the selecting cultivar; a temporary ripple in a semi-wild patho-system, but of real economic importance.

The case of the rice brown spot is by no means unique. There is a similar case of *Helminthosporium teres*, causing net blotch on barley, in Israel. Barley varieties, originally clean, succumbed one after the other to net blotch. The existence of typical races was not established.

Another case comes from the Netherlands. The Zuyder Zee Development Authority, managing the largest farm of the Netherlands, can test and grow new cultivars before they have been officially registered. During their life history these cultivars usually yield highest before registration; after registration cultivars gradually decline in yield by over 1% per year (Fig. 1; De Jong, 1981). The cause of the phenomenon is but partially known. Registration only marks the time. Loss of hybrid vigor through successive generations is a possible but improbable cause. Gradually increasing fungal infection of the foliage and possibly of the roots is certainly a cause, obvious for *Puccinia striiformis* and *P. recondita*, but elusive for many less specialized fungi.

How should we call this 'unnamed interaction phenomenon'? How frequent is this type of interaction between host and pathogen populations? Probably, it is very frequent but very difficult to prove beyond doubt.

THE SMALL INTERACTION PHENOMENON, 'TRAINING'

There exists a category of adaptations of fungal populations to their hosts that is better documented than the unnamed interaction phenomenon, though the cases are not very convincing. Pieters and Van der Graaff (1980) studied resistance to *Gibberella xylarioides* in Coffea arabica. In a germination test they determined the percentage of germinated conidia on the secondary wood of branch internodes for various coffee genotypes and pathogen isolates. Re-analysis of their data (Table 4) shows significance of main effects and, contrary to the authors' conclusion, a significant interaction can be found. Pieters and Van der Graaff conclude that their data is evidence for the horizontality of the coffee-vascular wilt patho-system. Horizontality certainly predominates but a rudiment of verticality is present, unless the significance is only accidental.

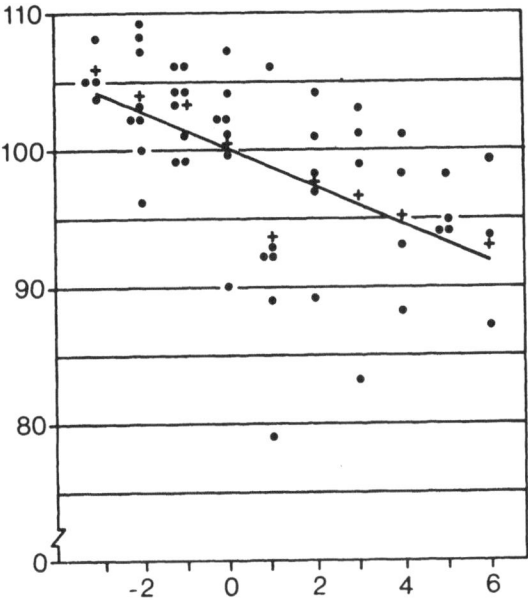

Fig. 1. Yield decline in winter wheat. Relation between relative
         yield of a variety in per cent (vertical) and age class
         in years of that variety (horizontal). Data from variety
         trials on the farm 'De Schreef', 1959-1969. The yield of
         a variety in any particular year was determined relative
         to the mean yield (100%) of the variety trial in that
         year. The age class of the variety was expressed in years
         from registration (year 0). The regression is -1.39% per
         year, r = 0.54, n = 53. ● : Yields per variety per age class
         + : Mean yield of varieties per age class.

     Another example comes from studies on *Armillariella mellea* and
forest trees. Guillaumin and Pierson (1978) studied the pathogenicity of various isolates to different tree species in a factorial
experiment; they found only main effects and concluded that no
physiologic specialization occurred. Re-analysis of their data by
ANOVA (Table 5) as well as by non-parametric techniques showed a
significant interaction between isolates and host species, suggesting
at least a certain specialization of an isolate on the host species
of origin.

     The interactions indicated in these two examples could be con-
sidered as epidemiological evidence for some degree of verticality
in predominantly horizontal pathosystems. Admittedly, other expla-
nations of the interaction can be thought of, but the *Armillariella*
example is rather convincing. The analysis does not tell us whether
the specialization is temporary or not, nor what the genetic background, if any, is. Unfortunately, the literature provides few

Table 4.   Test matrix for *Gibberella xylarioides* on
lines of *Coffea arabica*, after Pieters and
Van der Graaff, 1980, Table 3. Entries are
recalculated data: (observed - expected) * 100.

Coffee lines	Isolates			
	a	b	c	d
F 20	16	31	-59	12
F 30	-19	-22	50	-10
S 947	-26	-14	75	-35
F 54	-32	21	- 9	17
F 21	- 2	7	3	-10
S 288	58	-25	-62	26

LSD (p < 0.05)*100 = 65

Table 5.   ANOVA of field results with *Armillariella mellea*
based on data by Guillaumin and Pierson, 1978.
4 pathogen isolates, 4 host species, 20 trees
per species per isolate.

Source	SS	df	MS	F	p
isolates	91	3	30	39	***
host species	405	3	135	172	***
isolates X species	37	9	4	5	***
residual	213	272	0.8		
total	746	287			

hosts: peach   *Prunus persica*
       walnut *Juglans regia*
       oak   *Quercus pubescens*
       fir   *Abies pectinata*

data suitable for re-examination. Nonetheless, the 'small interaction phenomenon' is an established fact, whatever it may mean. As
long as the genetics of the phenomenon are not known, the authors
suggest the word 'training' for the small interaction phenomenon.
B. Clifford (personal communication) has literally 'trained' isolates
of *Puccinia hordei* to some host cultivars including the partially
resistant Vada. The training took six generations and its effect, a
slight but significant increase in spore production, remained after
six passages over a universal suscept.

Parlevliet (1978) has shown convincingly that the small inter-action phenomenon can have a genetical background. He studied the genetics of the latency period in the barley leaf rust pathosystem (*Puccinia hordei*). Several host genes are involved. At least one rust isolate could overcome the latency prolonging effect of a host gene. Parlevliet and Zadoks (1977) consider such an effect to be a form of physiological specialization at the minor gene level.

TYPICAL INTERACTION PHENOMENA

Typical interaction phenomena are well-known under the name of 'physiologic specialization'. The race nurseries for yellow stripe rust (*Puccinia striiformis*) provide data from the field (Zadoks, 1961). They have to be interpreted with care.

An experiment with 28 race nurseries spontaneously infected by a single race causing a local epidemic showed a significant cultivar effect, as expected, and also a significant nursery effect (Table 6). The nursery effect is due to inevitable differences in soil, planting, and growing conditions. The F-ratio of the mean squares for cultivars and fields was large and highly significant. The cultivar is what matters.

A test on 6 race nurseries at 6 localities in Switzerland ex-posed during 2 years to the local rust population shows no signifi-cant difference between years, a significant effect of fields and a highly significant effect of cultivars (Table 7). As the variance ratio of cultivars over fields is 29, the conclusion is as before: though fields differ somewhat, there is only one race discernible.

Table 6. ANOVA for data from race nurseries of *Puccinia striiformis* on wheat, Triumph race, after Zadoks, 1961, Table A24.33. 1 race, 41 cultivars, 28 nurseries, 1 year.

Source	SS	df	MS	F	p
cultivars	592605	40	14815	308	***
nurseries	3160	27	117	2.4	***
residual	51990	1080	48		
total	647755	1147			
cultivars			14815	126	***
nurseries			117		

Finally, data from 6 easily recognizable field races with two isolates each and tested in one field were re-analyzed (Table 8). The main effects cultivar and race are highly significant as is their interaction. The essence of physiologic specialization, expressed in statistical terms, is interaction. In extreme cases, the main effects cultivar and race could be non-significant, but interaction must be significant. With the yellow rust example, we are on firm grounds. The existaence of field races is an established fact, even though genetic experiments with yellow rust are difficult.

Table 7.   ANOVA for data from trap nurseries of *Puccinia striiformis* on wheat, Probus race, Switzerland, after Zadoks, 1961, Table A24.42.

Source	SS	df	MS	F	p
cultivars	164352	23	7146	55	***
nurseries	1304	4	326	2.5	*
years	6.	1	6	0.05	
cv X nu	10541	92	115	0.9	NS
cv X yr	1574	23	68	0.5	NS
na X yr	1132	4	283	2.2	NS
residual	11848	92	129		
total	190757	239			

Table 8.   ANOVA for race nurseries of *Puccinia striiformis* on wheat, after data from Table A24.42 in Zadoks, 1961. 6 races, 2 nurseries per race, 39 cultivars, 1 year.

Source	SS	df	MS	F	p
races	751	5	150 ⌐	1	NS
betw. n. within r.	714	6	119 ⌐		
nurseries	1465	11	133 ⌐	3	*
cultivars	231853	38	6101 ⌐	139	****
race X cv	82497	190	434 ⌐	10	***
residual	9970	228	44 ⌐		
total	325785	467			

Sometimes, field races do manifest themselves in monocyclic tests on seedlings in the greenhouse. Under such conditions, Johnson (1972) found differences in sporulation intensity, which is one of the components of the components analysis mentioned in the introduction. Interaction at the polycyclic level in the field and interaction at the components level in the greenhouse can corroborate each other.

CONCLUDING REMARKS

Natural pathosystems seem to have a tendency towards horizontality, agricultural pathosystems tend to develop towards verticality. A return to horizontality in agricultural pathosystems may not be impossible as the history of resistance breeding against potato late blight suggests. When the attempts will be successful, host-pathogen interaction phenomena will not disappear but decrease in strength.

Physiologic specialization, interpreted as strong interaction at the monocyclic level, due to specific genes in host and parasite, will recede but mild interaction at the components level will become more prominent. As long as the interaction is due to specific genes, there is no reason not to call the latter phenomenon physiologic specialization, but it is a physiological specialization at a low level of integration: the components level. The methodology of analysis is only just developing; many new results may appear in the next few years.

Avoidance of epidemics by stabilization of pathosystems is believed to be possible for three reasons. (i) When genes with minor but measurable and different effects are combined, a pathogen may need longer to adjust by mutation. (ii) Host-pathogen interaction may be more sensitive to environmental effects, especially weather effects, the more horizontality becomes prominent. Experience with the *Uromyces-Limonium* pathosystem suggests that weather can have a moderating influence. Evidence of three-way interaction between cultivar, race and temperature, which enforces this argument, has been provided by Katsuya and Green (1967) for *Puccinia graminis* and Chandrashekar and Heather (1981) for *Melampsora laricis-populina* (Table 9). (iii) There is good evidence that in case of an emergency a single treatment with a systemic fungicide will be highly effective.

Extrapolating the present train of events the authors expect research to penetrate more deeply in the dark area between pure verticality and supposed horizontality, probably to find a verticality at lower levels of integration which is more or less confounded with environmental interactions. The number of races may even increase as the significant three-way interaction between cultivar, isolate and environment creates many ecological niches for these races; at the same time the three-way interaction stabilizes these races because the environment does not remain constant.

Extrapolating the present train of thought in an attempt to foresee future trends, the authors expect that, with increasing horizontality of an agricultural pathosystem, epidemics will still occur but will be less frequent, slower to appear, and, when appearing, slower to develop and easier to control by fungicidal treatment. If so, resistance will be more durable.

Table 9. Three-way interaction between host cultivar, pathogen isolate and environment (temperature) in *Melampsora laricis-populina* on poplars (Chandrashekar and Heather, 1981). 4 races, 4 cultivars, 3 temperatures. The criterium was the number of uredia per leaf disk.

Source	SS	df	MS	F	p
temperature	74	2	37	183	***
race	63	3	21	105	***
cultivar	27	3	9	46	***
te X ra	132	6	22	111	***
te X cv	42	6	7	33	***
cv X ra	9	9	1	4.7	***
te X ra X cv	36	18	2	10	***
residual	133	665	0.2		
total		712			

REFERENCES

Chandrashekar, M., and Heather, W.A., 1981, Temperature sensitivity of reactions of *Populus* spp. to races of *Melampsora laricis-populina*, *Phytopathology*, 71:421-424.

Drandarevski, C.A., and Schicke, P., 1975, Spore formation and germination of apple scab and powdery mildew following curative treatment with triforine, *In:*"Rep. Inf. VIII Int. Plant Prot. Congress Moscow, Sect. III Chemical Control":211-213.

Guillaumin, J.J., and Pierson, J., 1978, Etude du pouvoir pathogène de quatre isolats d'armillaire, *Armillariella mellea* (Vahl.) Kast., vis-à-vis de quatre espèces-hôtes, *Ann. Phytopathol.*, 10:365-370.

Johnson, R., 1972, Minor genetic variations for virulence in isolates of *Puccinia striiformis*, *In:*"Proc. Eur. Mediterranean Cereal Rusts Conf., Prague":141-144.

Jong, G.J. de, 1981, Het beleid ten aanzien van de rassenkeuze bij wintertarwe op het grootlandbouwbedrijf van de Rijksdienst voor de IJsselmeerpolders. *In:* Anonymus: 50 jaar onderzoek door de Rijksdienst voor de IJsselmeerpolders. Rijksdienst, Lelystad. Vol. IIB:247-254.

Katsuya, K., and Green, G.J., 1967, Reproductive potentials of races 15B and 56 of wheat stem rust, *Canad. J. Bot.*, 45:1077-1091.

Klomp, A.O., 1977, Early senescence of rice and *Drechslera oryzae* in the Wageningen Polder, Surinam, *Agric. Res. Rep.*, Wageningen 859:97 pp.

Nilsson, H.E., 1969, Studies of root and foot rot diseases of cereals and grasses, *Lantbrukshögsk. Ann.*, 35:275-807.

Nilsson, H.E., 1972, The occurrence of lobed hyphodia on an isolate of the take-all fungus '*Ophiobolus graminis* Sacc', on wheat in Sweden, *Swed. J. Agric. Res.*, 2:105-118.

Parlevliet, J., 1978, Race-specific aspects of polygenic resistance of barley to leaf rust, *Puccinia hordei*, *Neth. J. Pl. Path.*, 84:121-126.

Parlevliet, J., and Zadoks, J.C., 1977, The integrated concept of disease resistance; a new view including horizontal and vertical resistance in plants, *Euphytica*, 26:5-21.

Pieters, R., and Van der Graaff, N.A., 1980, Resistance to *Gibberella xylarioides* in *Coffea arabica:* evaluation of screening methods and evidence for the horizontal nature of the resistance, *Neth. J. Pl. Path.*, 86:37-43.

Zadoks, J.C., 1961, Yellow rust on wheat, studies in epidemiology and physiologic specialization, *T.o.P. (Neth. J. Pl. Path.)*, 67:69-256.

Zadoks, J.C., and Schein, R.D., 1979, "Epidemiology and plant disease management", Oxford University Press, New York:427 pp.

DISCUSSION

VAN DER GRAAFF: The small interaction found by you in the table of Pieters and Van der Graaff (1980) warrants two remarks: (a) analysis for a suspiciously large value requires a lower P value to conclude significance; (b) I believe that, most likely, the specific interaction is caused by non-additivity in the scale we used. Only repetition of the experiment can solve whether this is a specific host-parasite interaction or a statistical interaction.

ZADOKS: The interaction found is small and it may be due to the inevitable 1:20 chance of a wrong conclusion. If such an interaction is found, the experiment must - indeed - be repeated. Scaling effects could be eliminated by using non-parametric tests.

VAN DER GRAAFF: The statement that agricultural pathosystems tend to verticality is only relevant in relation to a small number of crops (inbreeding cereals). In, for example, clonal crops and outbreeding crops, this statement is not valid.

ZADOKS: My words reflect my hesitation. Verticality is common in cereals, in many horticultural crops, in clonal crops such as potatoes (late blight) and perennial crops such as coffee (leaf rust).

ØSTERGAARD: Life tables are founded on proportions of infection units surviving to a certain stage of the life cycle. To include the influence of infection density on survival, it is essential to measure absolute numbers instead of proportions. Then it also becomes possible to estimate the density dependence parameter and the basic death rate independently (as was discussed in my paper).

ZADOKS: Proportions have been used to eliminate the influence of infection density and to compare data based on variable infection densities. If your objective is to study density-dependent effects, the classical life table technique using absolute numbers has to be applied.

ØSTERGAARD: Analysis of variance on proportions of infected leaf area has to be performed with care. In many cases a logarithmic transformation will provide a better clarification of data; a possible interaction might disappear because differences between proportions are relative instead of absolute. Another reason for the logarithmic transformation is that of scaling (compare with Dr. Putter's reading of Prof. Vanderplank's second paper).

ZADOKS: Among statisticians the opinions about the usefulness of transformations vary. In my yellow rust work, I applied one transformation expressing all data as percentages of the most severely infected cultivar. This transformation eliminates variation due to environmental differences and to differences in infection load. Homoscedacity can be improved by means of the arc-sinus transformation (not the logarithmic transformation) but that hardly affects the final conclusions.

ROBINSON: You have referred to "verticality" and horizontality" with reference to the presence or absence of a differential interaction. There are many kinds of differential interaction. I would like to suggest that the word 'vertical' be used only in connection with the Person differential interaction; that is, the gene-for-gene relationship.

ZADOKS: The terms 'vertical' and 'horizontal' have been used here as epidemiological descriptors without genetical connotation, conforming to Robinson, Rev. appl. Mycol. 1969. I am ready to accept any other suitable terminology, provided it is consistent.

DINOOR: You allowed for two pustules to develop from each infection (both sides of the leaf). I wonder how you will account for

second generation uredia (especially in yellow rust), where the
establishment of a sporulating colony does not depend on germination, appresorium formation, penetration, etc.
2. You stressed that the most important stage in the expression
of resistance is the establishment of a vesicle from an appresorium.
This might be a case where structural resistance is involved
(small stomata), which is not typical for rusts. Resistance
cannot be manifested before contact between host and parasite
(within the tissue at least) has been attempted.
3. The data from Israel on net blotch of barley do not indicate
that physiologic races are not present. They only indicate that
pathogen cultures are not stable in their performance and this
could well be due to heterokaryosis.
4. It was agreed by most participants of the study institute that
the operation of a gene-for-gene system cannot be claimed without
genetic analysis of the host and the parasite. I wonder how you
can conclude a gene-for-gene system for wheat and yellow rust.
5. You have claimed that the tendency in the wild is for horizontality. Therefore, you would expect that epidemics will be rare,
slow and will not reach a high level of disease. I think that there
is not enough evidence to support horizontality in the wild. More
than that, we have experienced many cases of very common epidemics in the wild, with a high rate and a high disease level.
6. You mentioned that there are more prospects for environmental
conditions to modify horizontal, polygenic resistance. There is
no reason why the environment will not modify susceptibility to
look like complete or at least partial resistance.

ZADOKS: As long as the pustules can be identified as separate
units and counted there is no problem in establishing life tables.
In more complex cases (yellow stripe rust) I prefer simulation
techniques.
2. What might suggest structural resistance is an artifact due to
the inadequate timing of the counts.
3. I only stated that the existence of physiologic races has not
been proven.
4. Sexual hybridization of yellow stripe rust is not yet possible.
The conclusion about the operation of the gene-for-gene system is
drawn by analogy and has been validated by experiments on heterokaryosis and by mutation studies of the fungus.
5. You may be right. I suggested a tendency, implying therewith
that exceptions may occur. I would like to compare intensity and
extent of natural and agricultural epidemics.
6. Agreed in principle. My personal experience is that intermediate levels of resistance vary more than extremely low or high
levels of resistance in response to varying environmental conditions.

KRANZ: Component analysis, including life tables, is a very
useful tool in epidemiology. However, can you really assess the
impact of individual components on the dynamics of the entire

epidemic by means of that tool?

ZADOKS: As long as internal and external conditions of plant and pathogen remain constant, the results obtained by component analysis will be valid for the entire duration of an epidemic. However, inoculum load changes, environment changes, the host develops and its physiology changes. Therefore, component analysis has more value for comparison of cultivars at one moment in time than for predictive purposes. For predictive purposes, component analysis should be repeated at successive stages of the epidemic.

EPIDEMIOLOGICAL PARAMETERS OF PLANT RESISTANCE

Jürgen Kranz

Phytopathologie und angewandte Entomologie des
Tropeninstitutes der Justus Liebig-Universität
6300 Giessen, Federal Republic of Germany

SUMMARY

The term parameter and its use in epidemiology characterize
processes, states, and their changes. Qualitative parameters of
epidemics and their scope are outlined in relation to plant resistance. Quantitative parameters for components of the infection
cycle, widely used in resistance screening, are briefly reviewed.
Their choice as selection criteria cannot always be related to
their actual epidemiological bearing nor do they yet indicate durability of resistance. The effects of resistance on pathogen populations through selection and inter- as well as intra-specific
competition are epidemiologically of greater interest, fitness in
particular. But too little is still known to monitor changes in
fitness to facilitate resistance management. Also longevity of
cultivars cannot yet be assessed and predicted.

INTRODUCTION

Epidemiology of plant disease studies the dynamics of populations. Resistance in the host population, whatever type it may be,
is one of the most powerful driving forces in the development of
pathogen and disease populations on them. Knowledge about these
interactions are of use in resistance breeding as well as in the
management of diseases by means of resistance. Hence, our emphasis
in the following pages is on the use of epidemiological parameters
to achieve these objectives.

AN EPIDEMIOLOGICAL PARAMETER - WHAT IS THAT?

What are parameters?

Surprisingly the word 'parameter' rarely appears in the indices of important epidemiological books. Nevertheless, it is used in the texts, usually in connection with mathematical models. The mathematical concept of parameters, however, may be too narrow for usage in epidemiology. On the other hand, the term parameter tends to be used rather loosely nowadays. Very often it is confused with 'parameter value'. It is, therefore, pertinent to clarify its usage in this paper.

In general, parameters are features or attributes that characterize processes and/or states. Most authors treat parameters like constants as does Webster's Dictionary. Typical examples are the constants a and b in linear regression equations or the limits, infection points and rate constants in logistic curves. However, these constants may attain varying parameter values.

Apart from these quantitative parameters which require appropriate dimensions, qualitative parameters are also needed to characterize processes and states. For instance, a pathogen may be a k- or r-strategist, a disease monocyclic or polycyclic, or the epidemic may follow a given pattern of typical disease progress curves. Thus parameters are more than just 'measurement terms' (Butt and Royle, 1980).

Which parameter eventually is defined depends on the specific objective, or emphasis, and on personal discretion. Whenever possible a parameter should be operational, i.e. to be used as a tool, and not merely descriptive. Essential criteria for their optimal choice are unambiguity, adequate sensitivity as well as little interdependence (Plinston, 1972). Sensitivity here means that changes in parameters as conditioned by external factors are constantly proportional to the size of stimulus. This, for instance, is ensured by the x in linear regression equations.

The liberty to choose parameters is evident from the 3-page-list of abbreviations in the book by Zadoks and Schein (1979), and the often numerous definitions of state, rate and driving variables in simulators (Waggoner, 1978). However, practically there are rather a few essential ones.

Variables, like x or t (for time), actually represent real figures or magnitudes, or statements. They can assume different values within the domain they have been defined. Hence, variables used in experiments may yield parameter values after evaluation of the results.

Epidemiological parameters

In epidemiology parameters characterize processes and states that result from interactions of host and pathogen (or disease) populations with environment and human activities (Kranz, 1974). These processes occur at various systems levels or integrate different levels, thus creating systems with entirely new structures and behavior (e.g. pathogen and host $\longrightarrow$ disease) (Kranz, 1978). Behavior of the resulting epidemics differs by virtue of their inherent variability, and varying determining factors. During current processes behavior expresses itself in transient states and rates. But behavior can become manifest as final states. This is the case with past epidemics and the pattern of their disease progress curves (Kranz and Hau, 1980) and/or gradients.

Epidemiological processes are, for instance, infection, pathogenesis, sporulation, spread, and so on. Their states may be disease intensity, latency, inoculum or similar ones. As their parameters we may choose percent disease severity, latent period, spores in m^3/day, pattern of disease progress curve, apparent infection rate r, etc. Each of these parameters can assume different values.

Preferably each process or state should be characterized by more than one parameter (and parameter values as examples). For barley powdery mildew epidemics these could be: disease progress curve: bilateral; optimum disease progress: at mean daily temperature $18^{\circ}C$; infection rate during progressive phase: up to r = 0.40; mean latent period at optimum temperature: 10 days.

Parameter estimation

Once quantitative parameters have been defined their parameter values are usually estimated from samples of populations or from experiments. Parameter estimation has been extensively treated elsewhere (Kranz, 1974). Here we shall confine ourselves to a brief reminder.

Curve fitting, including the linear regression lines, is one of the classical approaches. Curve fitting also gives clues both to qualitative parameters such as patterns of disease progress curve, and to the phase for which meaningful r-values can be computed. Various multivariate statistical methods are more commonly used at present for parameter estimation from experiments. Certainly, the best research procedure for parameter estimation in epidemiology is systems analysis, where data is recycled and tested between experiment and model (Kranz and Hau, 1980). Usually these techniques result in parameter values expressed as consants or coefficients.

EPIDEMIOLOGICAL PARAMETERS RELEVANT IN CONNECTION WITH PLANT RE-
SISTANCE

Epidemic patterns in relation to types of resistance

Disease progress curves are the temporal graphs describing
epidemic processes determined by factors from the disease square.
These patterns of behavior used as qualitative parameters (Kranz,
1978) may indicate the type of resistance involved or needed. For
instance, juvenile or adult plant resistance can be distinguished
from, perhaps, the yellow rust, but certainly from the potato late
blight curve, and from the ones of powdery mildew and apple scab
curve, respectively. The disease progress curves in Fig. 2 elaborate this aspect in some detail for powdery mildew of barley. In
the resistance groups (Ordóñez, 1981) only young leaves are susceptible for *Erysiphe graminis* f. sp. *hordei*. Cultivars of group
3 - 5 have resistant flag leaves, whereas in group 1 and 2 no such
adult resistance can be observed.

From non-linearized disease progress curves take-off-levels
of disease intensity at which generalization of an epidemic starts
may also be assessed and related to the crop phase (age) dependent
resistance. This is of interest when late or low initial diseases
incidence is correlated with low terminal disease intensity (Kranz,

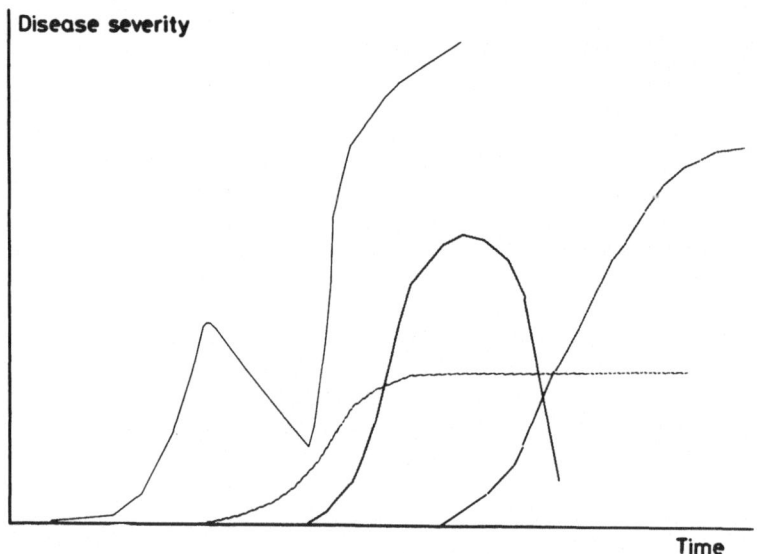

Fig. 1.   Epidemic patterns for the pathosystems wheat yellow rust
          (Zadoks & Schein, 1979) (——), barley powdery mildew (∿∿∿)
          apple scab (∿··∿) and potato late blight (⊢−−)

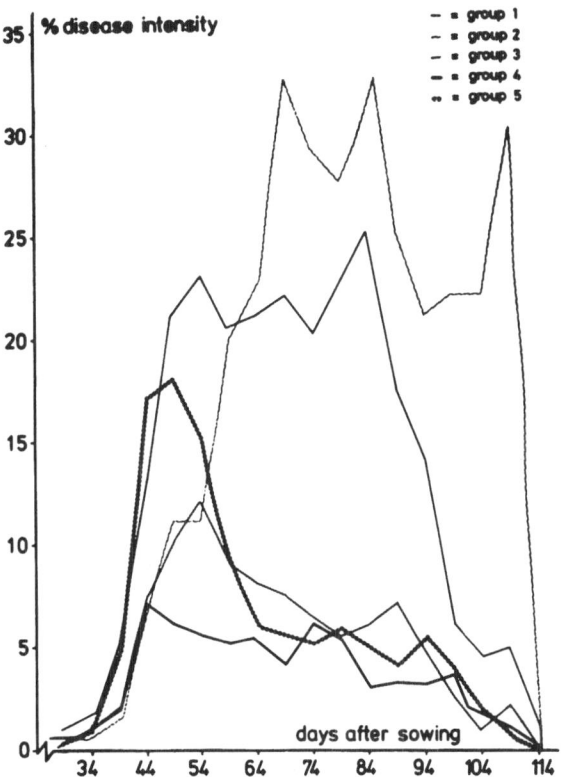

Fig. 2. Disease progress curves of five barley powdery mildew
         epidemics representing cultivars assigned to five resi-
         stance groups (after Ordóñez, 1981)

1968) which is not the same as slow rusting. From epidemic patterns quantitative parameters can also be derived (see Kranz,
1978), particularly position and rate parameters. Thus, for vertical resistance the delay in onset, Δt, i.e. position of the curve,
has already been suggested by Vanderplank (1963). Rates aptly
characterize speed of processes.

     Admittedly not enough is yet known about curve patterns and
their consistency for a given pathosystem (Kranz, 1974a). But epidemic patterns have already been used in relation to host plant
resistance (Campbell et al., 1980; Fried et al., 1979; Thompson
and Rees, 1979). Provided we know more about them they certainly
will be more elucidating than their linearized forms, so much used
in epidemiology nowadays. Furthermore, infection rates are not necessarily consistent over the whole epidemic (Table 2; Gassert,
1978; Kranz, 1975; McKenzie, 1976). Consequently, rates for entire
epidemics should be used only when, after transformation, linearity for the whole duration has been proven statistically. Other-

wise, rates for parts of the epidemic should only be used. Certain peaks might well have a biological significance. In addition, different epidemic patterns may require transformations other than logits to enhance precision.

Vanderplank (1963) when introducing linearization of disease progress curves by means of logits did not imply that this is essential, nor did he institute the infection rate r as the only possible epidemiological rate parameter. He used both tools to convey more important issues. Unfortunately, the infection rate r tends to vary greatly from year to year and site to site within the same pathosystem (Table 1; Kranz, 1968; Wilcoxson et al., 1975). This is either due to different degree of resistance, weather regimes, fertilizer dose (Shaner and Finney, 1977), growth stage (Parlevliet, 1979) or even stage of epidemic (Gassert, 1978). The widespread habit of only interpreting linearized curves can indeed be seriously misleading. It often obscures essential biological facts (Kranz, 1978).

Nevertheless, r is useful and very convenient in direct comparisons within a given experiment (Kranz, 1980; McKenzie, 1976) and with few measurements. When used to characterize disease progress between years and sites, rates as parameters should only be given in classes or in ranges (Kranz, 1980), e.g. with r = 0.47, to 0.55 compared with 0.10 to 0.13 for fast and slow-rusting cultivars, respectively.

Because of these reservations other or additional quantitative parameters are needed to characterize an entire epidemic. The area under a disease progress curve (Kranz, 1968; Vanderplank, 1963; Wilcoxson et al., 1975) seems much more elucidating than its

Table 1. The apparent infection rate r of some wild plant pathosystems studied simultaneously during 2 years (from Kranz, 1968)

Pathosystem	Apparent infection rate r for the years	
	1962	1963
Kordyana celebensis - Commelina	0.15	0.19
Cercospora bidentis - Bidens	0.30	0.06
Cercospora timorensis - Ipomoea	0.23	0.08
Puccinia conclusa - Cyperus	0.15	0.30
Cercospora contraria - Dioscorea	0.08	0.08
Peronospora favargeri - Euphorbia	0.02	0.03
Cercospora fusimaculans - Chasmopodium	0.25	0.20
Puccinia versicolor - Andropogon	0.04	0.20

Table 2. Disease frequency (%) and apparent infection rate r in
          Coffee Berry Disease (CBD) with and without clean-
          picking (after Gassert, 1978)

Weeks after flowering	without 'mummies'[a]		with 'mummies'	
	% CBD	r	% CBD	r
13	0.09		0.62	
		0.68		0.50
14	9.24		17.10	
		0.10		0.11
15	17.11		30.35	
		0.04		0.05
16	21.80		38.90	

[a] shrivelled coffee 'berries' from last season

rate and with software like the AUDPC (Wilcoxson et al., 1975) its
computation is no longer a problem. However, an additional parameter will at least be needed to make this information more meaningful. This may be a 'curve pattern' (Kranz, 1978) or one of the
curve elements, such as the lag-phase of the disease progress curve
(a position parameter), e.g. the time the powdery mildew needed to
reach 10% disease severity (Shaner and Finney, 1977).

Parameters of epidemiological processes and states

     Resistance (1) removes infection sites and (2) causes quali-
tative and quantitative changes in the pathogen population. Removed
infection sites make an immediate impact on population dynamics;
disease progress is either inhibited or slowed down. Changes in
the pathogen population become effective only with the delay of
latent periods. Fewer propagules then become available or are competent to exploit accessible infection sites. Underlying all this
are a number of processes, often called components (Zadoks and
Schein, 1979). Complete vertical resistance does not seem to operate through them, but partial or horizontal resistance certainly
does. It is not relevant whether the latter resistance is achieved
by pedigree breeding, mass, or recurrent selection. Whether these
components have relevance in the progress of an epidemic, as component analyses may suggest, can only be observed in the course of
an entire epidemic, or by adequate simulators. Differential interactions of components with weather factors may well lead to different effects.

However, which component the breeder eventually chooses as selection criteria is essentially a matter of convenience. Mass or
recurrent selection would need a parameter for disease intensity.
Certainly, more standardization of the methods and inoculum (in
terms of quantity and quality) used in screening is needed. Also
simulators are valuable tools to assist in the classification of
cultivars according to their resistance. None of these components
or their parameters, however, appear yet to be indicative for durable resistance.

Infection

Infection is certainly a key process in epidemiology of most
plant diseases. The appropriate parameter is 'infection efficiency'
which to some extent deals with density effects (Hirst and Schein,
1965; Rouse et al., 1980) expressed per unit of inoculum per unit
area of susceptible host tissue. It implies an accomlished penetration and is an excellent criterion for hypersensitive resistance,
but would be too tedious in routine screening.

Disease efficiency (Royer and Nelson, 1981) defined as the
number of manifest lesions per unit of inoculum per unit area of
susceptible host tissue is a more valid parameter for host resistance and its impact on further disease progress, and it is more
convenient with natural infection. This parameter will have to be
used relative to some standard cultivar, or to relevant spore cat-
ches.

However, infection types are useless criteria in epidemiology,
unless they refer to a defined sporulation intensity.

Pathogenesis or latency

The parameter of this process is the 'latent period' with appropriate time units chosen as dimensions, e.g. when 50 percent of all
lesions sporulate. Another useful paramter of latency could be
'variance of latent periods' in appropriate time units.

The great importance of latent pariods (Vanderplank, 1963) for
the progress of epidemics has been proven with simulators (Aust et
al., 1982; Kranz et al., 1973; Rapilly, 1977; Shaner and Hess,
1978; Zadoks, 1971). The latent period determines the number of pathogen generations during the period of host susceptibility, thus
also the delays within the system, and consequently, the speed of
disease progress.

Variability of latent periods is high, largely due to parabo-
lic relationships. Temperature has the greatest effect, inoculum
density, growth stage and age of host follow next, but contribute
considerably less (Aust et al., 1978; Kranz, 1968; Pauvert and

Tullaye, 1977; Schuh, 1980; Shaner et al., 1978; Zadoks and Schein, 1979). Races may or may not affect latent period duration (Evers-meyer et al., 1980; Sztejnberg and Wahl, 1976). This proneness to variability of latent periods makes very consistent methods in screening mandatory. Less favourable conditions, like lower tempe-ratures (Eversmeyer et al., 1980; Johnson, 1980; Ohm and Shaner, 1976; Schuh, 1980) and inoculum doses (Schuh, 1980) may ensure more clearcut differences between cultivars.

The importance of latent periods as components of resistance could be overrated. This may be so in areas where favorable infec-tion periods are infrequent as in dry seasons. Here it would be more relevant epidemiologically for the pathogen to maintain in-oculum over a long period or to react quickly to favorable condi-tions for sporulation once they occur.

Lesion growth

Parameters such as 'mean rate of growth', 'mean duration of growth' and 'final mean lesion size' relate to lesion growth and development. More susceptible cultivars may either have more and/ or larger lesions than more resistant ones (Forche, 1981).

Reproduction and multiplication

From the epidemiological point of view we distinguish sporula-tion (fission, replication) and infectiousness although both result in inoculum production.

Sporulation

There is plenty of good evidence that sporulation intensity is a safe parameter to assess relative differences in resistance. The parameter 'sporulation intensity' must be related to some area and a time dimension. Thus sporulation intensity could be defined as number (weight) of propagules produced per mm^2 of lesion - or even whole lesion - per day. It is essential to define age or stage of lesion during which sporulation intensity should be measured.

The infectiousness of lesions

Vanderplank (1963) and later Zadoks (1971) have shown the in-fluence of infectiousness on the rate of disease progress. Its pa-rameters are: 'mean infectious period' i (Vanderplank, 1963), i.e. its duration (Table 3), 'mean number of spore crops' per i, per mean lesion and 'mean amount of propagules per mean lesion per spore crop'.

Inoculum production

Inoculum production results from the foregoing two processes.

Table 3.   Duration of infectious periods i of *Mycosphaerella*
           musicola (after Kranz, 1968a)

| Leaf spots observed | | | infectious periods in days | |
from	-	to	$\bar{x}^a$	v^b
08.10.	-	23.10.	8	4-12
30.10.	-	30.11.	18	8-23

[a]Means from 15 individual leaf spots checked daily *in situ*
by means of a binocular.
[b]v = variance; i.e. individual leaf spots sporulated 4 to
12 days.

The parameter 'inoculum' defines the amount of infective propagules in appropriate units on the host (either in state of formation or landed and ready for infection), or in transport. Inoculum may also be estimated in absolute terms from the parameters
'disease severity' transformed into cm^2 lesions per hectare, and
'age frequency distribution'. The latter corrects the absolute
population size of lesions for their mean sporulation intensity.

Inoculum production early in an epidemic enhances further
disease progress. It may be influenced by the survivability of a
pathogen.

Aging of pathogen populations

As a lesion's spore crops, when disseminated, lead to daugh-
ter lesions inoculum spreads disease and increases its intensity.
One parameter for this relationship is Vanderplank's (1975) pro-
geny/parent ratio. This ratio not only implies spore-production,
but also the ability to disseminate as well as the spore-catch
capacity of the host, and some of the fitness 'attributes'. As
several spore crops of one particular lesion bring about overlapping generations new lesions enter this process whilst others are
removed. These processes result in various states of age frequency distribution in the course of an epidemic.

Age frequency distribution of lesions may be particularly relevant when peaks of fertile lesions coincide with susceptible
phases of a host plant, e.g. growth flushes of new and susceptible
leaves. Both progeny/parent ratio and age frequency distribution

are still neglected as epidemiological parameters. They turn out to
be of great value also in relation to plant resistance as they are
apt to characterize dynamics. The former could be employed in
screening for durable resistance by means of mini-epidemics. The
latter seems more relevant to resistance management. The success in
breeding *Hevea* clones with shorter refoliation periods to control
powdery mildew (Populer, 1972) is an example. This cuts down the
number of pathogen generations, and thus increments in inoculum.
With continuous infection periods the parameter 'number of pathogen generations' is the ratio of 'length of susceptible phase' of
the host divided by 'mean latent period' of the disease. Resistance
can affect this parameter through both the length of susceptible
phases of the host, and the duration of latent periods.

Host characteristics in relation to resistance

Epidemics are not only determined by resistance genes *per se*.
From the point of view epidemiology, plant breeders appear sometimes rather mesmerized by them. They could definitely derive clues
from an holistic approach to support durability of resistance.
Furthermore, genes that regulate plant characteristics which are
primarily unrelated to resistance may well create conditions unfavourable to the pathogen, or provide for escape. Their effects
on durable resistance could be greater than 'direct' genetic resistance. For instance, growth habits of plants or leaves may affect
the microclimate (Russel, 1975) or the values of the parameters
'impaction efficiency' p and the 'net collection efficiency' (Aylor,
1978) of cultivars for fungal spores. Thickened walls of epidermal
cells, silicated cells, more trichoma and less stomata protect
plants from excessive evaporation, but confer at the same time
'adult resistance' (Aust, 1981; Dutzmann et al., 1981). Adult resistance substantially reduces inoculum (Aust, 1981). Less tillers
per unit area, as in modern barley cultivars, create open crop
stands and thus unfavourable microclimates (Neervoort and Parlevliet, 1978).

In brief, whatever strategy they adhere to, breeders care too
little about the fringe benefits which could enhance disease resistance. Epidemiologists however, fail to elaborate in more detail which host characteristics, how and to what extent help the
crop to defend itself against diseases.

THE EFFECTS OF RESISTANCE ON PATHOGEN POPULATIONS AND THEIR EPIDEMIOLOGICAL IMPLICATIONS

Effects of plant resistance can best be monitored by changes
in pathogen populations. Regulation in turn is by the use of appropriate host plant resistance.

Qualitative and quantitative changes in pathogen populations triggered by a gene-for-gene relationship are mainly due to selective removal of infection sites for non-matching virulences. Non-differential plant resistance ensures similar changes by competition amongst genotypes in the pathogen population. Competition here is the reaction to environmental factors *sensu lato*. In both cases this results in distributions, diversities and frequencies of genotypes with different virulence, or fitness. If conditions permit, these states may reach an equilibrium.

Selection and competition as processes for differential and non-differential resistance, respectively, relate to resistance-management. Their separation here, however, is arbitrary. Obviously, there can be no competition without virulence *sensu* Vanderplank. Similarly virulence genotypes may differ in their competitive ability. The study of these processes certainly still offers great challenges for epidemiologists, population geneticists and breeders.

Selection - the change in virulence patterns

Selection of pathogen genotypes by a gene-for-gene relationship has already been extensively covered in the literature, and in this volume. For diversity of virulences, or virulence spectrum (Wolfe and Schwarzbach, 1978), ecological literature offers some parameters (see Lebeda, 1982). One is 'genetic richness', defined as the total number of virulence genes divided by total number of individuals or races. The other is eveness or equitability J (Sheldon, 1969) where J is the relative distribution of virulences in their population.

Selection leads to particular states of frequency and distribution in time and space, respectively, or diversity in pathogen populations. As parameters of frequency 'relative frequencies', i.e. relative to other virulences, at either the beginning or end of epidemic, or any time in between, are adequate. Wolfe and Schwarzbach (1978) suggest an equation to calculate the final frequency f_x of a particular race. Østergard (this volume) proposes elaborate models for the dynamics of these changes.

Changes in the above states are described by rate parameters: e.g. selection rate (or coefficient), a rate of gene accumulation, and rate inter-virulence changes. The latter is determined by a number of factors that reflect also pleiotropic and background effects as well as all characteristics summarized as fitness.

Epidemiological parameters for forces driving the system are popularity *sensu* Vanderplank (1968), rates of reproduction and propagule exchange, advantage to the pathogen of a virulence gene matching the corresponding resistance gene, and the ability of a pathogen genotype to attack more than one host genotype (after

Leonard, 1977; Wolfe and Schwarzbach, 1978). Quantitative techniques are available to detect these changes resulting from differential interactions. But they have to be conducted over several pathogen generations (Johnson and Taylor, 1976). An example may elucidate this process. In isolated miniplots, after an exodemic start, epidemics of barley powdery mildew soon became endodemic (Forche, pers. com.). Some virulences increased more or less rapidly while others decreased during crop growth thus changing in proportion. Also their rates differed. However, none of these virulences had disappeared from the population at harvest time.

Once relative frequencies are known absolute magnitudes of virulences can be assessed based on some estimates of total propagules (as the pathogen population). Knowledge of absolute magnitudes is highly relevant. For a large increase in population size relatively rare virulence also increase in absolute numbers. Changes observed within one cropping season and one field obviously occur also on village, region, and country levels (Schwarzbach and Limpert, pers. com.). Their diversity parameter (i.e. relative frequency per locality) strongly reflects the popularity of current and past populations of resistance genes in a given area.

Competition

The intraspecific competitive ability of pathogenic pathotypes is fitness whereas 'epidemic competence' should stand for interspecific competition. Virulence *sensu* Vanderplank is its condition *sine qua non* but has a specific bearing within the same virulence genotype. For dynamics within a pathogen population confronted with non-differential resistance where specific virulence genes become important the effect of competition appears to be one of the driving forces of changes. Fitness as a term is still not used uniformly (Leonard and Czochor, 1980; MacKenzie, 1980; Nelson, 1979; Person et al., 1976; Wolfe and Schwarzbach, 1978; etc.). We consider fitness here at the population level, where it represents the mean of the pathogen genotypes. We also feel that fitness is not only determined by the virulence genes in a given population and their relative frequency, but also by a number of epidemiological factors as well as their interactions amongst themselves, and with host-specific factors.

Brodny (1980), for instance, reports that the race group 264 - 276 of *Puccinia coronata* var. *avenae* which is less aggressive than race 263, nevertheless for 26 years is consistently widespread and dominant on wild and cultivated oats in Israel. This is ascribed to the higher fitness of race 276, as measured in the six following fitness parameters, i.e. viability of uredospores, percent germination, incubation periods, number of infections/cm^2, period and rate of sporulation, duration of uredosori development. Gregory et al. (1981) and MacKenzie (1978) use the parameters 'di-

sease efficiency' and 'disease severity' to measure fitness, e.g.
lower fitness results in lower disease severity.

Fitness in its epidemiological sense can embrace the following abilities: quicker infection, shorter latency, greater first
spore crop, more spore crops and at shorter intervals during infectious periods, greater aggressiveness (i.e. higher infection
efficiency, or disease efficiency *sensu* Royer and Nelson, 1981),
greater survivability, probability of effective hits (in multilines and mixtures), less stringent requirements for environmental
factors (temperature, leaf wetness, etc.), greater compensatory
capacities (Aust et al., 1980), and greater antibiotic potentials
as against other parasites.

Their parameters are infection time, latent period, disease
efficiency, sporulation intensity, overseasoning ratio, net collection efficiency, and environmental sensitivity. The latter parameter could be defined in various ways to characterize the reaction of pathogen genotype to climatic factors, and similar.

Obviously not each of these elements contribute alike to
competitive ability and their effects may change with the genotype.
Aggressiveness *sensu* Vanderplank (1975) could be excluded altoge-
ther and maintained as a separate term particular to artificial
inoculations. Its inclusion, however, makes fitness more handy for
field use.

If absolute fitness of a pathogen population (in our case) is
1 then relative fitness is 1 - s, where s is the selection coefficient. For MacKenzie (1978) biological or absolute parasitic fitness expresses the fitness of an individual genotype (isolate, population). The infection rate r is their measure: The higher r,
the fitter the genotype. The parameter absolute parasitic fitness
is defined as $F - e^r$ or $lnF = r$ being the infection rate (with
+ r,r = 0, and - r for increasing, static and decreasing populations, respectively). It results from the compatibility and the
aggressiveness of the genotype etc. on a given host. Data to quan-
tify F can be obtained from the observed change in absolute numbers of individuals or lesion (and disease severity) on a host
(MacKenzie, 1980). Relative fitness determines the fitness of a
genotype (or population) relative to another, and is measured in
relative frequencies and days rather than in absolute numbers (as
with the absolute fitness) and generation time. There is no essential difference to the above definition of relative fitness.

It will remain a great challenge to epidemiology to detect
the effects of these various elements of fitness. A lot of work is
still needed before the above parameters can be monitored effectively.

Estimation of longevity of cultivar resistance

A number of gene management strategies have been proposed in the literature to enhance longevity of varietal resistance. Usually they are based on the concept to set more complex problems for the pathogen (Wolfe and Schwarzbach, 1978) through genetic diversity of the host plants. Strategies emanating from this concept, like multilines, mixtures, gene pyramiding and deployment etc. are essentially pragmatic. Their explanation in epidemiological terms is still fragmentary and sometimes contradictory. They are, of course, all based on gene-for-gene relationships. For durable resistance a better understanding of processes behind such strategies is mandatory for the design of resistance management systems as well as to monitor and sustain them.

Durable resistance may be assessed by a simple check. If b, the regression coefficient of a linear regression with x = years and y = mean disease intensity over its entire acreage to the cultivar (popularity) is ~0, or not significant from 0, then the cultivar shows durable resistance. This can be extended for any year x_n until b is significantly different from b = 0. Little is known about the mechanism which ensures longevity. It is certainly not the overseasoning ratio alone, as Kiyosawa (1977) suggests, though it may contribute a great deal.

Even less is known how to make estimates on the durability of resistance if no gene-for-gene relationships are involved. Theoretically resistance should lower the mean fitness of the entire pathogen population through reduced aggressiveness (Person et al., 1976). On the other hand it is not known whether 'kin selection' (Maynard-Smith, 1974) within a virulence eventually leads to a steady increase in fitness or aggressiveness. It may as well lower virulence and/or fitness if this ensures higher survival prospects.

It is obvious from the foregoing remarks that we still lack concepts and tools to test for durable resistance. It is essential to have clues about future trends of developments in pathogen population for their prediction.

Survivability of host specific pathogen genotypes may be an essential process affecting durability. The question, however, is do the virulence levels and spectra that occur in one season represent the initial population for the next season. This is the case only when conidia survive on pioneers and seeds or on perennial crops or when inoculum is supplied from a congenial neighbouring crop. But it must not apply for pathogens with a saprophytic phase between two seasons, and possibly also not for biotrophic pathogens with intervening sexual recombination. There is evidence that from cleistothecia of *Erysiphe graminis* on spring barley not only does a different virulence population pass onto winter barley,

but also further changes seem likely before spring barley is infected again. Obviously these processes could restore diversity in
the exodemic inoculum for alloinfection every summer season. However, after having ensured primary foci of matching virulences at
the onset of a vegetation period in a field, the rapid endodemic
selection process takes over (Forche, pers. com.). Thus a threshold frequency of matching virulence genes depending on absolute
population size and infectivity in an area might suffice. The
catch area would then be crucial. Both the size and distribution
of acreage cropped to matching resistance genes in a region determine the probability of effective hits. Other factors which, in
relation to overseasoning, could enhance durable resistance are
high juvenile and/or rate reducing resistance. Together with the
magnitude of spore catch it would effect y_o, the primary infection.
This in turn is correlated with the final disease intensity.

It is evident that beyond the effects of relevant host and
pathogen genes, weather and agricultural practices greatly determine the longevity of a cultivar. We need to know what the equilibrium of virulence genes pertaining to durable resistance has to
be in a given pathogen, and situation. We need to know what oscil-
lations of virulence genes - and the disease they cause - are
still within the stability region of this equilibrium. We need to
know when oscillation charges to an undesirable trend, a trend
which leads to changes in virulences or fitness of a pathogen population to an extent which no longer permits a host to remain resistant. For the assessment of such virulence shifts Trenbath's
(1977) equilibrium point may be a useful parameter.

REFERENCE LIST

Aust, H. J. , 1981, Über den Verlauf von Mehltauepidemien inner-
        halb des Agro-Ökosystems Gerstenfeld, *Acta Phytomedica,* 7:
        1-76.
Aust, H. J., Hau, B., and Mogk, M., 1978, Wirkung von Temperatur
        und Konidiendichte auf die Inkubationszeit des Gerstenmehl-
        taus, *Z. Pflanzenkrankh.*, 85:581-585.
Aust, H. J., Bashi, E., and Rotem, J., 1980, Flexibility of plant
        pathogens in exploiting ecological and biotic conditions in
        the development of epidemics, *in*: Comparative epidemiology,
        J. Palti and J. Kranz, eds., PUDOC, Wageningen.
Aust, H. J., Hau, B., and Kranz, J., 1982, EPIGRAM - a simulator
        of barley powdery mildew, *Z. PflKrankh. PflSchutz,* (in
        press).
Aylor, D. E., 1978, Dispersal in time and space: Aerial pathogens,
        in: Plant Disease. An advanced treatise. Vol. II, J. G.
        Horsfall and E. B. Cowling, eds., Academic Press, New York
        London.

Brodny, U., 1980, Studies on the nature of mechanisms determina-
    ting the composition of physiological races populations of
    Puccinia coronata Cda. var. *avenae* Fraser & Ledingham on
    Avena sterilis L. in Israel, PhD Thesis, Tel Aviv Univ.
    (Hebrew, Engl. Summary):118.

Butt, D. J., and Royle, D. J., 1980, The importance of terms and
    definitions for conceptually unified epidemiology, *in*:
    Comparative Epidemiology, J. Palti and J. Kranz, eds.,
    PUDOC, Wageningen.

Campbell, C. L., Madden, L.V., and Pennypecker, S. P., 1980,
    Structural characterization of bean root rot epidemics,
    Phytopathology, 70:152-155.

Dutzmann, S., Forche, S., Döll, G., and Kranz, J., 1981, Licht-
    und rasterelektronenmikroskopischer Vergleich von Blatt-
    oberflächen verschiedener Gerstensorten, *Z. Pflanzen-
    krankh.,* 88:518-524.

Eversmeyer, M. G., Kramer, C. L., and Browder, L. E., 1980, Effect
    of temperature and host: parasite combination on the
    latent period of *Puccinia recondita* in seedling wheat
    plants, *Phytopathology,* 70:938-941.

Forche, S., 1981, Anzahl und Größe der Kolonien von *Erysiphe gra-
    minis* f. sp. *hordei* als Kriterien für die Resistenz von
    Gerstensorten, *Z. Pflanzenkrankh.,* 88:435-441.

Fried, P. M., Mackenzie, D. R.. and Nelson, R. R., 1979, Disease
    progress curves of *Erysiphe graminis* f. sp. *tritici* on
    Chancellor wheat and four multilines, *Phytopath. Z.,* 95:
    151-166.

Gassert, W. L., 1978, Rinde und mumifizierte Kirschen als Quelle
    für das Primärinokulum der Kaffeekirschen-Krankheit in
    Äthiopien, *Z. Pflanzenkrankh.,* 85:30-40.

Gregory, L. V., Royer, M. H., Hyres, J. E., and Nelson, R. R.,
    1981, The evaluation of relative parasitic fitness of iso-
    lates of *Helminthosporium maydis* race T., *Phytopath. Z.,*
    71:354-356.

Hirst, J. M., and Schein, R. D., 1965, Terminology of infection
    processes, *Phytopathology,* 55:1157.

Johnson, R., and Taylor, A. J., 1976, Spore yield of pathogens in
    investigations of the race specificity of host resistance,
    Ann. Rev. Phytopath., 14:97-119

Johnson, D. A., 1980, Effect of low temperature on the latent pe-
    riod of slow and fast rusting winter wheat genotypes,
    Plant Disease, 64:1006-1008.

Kiyosawa, S., 1977, Development of methods for the comparison of
    utility values of varieties carrying various types of re-
    sistance, *in:* The Genetics Basis of Epidemics in Agricul-
    ture, P. Day, ed., *Ann. N.Y. Acad. Sci.,* 287:107-123.

Kranz, J., 1968, Eine Analyse von annuellen Epidemien pilzlicher
    Parasiten, *Phytopath. Z.,* 61:171-190.

Kranz, J., 1968a, Zur Konidienbildung und -verbreitung bei *Myco-
    sphaerella musicola* Leach., *Z. Pflanzenkrankh.,* 75:327-338.

Kranz, J., 1974, The role and scope of mathematical analysis and modeling in epidemiology, *in*: Mathematical Analysis and Modeling of Epidemics, J. Kranz, ed., Springer, Berlin - Heidelberg - New York.

Kranz, J.. 1974a, Comparison of epidemics, *Ann. Rev. Phytopathology*, 12:355-374.

Kranz, J., 1975, Beziehungen zwischen Blattmasse und Befallsentwicklung, *Z. Pflanzenkrankh.*, 82:621-654.

Kranz, J., 1978, Comparative anatomy of epidemics, *in*: Plant Disease, An Advanced treatise, Vol. II, J. H. Horsfall and E. B. Cowling, eds., Academic Press, New York - London.

Kranz, J., 1980, Comparative epidemiology: an evaluation of scope, concepts and methods, *in*: Comparative Epidemiology, J. Palti and J. Kranz, eds., PUDOC, Wageningen.

Kranz, J., and Hau, B., 1980, Systems analysis in epidemiology, *Ann. Rev. Phytopathology*, 18:67-83.

Kranz, J., Mogk, M., and Stumpf, A., 1973, EPIVEN - ein Simulator für Apfelschorf, Vorl. Mitt. *Z. Pflanzenkrankh.*, 80:181-187.

Lebeda, A., 1982, Measurement of genetic diversity of virulence in populations of phytopathogenic fungi, *Z. pflanzenkrankh.*, (in press).

Leonard, K. J., 1977, Selection pressures and plant pathogens, *in*: The genetic basis of epidemics in agriculture, P. R. Day, ed., The New York Academy of Sciences, New York.

Leonard, K. J., and Czochor, R. J., 1980, Theory of genetic interactions among populations of plants and their pathogens, *Ann. Rev. Phytopathology*, 18:237-258.

Mackenzie, D. R., 1976, Application of two epidemiological models for the identification of slow stem rusting in wheat, *Phytopathology*, 66:55-59.

Mackenzie. D. R., 1978, Estimating parasitic fitness, *Phytopathology*, 68:9-13.

Mackenzie, D. R., 1980, The problem of variable pests, *in*: Breeding plant resistance to insects, F. G. Maxwell and P. R. Jennings, eds., John Wiley & Sons, New York.

Maynard-Smith, J., 1974, Models in Ecology, Cambridge Univ. Press, Cambridge.

Neervoort, W. J., and Parlevliet, J. E., 1978, Partial resistance of barley to leaf rust *Puccinia hordei*, V. Analysis of the components of partial resistance in eight barley cultivars, *Euphytica*, 27:33-39.

Nelson, R. R., 1979, The evolution of parasitic fitness, *in*: Plant Disease Vol. IV., J. G. Horsfall and E. B. Cowling, eds., Academic Press, New York - London.

Ohm, H. W., and Shaner, G. E., 1976, Three components of slow leaf-rusting at different growth stages in wheat, *Phytopathology*, 66:1356-1360.

Ordonez, M. T., 1981, Epidemiologische und cytologisch-histologische Untersuchungen zum Verhalten von Sommergerste gegenüber Mehltau (*Erysiphe graminis* DC. f. sp. *hordei* Marchal) unter besonderer Berücksichtigung der Altersresistenz, Diss. Univ. Göttingen.

Parlevliet, J. E., 1979, Components of resistance that reduce the rate of epidemic development, *Ann. Rev. Phytopath.*, 17: 203-227.

Pauvert, P., and De la Tullaye, B., 1977, Etude des conditions de contamination de l'orge par l'*Oidium* et de la période de latence chez *Erysiphe graminis* f. sp. *hordei*, *Ann. Phytopath.* 9:495-501

Person, C., Groth, J. V., and Mylik, O. M., 1976, Genetic change in host-parasitic populations, *Ann. Rev. Phytopath.* 14:177-188.

Plinston, D. T., 1972, Parameter sensitivity and interdependance in hydrological models, *in:* Mathematical Models in Ecology, J. N. R. Jeffers, ed., Blackwell Scientific Publ., Oxford-London.

Populer, C., 1972, Les épidémies de l'*Oidium* de l'hévéa et la phénologie de son hôte dans le monde, *Serie Scientifique* No. 115 de l'INEAC, Brussels.

Rapilly, F., 1977, Recherche des facteurs de résistance horizontale à la septoriose du blé (*Septoria nodorum* Berk.), Resultats obtenus par la simulation (1), *Ann. Phytopath.* 9:1-19

Rouse, D. I., Nelson, R. R., Mackenzie, D. R., and Armitage, C. R., 1980, Components of rate-reducing resistance in seedlings of four wheat cultivars and parasitic fitness in six isolates of *Erysiphe graminis* f. sp. *tritici*, *Phytopathology*, 70:1097-1100.

Royer, M. H., and Nelson, R. R., 1981, The effect of host resistance of relative parasitic fitness of *Helminthosporium maydis* Race T, *Phytopathology*, 71:351-354.

Russell, G. E., 1975, Deposition of *Erysiphe graminis* f. sp. *hordei* conidia on barley varieties of differing growth habit, *Phytopath. Z.*, 84:316-321.

Schuh, W., 1980, Variabilität der Inkubationszeit und des Koloniewachstums beim Gerstenmehltau, Diplomarbeit, FB 16, Justus Liebig-Universität, Giessen.

Shaner, G., and Finney, R. E., 1977, The effect of nitrogen fertilization on the expression of slow-mildewing resistance in Knox wheat, *Phytopathology*, 67:1051-1056.

Shaner, G., and Hess, F. D., 1978, Equations for integrating components of slow leaf-rusting resistance in wheat, *Phytopathology*, 68:1464-1469.

Shaner, G., Ohm, H. W., and Finney, R. E., 1978, Response of susceptible and slow leaf-rusting wheats to infection by *Puccinia recondita*, *Phytopathology*, 68:471-475.

Sheldon, A. L., 1969, Equitability indices: Dependence on the species count, *Ecology*, 50:466-467.

Sztejnberg, A., and Wahl, I., 1976, Mechanisms and stability of slow rusting resistance in *Avena sterilis*, *Phytopathology*, 66:74-80.

Thompson, J. P., and Rees, R. G., 1979, Pattern analysis in epidemiological evaluation of cultivar resistance, *Phytopathology*, 69:545-549.

Trenbath, B. R., 1977, Interactions among diverse hosts and diverse parasites, *in*: The genetic basis of epidemics in agriculture, P. Day, ed., *Ann. N.Y. Acad. Sci.*, 287:107-123.

Vanderplank, J. E., 1963, Plant Diseases: Epidemics and control, Academic Press, New York - London.

Vanderplank, J. E., 1968, Disease Resistance in Plants, Academic Press, New York - London.

Vanderplank, J. E., 1975, Principles of Plant Infection, Academic Press, New York - London.

Waggoner, P. E., 1978, Computer simulations of epidemics, *in*: Plant Disease, An Advanced treatise, Vol. II, J. G. Horsfall and E. B. Cowling, eds., Academic Press, New York - London.

Wilcoxson, R. D., Skormand, B., and Atif, A. H., 1975, Evaluation of wheat cultivars for ability to retard development of stem rust, *Ann. appl. Biol.*, 80:275-281.

Wolfe, M. S., and Schwarzbach, E., 1978, Patterns of race changes in powdery mildews, *Ann. Rev. Phytopathology*, 16:159-180.

Zadoks, J. C., 1971, Systems analysis and the dynamics of epidemics, *Phytopathology*, 61:600-610.

Zadoks, J. C., and Schein, R. D., 1979, Epidemiology and Plant Disease Management, Oxford Univ. Press, New York - Oxford.

DISCUSSION

PARLEVLIET: Often the apparent infection rate 'r' is used in an experiment to compare the partial resistance of varieties. It seems that 'r' varies with the phase of the epidemic, becoming smaller as the epidemic advances. This would mean that varieties which differ in resistance will be in different phases of the epidemic at any given moment. Comparing 'r' over a certain time span as a measure of partial resistance would be invalid because the varieties are not in the same phase.

KRANZ: Yes, I agree with that.

DE MILIANO: You have pointed out many pitfalls that can be encountered when working with epidemiological parameters of resistance but I would like to know how these are related to yield, particularly in different agricultural systems; I have to make decisions

concerning plant breeding on the basis of yield. When I find a
change in pathotype associated with changes in disease severity
one year, should I throw away the cultivar with increased susceptibility, or should I continue to test the yield over several more
years, perhaps finding that relative yield may remain constant?

KRANZ: You are right; either stable or high yield (both in terms of
quality and quantity) are the most important objectives in plant
breeding. Resistance to diseases and pests serve these ends. Yield
loss caused by diseases and pests occur if their intensity reaches a
certain level at a certain growth stage (on certain plant organs).
Resistance breeding implicitly tries to avoid just this. For the
selection of progenies with such characteristics (i.e. low disease
intensity at critical stages) breeders rely on epidemiological
parameters which have been found to be correlated with cultivars
that have shown such behavior in the field. In the literature there
are, however, rather few references to the yield of, e.g. slow
rusting cultivars.

In field selection a progeny or cultivar with high yield should only
be discarded when disease intensity is distinctly higher on yield-
relevant plant organs (with a subsequent yield loss) than on other
progenies. Generally the disease intensity/crop loss relationship
should be studied for a number of years, yield (or its loss) should
always be relevant to disease intensity at certain growth stages
and on relevant plant organs.

DINOOR: If one is looking for epidemiological criteria or parameters
for selection for disease resistance, I think that one should rea-
lise that loss (caused by disease) is not constantly related to
disease level or to the extent of the epidemic. I think that the
multiple regression analysis would be much more suitable to expose
the critical stages of disease development and even previous stages
which will contribute to them. Then selection could be done at
those critical stages.

KRANZ: Disease assessment for the selection of resistance should
be done during relevant growth stages of the host, these are (1) stages
during which disease intensity at critical points is determined, (2)
stages which are most relevant to yield production, or sensitive to
yield loss. The former can more easily be determined from the course
of disease progress curves. This may be valid for the second aspect
as well, when it is known which yield component contributes most to
the eventual yield (e.g. number of tiller per ha, number of grains
per ear, mean grain weight). If it is not known multivariate
statistical methods certainly are the most suitable ones for parameter estimation.

DURABLE RESISTANCE IN CROPS : BIOMETRIC ANALYSIS OF RESISTANCE

J.E. Vanderplank

Plant Protection Research Institute
Department of Agriculture & Fisheries
PRETORIA, R.S.A.

THE HOST PLANT AS THE INDEPENDENT VARIABLE

With the host plant taken to be the only independent variable much attention has been given to alleged safety in polygenic inheritance of resistance. Polygenic resistance, it is supposed, is stable because the pathogen cannot simultaneously overcome many genes. The argument is confused. First, there is little or no evidence for true polygenic resistance. Second, polygenes have been confused with additive variance. Third, there are enough examples of host plants with many resistance genes which even in combinations are ineffective.

The confusion starts with quantitative inheritance, i.e. with continuous variation of the phenotype. A trait conditioned by a large number of genes each of small effect will vary continuously. It is the converse that is wrong. Continuous variation does not necessarily imply polygenic inheritance. This was pointed out by Thompson (1975). Continuous variation may even be conditioned monogenically provided that nonheritable (environmental) effects are large. Walker (1966) noted that resistance to cucumber mosaic virus was thought to be polygenic until tests in a better controlled environment showed it to be monogenic. Hoff and McDonald (1980) found that the quantitative resistance of *Pinus monticola* to *Cronartium ribicola* could best be explained by monogenic inheritance, nonheritable variance being as much as 64%.

Consider monogenic additive resistance, and a cross between a resistant and a susceptible parent. In the F2 generation there would be three phenotypes, with resistant, intermediate and susceptible types in the proportion 1:2:1. With nonheritable

variance of 50%, the distribution would be smoothed and easily
taken for continuous. With two loci the F2 generation would
reveal five classes. If unequal effects are added as a possibility,
there would be nine phenotypic classes and so far nonheritable variation has not been considered. Consider trigenic additive resistance.
If the resistance were conditioned by three genes of equal effect,
there would be seven phenotypes and if unequal effects and some
nonheritable variation is present, this would probably be taken
for continuous variation and trigenic resistance mistaken for
polygenic. The essence here is that the resistance in the examples
is additive. Had it been dominant, the seven phenotypes would have
shrunk to two, in the ratio 63:1 and the distribution would clearly
have been seen to be discontinuous and conditioned oligogenically.
In short, trigenic inheritance with some nonheritable variation
would probably be taken for polygenic if the resistance were additive; and oligogenic if it were dominant. Moreover, nonheritable
variance would probably act to increase the confusion, because as
a rule additive resistance is more affected by environmental factors
than dominant resistance.

The safety given by a large number of resistance genes has
often been overrated. A wheat cultivar or multiline with the eleven
genes, Sr5, 7a, 7b, 9b, 10, 11, 14, Tmp, and Ttl would be susceptible
to stem rust in Canada, because most isolates of *Puccinia graminis
tritici* in Canada are virulent for all eleven of them.

The moral of all this is that in probing durable resistance
we should stop stressing safety in numbers and start investigating
whence comes the virtue of additive variance. There may be a tie
with enzyme dose.

To avoid misunderstanding let it be said that the link between
durability and additive variance is frequent but not necessarily
universal. Universality is not implied. There are examples of
durability going with dominance variance, i.e., with dominant resistance or dominant susceptibility. Clearly more than one system is
involved.

HOST AND PATHOGEN IN A TWO-VARIABLE SYSTEM

If host and pathogen are the only two variables, there can be
two and only two sorts of resistance- related to main effects and
interactions respectively. Resistance found as a main effect (with
this term used in its biometrical sense), is horizontal. Resistance
as a host-pathogen interaction (in its biometric sense), is vertical.

Horizontal resistance and vertical resistance can coexist.
This was brought out when the definitions were first proposed and
illustrated by the coexiting resistances of the potato variety
Maritta to *Phytophthora infestans*. Maritta has the gene Rl. Against
race (0) of *P. infestans* and all other races without the number 1

in them, Maritta has vertical resistance. Against race (1) and all
other races with the number 1 in them, Maritta has some horizontal
resistance. All this is old ground which need not be covered again.
What needs to be interpolated here is that the concept of the co-
existence of horizontal and vertical resistance which is the concept
of the coexistence of main effects and interactions, has wrongly
been disputed; and Nelson (1978) has based a review on the mistaken
belief that main effects and interactions are mutually exclusive.

Both host and pathogen must vary, and the concept of horizontal
and vertical resistance implies that there must be at least two types
of hosts differing relevantly and two pathogen types differing relevantly. This is the minimum requirement for allowing at least
one degree of freedom for interaction.

Vertical resistance has often been assumed to occur when in
fact the minimum requirement is not met that could distinguish it
from horizontal resistance. Consider fusarium wilt of tomato caused
by *Fusarium oxysporum* f.sp. *lycopersici*. Two resistance genes,
called gene I and gene I-2, are known in tomato. A race of the
pathogen is known, called race 2 by Cirulli and Alexander (1966)
and race 1 by Gabe (1975). It attacks tomato varieties with the
gene I but not those with gene I-2. There is no degree of freedom,
and we are left in the dark about whether the resistance is horizontal
or vertical. Until such time as race 2 *sensu* Gabe is discovered
that attacks tomatoes with the gene I-2 but not those with the gene
I, we cannot assume the resistance to be vertical. A race that
overcame the resistance given by both gene I and gene I-2 could be
either race (1,2) *sensu* Gabe in a system of vertical resistance;
or it could be a more aggressive race in a system of discontinuous
horizontal resistance, with gene I-2 conditioning more horizontal
resistance than gene I. This matter is taken up again in the next
section, in relation to the third variable and pseudospecificity.

In terms of the topic of this conference, horizontal resistance
will be durable (although not necessarily adequate) in the two-
variable system for which it is defined, but can be endangered by
a third variable. Vertical resistance, because of the differential
interaction between host and pathogen will be unstable, because of
substantial mutation rates from avirulence to virulence (Person,
Groth & Mylyk, 1976; Parlevliet & Zadoks, 1977), unless stabilizing
selection intervenes and favors the avirulence over the virulence
allele. For durable vertical resistance, stabilizing selection is
a *sine qua non*.

THE THIRD VARIABLE : PSEUDOSPECIFICITY: ADEQUACY

Resistant crop varieties are there for the farmer to use, and
the farmer's response to them can conveniently be called the third
variable. A new variety with added resistance might allow him to
dispense with foliar fungicides or soil fumigation; it might allow

him to plant a crop where it was hitherto uneconomic to grow; and
so on. The use of resistance by the farmer, if it introduces a
third variable, places a strain on the resistance. For example,
the use of the gene I against fusarium wilt has allowed tomato
growers in warm climates to shorten rotations and shorter rotations
have led in some areas to a substantial accumulation of race 2
sensu Cirulli and Alexander. This accumulation is a response to
the third variable and the response could equally have occurred
irrespective of whether the resistance was horizontal or vertical.
The third variable has introduced a pseudospecificity which is
outside the definitions of horizontal and vertical resistance in
their two-variable frame.

The third variable explains some reports that horizontal resistance has "broken down". What is meant is that the quantity of
horizontal resistance in the reported instances was inadequate to
cope with a third (or other) variable when this is more favorable
to disease. For example, potato varieties rated in Europe or the
United States as adequately resistant to blight have a resistance
that proves to be quite inadequate when the same varieties are
planted in the Toluca Valley of Mexico. The resistance is the same;
its adequacy is not. The change from adequacy in one environment
to inadequacy in another, is not a special feature of horizontal
resistance. It can also be a feature of vertical resistance which
is often also only partial, i.e. adequate according to the circumstances.

A measure of homeostasis or stabilizing selection is needed to
maintain adequacy if the third variable changes adversely for the
host, that is, homeostasis is needed to restrict the development of
virulence if the resistance is vertical or aggressiveness if the
resistance is horizontal. Homeostasis is relevant to both horizontal and vertical resistance via the third variable.

We need not stop at three variables. There are many more.
The plant breeder's response is another. For example, maize breeders
now have available the convenient Ht gene for resistance to *Helminthosporium turcicum* and are less likely to pay attention to the
older "lesion-number" sort of resistance; varying amounts of the
older sort of resistance in the background would constitute a fourth
variable; and so on.

Durable resistance is therefore a term imprecisely defined in
terms of variables. It is defined only by invariability. It is
invariable with time. Beyond this invariability there is no precision at all.

REFERENCES

Cirulli, M. and Alexander, L.J., 1966, A comparison of pathogenic
    isolates of *Fusarium oxysporum* f.sp. *lycopersici* and different

sources of resistance in tomato, *Phytopathology* 56:1301-1304.

Gabe, H.L., 1975, Standardization of nomenclature for pathogenic races of *Fusarium oxysporium* f.sp.*lycopersici*, *Trans. Br. Mycol. Soc.*, 64:156-159.

Hoff, R.L. and McDonald, G.I., 1980, Resistance to *Cronartium ribicola* in *Pinus monticola:* reduced needle-spot frequency, *Can. J. Bot.* 58:574-577.

Nelson, R.R., 1978, Genetics of horizontal resistance to plant diseases, *Ann. Rev. Phytopath.* 14:177-178.

Parlevliet, J.E. and Zadoks, J.C., 1977, The integrated concept of disease resistance; a new view including horizontal and vertical resistance in plants, *Euphytica*, 26:5-21.

Person, C., Groth, J.V. and Mylyk, O.M., 1976, Genetic change in host-parasite populations, *Ann. Rev. Phytopath.*, 14: 177-178.

Thompson, J.N., 1975, Quantitative variation and gene number, *Nature*, (London), 258:665-668.

Walker, J.C., 1966, *In:* Plant Breeding, ed., K.J. Frey, pp.219-242. Iowa State Univ. Press, Iowa City.

EXPRESSION OF INCOMPLETE RESISTANCE TO PATHOGENS

Albertus B. Eskes

Food and Agriculture Organization of the United Nations
Instituto Agronomico of Campinas
Campinas, SP, Brazil

SUMMARY

A critical literature review is given on resistance terminology, on measurement of incomplete resistance and on factors that affect the host–pathogen relationship. Most reviewed research data refer to lesion causing leaf diseases and to obligate pathogens. The supposed relationship between durability and the expression of the resistance is questioned. The possible importance of environment in obtaining a balanced host–pathogen relationship is discussed.

INTRODUCTION

This paper deals with the expression of incomplete resistance. The expression of resistance is the result of a given host–pathogen–environment relationship. If reproduction of the pathogen occurs with an intensity less than maximal in the particular environment, the resistance is called here incomplete.

Many researchers focused attention on intermediate types of resistance, and indicated that this may confer a more durable protection than the resistance of the hypersensitive type (Caldwell, 1968; Vanderplank, 1968; Simons, 1972). However, the concepts on durable resistance continue to attract ample discussion (Nelson et al., 1970; Nelson, 1978; Parlevliet and Zadoks, 1977, Vanderplank, 1978).

In this paper, some of the literature on the expression of incomplete types of resistance is reviewed and is related to resist-

ance models. In this analysis, the complexity of the host–pathogen–
environment relationship is taken into account.

TERMINOLOGY

Terms for resistance describe either the inheritance of the
resistance, its phenotype, its durability, or the relationship between host and pathogen.

Genetic descriptors are, for example, polygenic/monogenic
resistance and minor/major gene resistance. Slow rusting, partial
resistance, incomplete resistance, resistance of a high or low
infection type, field resistance, hypersensitive resistance, quantitative resistance, adult plant resistance, resistance to infection
are phenotypic descriptors. Parlevliet (1979) used the term
partial resistance specifically to indicate resistance with a high
infection type, that can however be measured by its components,
such as infection frequency, latency period, sporulation frequency
and sporulation period.

Examples of terms used to describe the durability of the
resistance are durable/temporary resistance and stable/unstable
resistance. The term durable is to be preferred over stable
because its meaning is only related to time, whereas stable has
a much wider meaning.

Terms indicative for the host–pathogen relationship are for
example horizontal/vertical resistance, uniform/differential resistance and race–non–specific/race–specific resistance. These pairs
have generally been used in synonymity to indicate the absence or
presence of host–pathogen interactions. The terms are based on
the concept that host–pathogen relationships can be divided into
two groups. The validity of this concept has been questioned by
several authors, and, hence, the terminology is under discussion.

The terms may easily be used out of context. For example,
the expression of the resistance (slow rusting, incomplete resistance) can only be retrospectively or speculatively related to its
durability. Incomplete resistance or slow rusting has often been
titled horizontal resistance without giving any indication of race-
specificity. In some cases, this supposedly horizontal resistance
proved to be race-specific later (e.g., Browder, 1973). Although
non-specificity may give some indication of durability, horizontal
resistance may be eroded by an increase in horizontal pathogenicity.

The resistance terminology is still under wide discussion.
For example, Nelson (1978) postulated that, by definition, vertical
and horizontal resistance cannot coexist within the same host
genotype. The author proposed that horizontal resistance should

only be used in an epidemiological context, indicating rate-reducing resistance. However, Johnson (1979) pointed out that the nature of the resistance is related to the effect of individual genes and not to the effect of the entire genotype. A host genotype containing several genes for resistance may confer different types of resistance simultaneously. Nelson's proposal to use horizontal resistance in an epidemiological context is likely to increase the confusion on terminology. All types of incomplete resistance, specific or non-specific, will reduce the rate of epidemic growth. Also, a high level of non-specific resistance may delay the detectable onset of the epidemic (x_0 value). Ample evidence of this can be found in the literature (e.g. Rees et al., 1979; Parlevliet, 1979).

Each term describing host resistance can be matched with an equivalent term for pathogenicity. Horizontal/vertical pathogenicity have been proposed to indicate the absence or presence of host-pathogen interactions (Robinson, 1976). In this paper, the terms cultivar-specific and non-specific pathogenicity or aggressiveness and virulence (Vanderplank, 1968) are used.

INCOMPLETE RESISTANCE

Minor Genes

Incomplete resistance may be due to a varying number of resistance genes. If the resistance is polygenic, each gene is supposed to have only a small effect and can be called a minor gene. It has been suggested that non-specific resistance is mainly due to minor genes that do not interact with genes in the pathogen (Vanderplank, 1968). The expression of this resistance to different races of the pathogen should not be visualized as a horizontal line (equal disease by all races) as originally suggested by Vanderplank, if the pathogen is allowed to vary for aggressiveness (horizontal pathogenicity). A better interpretation of this host-pathogen relationship should be the additive model, as described by Parlevliet and Zadoks (1977), or the horizontal resistance model described by Robinson (1979). In these models no differential interaction between host and pathogen genotypes occurs, and, thus, there is a constant ranking of host and pathogen genotypes.

With the increasing number of research papers on incomplete types of resistance, however, also an increasing number of small interactions between host and pathogen have been reported involving minor genes (e.g. Parlevliet, 1979). With the present knowledge, we have to conclude that at least part of the quantitative variation in resistance may be due to minor genes for specific resistance.

Incomplete Vertical Resistance of Major Genes

Many known major genes for specific resistance give an incomplete protection to incompatible races of the pathogen. Even many host differentials chosen for their clear-cut reactions, may only give incomplete protection, which may be expressed by lesions of low infection type with some sporulation (e.g. Roelfs and Mc Vey, 1979). Furthermore, several authors have reported that the genetic background of the host may affect the level of resistance of major genes (Parlevliet and Kuiper, 1977b; Dyck and Samborski, 1968; Athwal and Watson, 1954). Major resistance genes that give only incomplete protection may be called genes for incomplete vertical resistance. As a complementary term for the pathogen, incomplete vertical pathogenicity of incomplete virulence may be used.

Statler and Jones (1981) presented examples of incomplete vertical pathogenicity: isogenic wheat lines, carrying different Lr resistance genes, were inoculated with two pathogen cultures that had different virulence spectra to these resistance genes. Segregation for virulence to eight Lr genes occurred in the F_2 of the pathogen. For seven of these genes, no good 3:1 fit was obtained. For some genes, the distribution for virulence even approached normality.

This wide distribution of virulence may partly be caused by incomplete dominance of the avirulence allele which results in an intermediate heterozygote. This may explain the existence of three levels of compatibility, for diploid or dicaryotic fungi. However, some authors have reported the existence of at least four levels of compatibility in relation to major resistance genes (Watson and Luig, 1968) even for haploid fungi (Schwartzbach, 1979). This indicates allelism for virulence or non-allelic interactions in the pathogen (Watson and Luig, 1968; Samborski, 1963). Thus, incomplete vertical resistance and incomplete virulence, especially when occurring together, may explain a considerable part of the observed variation for incomplete resistance.

HOW INCOMPLETE RESISTANCE IS MEASURED

Assessment Scales for Resistance and Analysis of Research Data

When incomplete resistance is to be measured, the problem of scaling arises. The ideal quantitative assessment scale needs to meet two basic requirements: (a) the relation between resistance and the scale values should be linear, and (b) the variance, due to non-genetical effects, should be similar for all host genotypes, at all resistance levels.

Most of the assessment scales for resistance do not meet,

or only partially meet, these basic requirements. Within the lower and upper limits of the scale, which represent respectively high resistance and high susceptibility, variation may be considerable but cannot be recorded. Therefore, the statistical or genetical analysis of data involving incomplete resistance should be handled with great care. For example, absence of linearity in the data of a host-pathogen relationship, which will show up as an interaction in the traditional statistical analysis, may not be related to a real host x pathogen interaction but may be due to a problem of scaling. This problem can partly be overcome by transformation of the data or by the use of non-parametric statistics. Interpretations of the inheritance of resistance also depends on the scale. For example, intermediate inheritance of a resistance factor may not be observable when dealing with extremely resistant or extremely susceptible parent material because the intermediate resistant type may then still fall within the lower or upper limit of the scale.

Another problem in the analysis of quantitative data is the often substantial effect of the environment on resistance. Therefore, the analysis always depends on the environmental conditions under which the experiment was performed, and care should be taken before extrapolating data obtained in confined environments to field conditions.

Infection Type

For many biotrophic diseases, such as rusts, the traditional method for assessment of resistance is based on the infection type (IT). The IT is generally related to (a) the size of the pustules or lesions, (b) the intensity of the sporulation and (c) the occurrence of chlorosis or necrosis. IT "0" indicates a necrotic or chlorotic fleck, without sporulation, and the IT's "1" to "4" indicate pustules with increasing sporulation and with decreasing chlorosis or necrosis around the pustule.

McNeal et al. (1971) proposed the use of a 0 to 9 scale for evaluating resistance. A "0" indicates immunity, "1" to "3" indicate variation for low IT's (resistant), "4" to "6" indicate variation for intermediate IT's, and "7" to "9" indicate variation for high IT's (susceptible). This scale has the advantage that the data on IT are better suited for statistical calculations. Furthermore, quantitative variation within the resistant group can be recorded and the variation for high IT's is increased in comparison with the 0 to 4 scale. Ideally, for each disease, the 0 to 9 scale should be defined in such a way that the variation from complete resistance to high susceptibility is linearly expressed by the scale values.

The 0 to 9 scale for IT is a first step towards a quantitative

scale for resistance measurements. However, for more accurate
quantitative observations, measurements on either the components
of resistance or on disease incidence and severity are necessary.
Johnson and Taylor (1976a) proposed that spore yield of pathogens
may be a more sensitive tool to measure small differences in
resistance than IT. The problem of scaling, as mentioned earlier,
can be illustrated by their data. The authors obtained a quadratic
relationship between spore yield per cm^2 and IT's of *P. striiformis*
in wheat. Also, the variation for spore yield was much higher for
high infection types than for low infection types. This shows that
infection type may not be a good measure for the reproductive
capacity of the pathogen and, hence, may be insufficiently accurate
to determine the genetics of the host–pathogen relationship or to
study small host–pathogen interactions.

Infection Type and Specificity of Resistance

Resistance of a low infection type (LIT) is generally
considered to be race-specific (Parlevliet, 1979) and resistance
combined with a high infection type (HIT) race non-specific
(Vanderplank, 1968; Robinson, 1976). However, as pointed out
by several authors, specific resistance is not only restricted to
LIT's. Parlevliet (1979) mentioned resistance of sugarbeet to
Cercospora beticola , for which great variation in lesion number
but not in lesion type was found. This resistance proved however
to be race-specific. It has also been reported that by intercrossing
host genotypes with a high IT, dramatic shifts in resistance can be
obtained. Umaerus (1973) reported that through selection for field
resistance to late blight in the susceptible *Solanum tuberosum*
subsp. *andigena* some clones were obtained that showed a hyper–
sensitive reaction to late blight. He suggested that the inheritance
of this resistance is polygenic. Pope (1968) described that through
inter–crossing wheat cultivars with IT "4" to stripe rust, progenies
were obtained which gave a "0" IT. These examples suggest that
the supposed relation between IT and the genetics or the specificity
of a resistance may not always be valid.

Heterogeneous Infection Types

Heterogeneous or mesothetic IT's are those in which a mixture
of sporulating and non-sporulating lesions are present. This reac–
tion has been associated with a mixture of incompatible and com–
patible pathotypes ("X" type of infection). However, heterogeneous
IT's may also reflect the expression of a certain host–pathogen
relationship and should then be considered as a distinct infection
type.

Many single genes for specific resistance may express

themselves as a heterogeneous IT. Parlevliet (1976b) described
the expression of the Pa-6 gene for resistance to *Puccinia hordei*
in the barley cultivar Bolivia as an X infection type in the seedling
stage and as a less extreme heterogeneous type, including lesions
of o_n, $1-3^+$, in the flag leaf stage. Vanderplank (1978) indicated
that temperature-sensitive monogenic resistance is often expressed
by a mesothetic IT in the transition phase between resistance and
susceptibility. In these examples, heterogeneous IT's are caused
by genes for incomplete vertical resistance. Watson and Luig (1968)
also reported heterogeneous IT's as a result of incomplete vertical
pathogenicity.

Heterogeneous infection types may also result from polygenic
resistance. In our research in Brazil we found many coffee plants
with either incomplete vertical resistance or apparent polygenic
resistance to coffee leaf rust; both categories showed heterogeneous
IT's. The IT's also depended on the environmental conditions, on
the leaf age and on productivity. Even in normally compatible
combinations, low IT lesions may be observed. No indications
were found that mixtures of races of *H. vastatrix* were involved.
In other pathosystems, heterogeneous infection types have also
been associated with compatible host-pathogen combinations.
Clifford and Roderick (1978) made histological observations on
barley infected with brown rust. The most susceptible cultivar,
which was compatible to the two races of the pathogen used, showed
some necrotic lesions with one race but not with the other. Successful infections varied from 38 to 63 percent in the experiment.
The authors concluded that a lesion may apparently stop its growth
at any moment during development. Ashagari and Rowell (1980)
mentioned that low receptivity (number of visible infections per
cm^2) of wheat to *P. graminis* f.sp. *tritici* is the major feature of
the slow rusting character. They observed that low receptivity
was related to a lower infection type. The infection types for adult
wheat plants varied from 0 to 3 with reactions such as 1-3, 2-3,
3-4, etc. often being obtained. The compatible infections were
also significantly smaller in cultivars with low receptivity than in
those with high receptivity, and incompatible infections were associated with collapsed host cells. In fact, the authors described a
heterogeneous IT in which not all lesions sporulate. Carver and
Carr (1980) described the reaction of oat cultivars with specific
or general resistance to oat mildew. On all hosts tested, a proportion of germinated conidia failed to achieve primary penetration;
even in compatible combinations less than 30 percent were successful. On cultivars with a high level of race non-specific resistance,
the proportion of failures was higher. The authors concluded that
both race specific and non-specific resistance shifts a series of
reactions of a similar type from a susceptible to a resistant extreme. Milus and Line (1980) described the resistance of six wheat
cultivars to leaf rust. At low temperatures, all cultivars showed
HIT's in the seedling stage. However, at higher temperatures and

in the adult plant stage, the infection types shifted towards heterogeneous reactions. The authors stated that "the range of infection
types observed is similar to the mesothetic reaction, but it may
not consist of the whole range of infection types".

The data presented above indicate that heterogeneous IT's:
(a) may be a common expression of incomplete resistance, (b) should
be accepted as intermediate infection types for those pathogens
where they are observed with a certain frequency, (c) probably
depend on either major or minor resistance genes, (d) may be
affected by environment or development stage of the plants. It is
not known how far the traditional intermediate IT's have already
factually been used to indicate heterogeneous reactions. In coffee,
the 1 to -3 infection types were generally used to describe the
maximum lesion type instead of the average lesion type. In
0 to 9 scale for infection types for coffee leaf rust, proposed by us,
heterogeneous infection types with a smaller or wider range have
been included (Eskes and Toma Braghini, in press).

In this context, some remarks should be made on the resistance of rice to rice blast. In a review article on rice blast resistance, Ou (1980) described the difficulties he and his co-workers
found in race identification of P_o $oryzae$. Races were often unstable
and monoconidial isolates generally resulted in a wide range of new
races. However, individual lesions of a distinct type were considered to be the result of genetically controlled interactions
between host and pathogen. Often, resistant (R), moderately
resistant (M) and susceptible lesions (S) types were found in
close proximity on leaves inoculated with single conidial cultures.
According to Ou, susceptible plants generally show S type lesions,
mixed with a few M or R lesions, and the proportion of R and MR
lesions increased on the more resistant plants.

It may be assumed that Ou is explaining the mesothetic reaction of rice cultivars erroneously as being only the result of different interactions between host and pathogen genotypes. More probably, the mesothetic reaction to rice blast may also be an expression of intermediate resistance, as it is for many other diseases. Furthermore, it has been shown that the monogenic resistance of differential rice cultivars is affected by temperature
(Manibhushanrao and Day, 1972). This may explain heterogeneous
IT's also as a result of the environmental conditions.

Components of Partial Resistance

Parlevliet (1979) has given an excellent review on the components of partial resistance. These components include infection
frequency, latency period, size of the lesion, spore production and
duration of sporulation. No detailed review of these components

will be given here.

Relation Between Infection Type and Components of Partial Resistance

Reports on simultaneous observations on infection type and the components of resistance are rare. Parlevliet (1976a) suggested that extremely long latency periods may be related to a slightly lower IT. Milus and Line (1980) made simultaneous observations on infection types of slow leaf-rusting wheat cultivars, the number of uredia, the latency period and the sporulation intensity. They found high IT's in the seedling stage of slow rusting cultivars, but in the adult plant stage a shift towards lower IT's and also towards heterogeneous IT's occurred. All the components of partial resistance were correlated with the IT. For coffee leaf rust, we have also often observed a relation between infection type and the components of resistance.

The relationship indicated above can easily be explained. In fact, the components of partial resistance are a quantitative extension of the scale used for IT. The components of partial resistance, as well as the IT's relate to the same basic criteria, which are lesion size, intensity of sporulation and the occurrence of chlorosis or necrosis. The latency period is related to lesion size: when fungal growth is slow, the sporulation will generally be delayed and the lesions will be smaller. In IT's, chlorosis or necrosis of the host tissue will reduce the spore production and/or the duration of sporulation.

The only component that, at first sight, is not directly related to IT is the infection frequency (total number of visible infections). However, several authors seem to relate this component to the number of sporulating lesions only, excluding the incompatible ones. In such cases, infection frequency may also be a measure of the heterogeneous infection type. In our selection work for coffee leaf rust resistance, we looked for plants showing a low infection frequency but having a high IT. Five to ten fold differences in infection frequency were often observed. However, plants that show a low infection frequency under certain conditions, may also show a LIT or a heterogeneous infection type under different conditions. A variation may occur, within the same plant, from high resistance expressed by near immunity to intermediate resistance expressed by a low infection frequency and by a LIT, towards normal susceptibility expressed by a high infection frequency and a a high infection frequency and a high infection type. This variation depends on leaf age and environmental conditions.

It is considered that a low infection frequency may, in certain cases, represent a heterogeneous infection type that may vary in

range from immunity, or near immunity, to normally sporulating
lesions. As already indicated earlier (also see Carver and Carr,
1980; Parlevliet, 1979; Rowell and McVey, 1979), the resistance
expressed by a low infection frequency may also be specific and
may either be inherited monogenically, oligogenically or polygenically (Brennen, 1977). Therefore, selection for low infection
frequency may not necessarily be an extra guarantee that we are
selecting for non-specific resistance.

It is concluded that both resistance expressed by a low
infection type and by the components of partial resistance may
not necessarily represent different resistance mechanisms. The
difference between both assessment methods may consist, basically, of the accuracy with which small differences in resistance
can be observed.

Field Assessment of Resistance

The environment influences the expression of incomplete
resistance; therefore, this resistance has to be tested under
field conditions to determine its value for the protection of the
crop.

Epidemiological parameters.   Many researchers have only
studied incomplete resistance under field conditions. It has been
suggested that non-specific resistance is expressed by a slow development of the epidemic, which can be measured by the apparent
infection rate, and that specific resistance should be characterized
by a delay of the onset of the epidemic (Vanderplank, 1968). However, as several authors have pointed out (Parlevliet, 1979;
Johnson, 1979), all types of incomplete resistance, including
specific resistance, will reduce the apparent infection rate.
Similarly, the detectable onset of the epidemic may also be
delayed by any form of incomplete resistance (Rees et al., 1979;
Parlevliet, 1979).

The value of the apparent infection rate for measuring
differences in incomplete resistance between cultivars is questionable. The apparent infection rate only has practical value
for evaluating resistance if (a) host resistance does not change
during the experiment, (b) environmental conditions for disease
development do not change, and (c) relative amounts of green host
tissue do not change among the cultivars. These conditions are
normally not fulfilled. Incomplete resistance may depend on the
development stage and the maturing date of the host (see further
on). If this occurs, deviations from the logit curve will be obtained.
A cultivar that is relatively resistant during the active growth
phase but susceptible during the ripening phase will delay the
phase of exponential increase of the epidemic, but may show a

rapid increase of disease incidence when approaching maturity. Equally, when the climatic conditions during the second part of epidemic development become more favourable for disease, the apparent infection rate of a more resistant cultivar may be similar to the apparent infection rate of a more susceptible cultivar earlier in the season. The occurrence of late rusting, in addition to slow rusting, has been recognized by Rees et al. (1979), but it could not be related to a change in the pathogen population. The authors concluded that infection rate is not a valuable parameter for measuring field resistance to stem rust of wheat.

AUDPC.   The area under the disease progress curve may be a better parameter for evaluating field resistance than the apparent infection rate. This parameter does not show the disadvantages of the infection rate pointed out above. The average disease incidence or severity during the epidemic may be equally valid.

Yield.   The final objective of disease resistance is to avoid losses in yield. It has been proposed that yield can be used to select for resistance (Robinson, 1976). This parameter, however, not only measures resistance but also tolerance (not always a desirable trait). Furthermore, yield may be positively correlated with disease incidence in the field. Simons (1975) measured the relative yield differences between sprayed and unsprayed plots of breeding lines of oats. He found a positive correlation between total yield and susceptibility indicating that productive plants suffered more from disease than unproductive ones. This also indicates that it is necessary to select simultaneously plants for both yield and resistance.

Some factors affecting field resistance.   Some pitfalls can be recognized: quantitative resistance in the field may be due to genes for either specific or non-specific resistance. Even high levels of race-specific resistance may be expressed as quantitative resistance, if the matching pathotype is rare (Ou et al., 1975) or if the pathotype is only incompletely virulent. Furthermore, allo infection from severely diseased plots or plants may increase the disease incidence of plots with more resistant plants, that are located nearby (interplot and interplant interference) (Parlevliet and Van Ommeren, 1975; James et al., 1976; Vanderplank, 1968).

FACTORS AFFECTING INCOMPLETE RESISTANCE

Development Stage of the Plant

Plant age.   The development stage of the plant may affect the resistance of a high infection type as well as that of a low infection type. The effect of development stage has been studied

for the potato-late blight relationship. Umaerus (1970, 1973)
recognized three growth phases for resistance: the young plants
were susceptible, the adults were more resistant and the old plants
again become more susceptible. He related the latter transition
to tuber formation. For rusts, seedlings are also generally more
susceptible than adult plants (Hooker, 1967). For barley/*Puccinia*.
hordei, this increase in resistance in adult plants may be expressed
by a lower infection type, for both monogenic and polygenic resistance. Parlevliet (1975, 1976a, 1976b) described how several types
of resistance to brown rust can be detected in barley: (a) low infection types (0 to 3) in all growth stages, (b) infection type 4 in seedlings and 0-2 in adult plants, or (c) an infection type 4 in seedlings
and 3-4 in adult plants. The effect of the development stage can
also be expressed by the components of resistance, e.g. latency
period (Parlevliet, 1975), infection frequency (Parlevliet and
Kuiper, 1977a) or receptivity (Rowell and McVey, 1979).

For perennial crops less information on the relation between
plant age and susceptibility was found. Rowan and Steinbeck (1977)
report that susceptibility of loblolly pine to fusiform rust increases
with the age of the seedling plants.

Age of plant organs. Young potato leaves are generally more
susceptible to late blight than adult leaves. Susceptibility again
increases when leaves are ageing or are detached from the plant
(Umaerus, 1970, 1973). Susceptibility (infection frequency) of
barley leaves to *Puccinia hordei* increases with age (Parlevliet
and Kuiper, 1977a). Adult leaves of *Coffea canephora* cv. Kouillou
show a higher resistance to coffee leaf rust than very young or
old leaves. Sharma et al. (1980) mentioned that more mature
leaves and older shoots of *Populus* species are more resistant to
Melampsora larici-populina than younger ones. This is also the
case for the resistance of *Malus* to *Venturia inaequalis* (Williams
and Kuc, 1969).

Difference in resistance among plant organs. Hooker (1967)
indicated that resistance to stripe rust in wheat may vary between
primary leaves, stem leaves, stems and ears. Resistance of
potato to late blight is greater after inoculation on the upper leaf
surface than on the lower leaf surface (Malcolmsen, 1969). These
examples show that different host tissues may have different levels
of resistance.

Interactions between development stages and cultivars. The
effect of development stage or age of the tissue on resistance is
not equal for all host genotypes. Normally, highly susceptible
plants are so in all growth stages and at all tissue ages. More
resistant plants tend to express their resistance more in one
development stage than in others. Significant host cultivar x
development stage interactions have been reported for many

diseases (Parlevliet, 1976; Parlevliet and Kuiper, 1977a; Milus and Line, 1980; Rowell and McVey, 1979).

Productivity and Date of Maturity

Relatively few reports exist on the effects of productivity on incomplete resistance. As indicated earlier, the increase of susceptibility of the potato to late blight, when plants age, has been associated with tuber formation. This could partly explain the correlation found between late maturing of potato cultivars and field resistance to late blight (Toxopeus, 1959). A similar relationship has been suggested between maturity date and resistance of oats to crown rust (Luke et al., 1975). The early ripening varieties did not rust faster than the late ripening varieties at the onset of the epidemic, but rusted rapidly when approaching maturity. Simons (1975) observed a negative correlation between total yield and resistance. Shaner and Dana Hess (1978) mentioned that the hypothetical disease progress curves calculated by them for slow leaf rusting resistance in wheat did not fit well with the real data. They suggested that this may be due to an increased susceptibility of the slow rusting wheat cultivars when the ripening phase began.

The effect of productivity on incomplete resistance has also been confirmed in our work on the coffee/*H. vastatrix* relationship. Leaves of branches with developing berries are more susceptible to coffee leaf rust than those of the same plants without berries; this was determined by an increased infection frequency and a decreased latency period. Parlevliet (personal communication) also found a higher resistance to brown rust of flag leaves of barley plants without ears than of those with ears.

Effect of Environment

As already indicated, the environment greatly affects disease and disease resistance. The individual environmental factors may change the fitness of the pathogen, the resistance of the host or may change the host-pathogen relationship. The effect of environment on resistance may or may not be linear. If the effect of environment is not linear, two or three factor interactions between host genotype, pathogen genotype and environment may be observed (see Wilcoxson et al., 1975; Stubbs, 1980).

The environmental factors involved in the modification of resistance may be numerous. Examples of how some individual factors may affect the resistance are given hereafter.

Temperature.   Both complete as well as incomplete

resistance may be affected by changes in the air temperature.
Roelfs and McVey (1979) mentioned that, although the resistance
of wheat differentials with Sr genes is generally stable, the genes
Sr 6, Sr 10, Sr 15 and Sr 17 become less effective, whereas other
genes such as Sr 13 and Sr 9-b become more effective at higher
temperatures. Temperature also affects the Lr 17 and Lr 18 genes
for leaf rust resistance in wheat (Dyck and Samborski, 1968).
Armstrong and Armstrong (1978) reported that the resistance of
Cucurbitaceae to *Fusarium oxysporum* is best expressed at 30°C
and that many resistant hosts become susceptible at 16°C, whereas
the optimum growth of the fungus occurs at 26-28°C.

Brown and Sharp (1969) reported that "minor" genes for
resistance of wheat to stripe rust may be extremely temperature
sensitive. In critical phases, short exposures to contrasting
temperatures (as short as four hours) resulted in significant
changes in infection types.

Sharp et al. (1976) also reported that certain additive genes
for stripe rust were only effective in susceptible wheat cultivars
at a high temperature. Kochmann and Brown (1975) showed that
the resistance of oats to *P. graminis* , as expressed by latency
period, pustule size and uredospore production, was lower at 30°C
or 35°C than at 20 or 25°C, and that the cultivar with the highest
level of resistance was more affected by a change in temperature
than others.
These examples indicate that the effect of temperature is not
uniform: the resistance may either be more or less effective at
high or low temperatures. This is at variance with the conclusion
of Vanderplank (1978) that resistance is generally lost at higher
temperatures.

Temperature may not only affect host resistance, but also
the fitness of the pathogen. Katsuya and Green (1967) reported
that race 15 B of wheat stem rust predominated when grown in
mixtures with race 56 at low temperatures, but at high temperatures the reverse took place. However, this may not be a general
phenomenon for rust fungi: Eyal and Peterson (1967) reported
only slight differences in spore production of five races of *P.
recondita* at different temperatures.

Light intensity. The effect of light intensity is less well
studied. Stubbs (1967) observed that certain differential wheat
varieties developed a higher infection type to *P. striiformis* under
reduced light intensity. The reverse effect was observed for a
barley variety. Intermediate reaction types were generally heterogeneous. Field resistance of potato, measured by the infection
frequency or lesion size, may be more strongly expressed under
high light intensities (Victoria and Thurston, 1974; Schumann and

Thurston, 1977). There are similar reports for the effect of day
length on the resistance of wheat and potato (Thurston, 1971; Stubbs,
1967).

In the coffee/*H. vastatrix* relationship high light intensity
before inoculation increases the susceptibility (infection frequency)
of the leaves. The infection type of more resistant plants, including plants with incomplete vertical resistance, may also be affected.

The opposite effect of light intensity on coffee and wheat may
be related to the natural environment in which these crops grow.
In the natural pathosystem, coffee is normally shaded whereas
wheat grows in the full sun.

Other environmental factors.   The level of nutrition may
affect the expression of resistance. Excessive N-fertilization of
potatoes increases the lesion size of late blight under field conditions (Main and Gallegly, 1964). The effect of fertilizers on
resistance expression has also been observed in certain other
pathosystems (Rowan and Steinbeck, 1977; Figueiredo et al.,
1976).

Not many studies have been found on the effect of relative
air humidity. In experiments, the effects of relative humidity on
the host cannot be separated from its effects on the pathogen.
Concerning soil moisture, Van der Wal et al. (1974) reported that
a susceptible wheat variety had the lowest level of leaf rust at
average levels of soil moisture.

Host-Pathogen-Environment Interactions

The environment may not only interact with the resistance
or the pathogenicity but may also influence the host-pathogen
combination. Most of the reported interactions relate to temperature.

Zimmer and Schäfer (1961) tested two resistant oat cultivars
with three races of *Puccinia coronata* at three constant temperatures. In one race-cultivar combination, the infection type changed
to a higher type, while in the other cases the expression of the
resistance did not change. Eversmeyer et al. (1980) reported
that significant differences were found at low temperatures in the
latency periods of *P. recondita* races on wheat cultivars. At higher
temperatures, these differences disappeared. Some incompatible
reactions were also observed at low temperatures only. Johnson
and Taylor (1976b) reported that the genes for resistance of Hybride
de Bersée to *P. striiformis* were effective at different temperatures.
In the combination wheat and leaf rust, Milus and Line (1980) found

significant differential interactions between host genotype, development stage, pathogen genotype, and temperature. Chandrashekar and Heather (1981a), working with *Populus* and *Melampsora laricipopulina* found significant two, three, and four factor interactions between host cultivars, pathogen genotypes and pre- and post-inoculation temperatures.

Genetic Background

Monogenic major gene resistance may be affected by other genes in the host. Parlevliet and Kuiper (1977b), Roelfs and McVey (1979), Dyck and Samborski (1968) and Athwal and Watson (1954) showed that the level of resistance conferred by monogenes may vary according to the genetic background of the host. Gerber and Green (1980) reported the effect of a suppressor gene on resistance.

A combination of minor genes for resistance may produce an unexpectedly high level of resistance (measured by infection type). Thurston (1971) mentioned that by selection for field resistance to late blight in the susceptible *S. tuberosum* subsp. *andigena* clones were obtained that showed hypersensitive resistance. In wheat/stripe rust, Pope (1968) observed dramatic increases in resistance in crosses between susceptible cultivars.

In the coffee population Icatu, a new race of *H. vastatrix* has been identified by us that breaks, simultaneously, the resistance of certain R, MR, MS and S plants. It seems plausible that the same resistance factor is involved, which apparently operates at different levels of efficiency in different plants.

The Effect of Major Genes on Incomplete Resistance

Several times it has been suggested that major genes, when overcome by the matching pathotype, may have some residual effect on resistance. Such a hypothesis can only be tested if fully isogenic lines of the host and a susceptible control are inoculated with compatible races of the pathogen. Available information indicates that direct effects of broken resistance genes have generally not been detected (e.g. Skovmand et al., 1978; Mantin and Ellingboe, 1976). Nazareno and Roelfs (1981) did not find a relationship between independent Sr genes for seedling resistance and the adult plant resistance of Thatcher wheat to stem rust. Also, most of the combinations of Sr_5, Sr_{12}, Sr_{16} and Sr Tc genes generally did not increase the adult plant resistance. Only some progenies with the combination of the Sr_{12} and Sr Tc genes were similar in resistance to Thatcher.

Apparent effects of major genes may have several causes.

Three important ones are indicated here:

(a) The residual effect may not be due to the major gene itself, but rather to hidden minor genes in the same host genotype. A clear example of this has been given by Parlevliet and Kuiper (1977b). These minor genes may be closely linked to the major gene. In segregating populations the effect of the linked minor genes may then be counfounded with the effect of the major gene. Skovmand et al. (1978a, 1978b) have suggested that effect occurs for some stem rust resistance genes in wheat.

(b) Incomplete vertical pathogenicity may also produce a residual effect. The pathogen may need more than one genetic modification to achieve full compatibility towards the matching resistance gene.

(c) The virulence genes needed to match the resistance gene may have negative pleiotropic effects on the survival of the pathogen (stabilizing selection, Vanderplank, 1968). This negative effect of the virulence gene may also be classified as a residual effect of the major resistance gene.

Self Protection and Cross Protection

Resistance can be induced by pre-inoculation of compatible pathotypes, incompatible pathotypes or even by a different parasite species. This "acquired resistance" was recognized a long time ago (Chester, 1933).

Kuc and Richmond (1977) observed the occurrence of systemic protection of cucumber against *Colletotrichum lagenarium* by pre-inoculation of a compatible race of the fungus on the cotyledon or on the first leaf. The protection was expressed by a reduction in lesion size and number of lesions, and was efficient for four to five weeks. Differences in protective responses among cultivars and to pathotypes were observed. Tjamos (1979) showed that resistance to *Verticillium* wilt in cucumber could be induced by filtrates of the fungus, which reduced the extent of vessel colonization following subsequent inoculations. Self protection has also been demonstrated for perennial plants (Partiot et al., 1977; Moraes et al., 1976; Randall and Helton, 1977).

Striking levels of cross protection involving different species of pathogens have also been obtained. McIntyre et al. (1981) demonstrated how localized infections of TMV virus induced systemic and long lasting resistance in *Nicotiana tobacum* against reinoculation with a compatible strain of TMV, and also against *Phytophthora parasitica*, *Pseudomonas tabaci* and *Peronospora tabacina*. Furthermore, the TMV infections reduced the reproductive capacity of *Myzus persicae*. Fernandez et al. (1975) also demonstrated that potato plants infected with different types of virus were more resistant to late blight.

Thus, cross protection and self protection can be important factors in determining incomplete resistance towards some fungal pathogens. This interesting aspect of resistance has been studied in only a few pathosystems.

CONCLUSIONS IN RELATION TO THEORETICAL RESISTANCE MODELS

The Gene-for-Gene Relationship

Some doubts have appeared in the relation to the gene-for-gene hypothesis. The occurrence of modifier genes, and even suppressor genes, of race-specific major gene resistance indicates that more than one gene may be involved before resistance can be fully expressed. The occurrence of several levels of incomplete virulence, in relation to one resistance gene, also indicates genetic background effects on virulence or the existence of alleles with different levels of virulence.

The Concept of Horizontal Resistance

The available information on resistance expression does not permit a strict separation of resistance into specific or non-specific types. Neither the phenotypical expression or the inheritance of resistance, nor the expression of the host-pathogen relationship are proof of the non-specificity of the resistance.

The Integrated Concept of Horizontal and Vertical Resistance

In this concept, as presented by Parlevliet and Zadoks (1977), the gene-for-gene relationship is extended also to polygenic resistance. Although it is extremely difficult to prove a gene-for-gene relationship for minor genes, it is considered that some evidence exists to support this hypothesis (Parlevliet, 1979).

However, one reservation to the integrated concept can be made in relation to non-specific pathogenicity. As the concept suggests that all host-pathogen relationships are a consequence of specific interactions, no variation in pathogenicity should be observed on hosts that do not carry resistance genes. This is not always the case: the variation in pathogenicity may even be greater on hosts without resistance genes than on those carrying several resistance genes. However, Parlevliet (1976a) detected significant differences in latency period between brown rust cultures on the most susceptible barley cultivar L94, whereas this cultivar apparently does not possess any gene for longer latency periods (Parlevliet, 1978). This indicates the existence of non-specific pathogenicity, which, however, is no proof of non-specific resistance. Thus, variation for pathogenicity seems to exist that is unrelated to resistance genes in the host.

Horizontal Resistance Genes are the Same as Vertical Resistance Genes

This idea was suggested by Nelson et al. (1970) and Nelson (1978). His hypothesis may be supported by the similarity of the expression of specific and non-specific resistance. Furthermore, specific major resistance genes, when broken, may function as minor genes for resistance. It is considered that this has not been proved sufficiently. Published data indicate that the presence of one or more broken vertical resistance genes normally does not improve the level of resistance.

The origin of Nelson's concept goes back to his 1970 paper (Nelson et al., 1970). However, after careful reading of this paper, it was concluded that the authors interpreted horizontal resistance as a horizontal line (equal resistance to all races). They concluded that horizontal and vertical resistance genes were the same, because the absolute variation in pathogenicity of *H. turcicum* isolates was high on the most susceptible maize inbred, but low on the most resistant inbred, and not all isolates were able to infect the resistant inbreds. The authors deduced from this observation that the more resistant genes there are present, the more "horizontal" the nature of the resistance is. However, the deductions of the authors apparently do not agree with the data they present. In fact, their data could be much better explained by applying a model for non-specific resistance.

Models for Durable Resistance Related to the Host-Pathogen-Environment Relationship

Most of the resistance models refer only to the host-pathogen relationship. However, for a full appreciation of how a host-pathogen balance can be achieved, the effect of the biotic and abiotic environment will also have to be considered.

Firstly, the factors affecting the expression of incomplete resistance may be important for the durability of resistance because they increase discontinuity of the contact between the resistant host and the virulent pathogen. As Robinson (1979) suggested, the relative value of race-specific resistance for pathosystem balance is related to the degree of discontinuity of the contact between host and pathogen. Incomplete resistance that is expressed only in the adult plant stage, for example, will exert no selection pressure on the pathogen when the plant is young. Temperature labile resistance genes will only exert selection pressure for virulence under certain environmental conditions. Resistance expressed by heterogeneous reaction types does allow for reproduction of incompatible pathogen races. When several host and environmental factors simultaneously affect resistance, it is unlikely that selection pressure for increased virulence will be exerted for more than a few reproductive cycles of the pathogen. Generally, mutants for increased virulence are initially weak growers. They may become less competitive if the selection pressure for virulence is not constant.

Secondly, the occurrence of differential interactions between pathogen genotype and host genotype, development stage of the host and environment can be an additional stabilizing factor on the pathogen population (Chandrashekar and Heather, 1981a). These interactions indicate that certain virulence genes may receive negative selection pressure for one host/environment combination but positive selection pressure in another combination.

These arguments may be an additional explanation for the durability that is often observed for polygenically inherited resistance, especially if interactions with environment and development stage of the host occur, even if the individual genes can be specifically overcome by the pathogen.

Stabilizing selection. The occurrence of pathogen-host-environment interactions suggest that the forces regulating stabilizing selection, sensu Vanderplank (1968), may be specific rather than general ones. This could explain why stabilizing selection is often not operative in the agricultural pathosystem, as pointed out by Parlevliet (1981). The experiments on stabilizing selections are generally carried out under restricted environmental conditions and with a restricted host population. If stabilizing selection is indeed governed by specific forces, then more detailed tests, including a wide host range and varying environmental conditions, will be necessary to detect these forces.

REFERENCES

Armstrong, G.M., and Armstrong, J.K., 1978, Formae speciales and races of *Fusarium oxysporum* causing wilts of the Cucurbitaceae, *Phytopathology*, 68:19-28.
Ashagari, D., and Rowell, J.B., 1980, Post penetration phenomena in wheat cultivars with low receptivity to infection by *Puccinia graminis* f. sp. *tritici* *Phytopathology*, 70:624-627.

Athwal, S.D., and Watson, I.A., 1954, Inheritance and the genetic
    relationship of resistance possessed by the two Kenya wheats
    to races of *Puccinia graminis*, *Proc. Linn. Soc. N.S.W.*,
    79:1-14.
Brennen, P.S., 1977, Non-hypersensitive resistance to wheat-
    stem rust and its race specificity. *Proc. 3rd Int. Congr.
    Soc. Adv. Breed. Res. in Asia and Oceania (SABRAO)*,
    Canberra, Australia , 4(a):3-5.
Browder, L.E.. 1973, Specificity of the *Puccinia recondita* f. sp.
    tritici: Triticum aestivum "Bulgaria 88" relationship,
    Phytopathology, 63:524-528.
Brown, J.F., and Sharp, E.L., 1969, Interaction of minor host
    genes for resistance to *P. striiformis* with changing tempera-
    ture regimes, *Phytopathology* , 59:999-1001.
Caldwell, R.M., 1968, Breeding for general and/or specific plant
    disease resistance, *Proc. 3rd Int. Wheat Genetics Symp.*,
    Canberra, 263-272.
Carver, T.L.W., and Carr, A.J.H., 1980, Some effects of host
    resistance on the development of oat mildew, *Ann. Appl.
    Biol.*, 94:290-293.
Chandrashekar, M., and Heather, W.A., 1981a, The effect of pre
    and post inoculation temperature on resistance in certain
    cultivars of poplar to races of *Melampsora larici-populina*
    Kleb., *Euphytica* , 30:113-120.
Chandrashekar, M., and Heather, W.A., 1981b, Temperature
    sensitivity of reactions of *Populus* spp. to races of *Melampsora
    larici-populina* Kleb., *Phytopathology*, 71:421-424.
Chester, K.S., 1933, The problem of acquired physiological
    immunity in plants. *Q. Rev. Biol* ., 8:129-154, 275-324.
Clifford, B.C., and Roderick, H.W., 1978, A comparative his-
    tology of some barley/brown rust interactions, *Ann. Appl.
    Biol.*, 89:295-298.
Dyck, P.L., and Samborski, D.J., 1968, Host-parasite interactions
    involving two genes for leaf rust resistance in wheat, *Proc.
    3rd Int. Wheat Genetics Symp., Canberra* , 245-250.
Eskes, A.B., and Toma-Braghini, M., 1981, Assessment
    methods for resistance to coffee leaf rust (*Hemileia vastatrix*
    Berk. et Br.), *Plant Prot. Bull.* , 29, in press.
Eversmeyer, M.G., Kramer, C.L., and Browder, L.E., 1980,
    Effect of temperature and host: parasite combination on
    latent period of *Puccinia recondita* in seedling wheat plants,
    Phytopathology , 70:938-941.
Eyal, Z., and Peterson, J.L., 1967, Uredospore production of five
    races of *P. recondita* as affected by light and temperature,
    Can. J. Bot. 45:537-540.
Fernandes de Cubillos, C., and Thurston, H.D., 1975, The effect
    of viruses on infection by *Phytophthora infestans* (Mont.) de
    Bary in potatoes, *Am. Potato Journal* , 52:221-226.
Figuereido, P., Hiroce, R., and Oliveira, D.A., 1976, Estado

nutricional e ataque da ferrugem do cafeeiro (*Hemileia vastatrix* Berk. et Br.), *O Biologico*, 42:164-167.

Hooker, A.L., 1967, The genetics and expression of resistance in plants to rusts of the genus *Puccinia, Ann. Rev. Phytopath.* 5:163-182.

James, W.C., Smith, C.S., Hodgson, W.A., and Callbeck, L.C., 1976, Representational errors due to interplot interference in field experiments with late blight of potato, *Phytopathology*, 66:695-701.

Johnson, R., 1979, The concept of durable resistance, Letter to the editor, *Phytopathology*, 69:198-199.

Johnson, R., and Taylor, A.J., 1972, Isolates of *Puccinia-striiformis* collected in England from the wheat varieties Maris Beacon and Joss Cambier, *Nature*, 238:105-106.

Johnson, R., and Taylor, A.J., 1976a, Spore yield of pathogens in investigations of the race specificity of host resistance, *Ann. Rev. Phytopath.*, 14:97-119.

Johnson, R., and Taylor, A.J., 1976b, Genetic control of durable resistance to yellow rust *(Puccinia striiformis)* in the wheat cultivar Hybride de Bersee, *Ann. Appl. Biol.*, 81:385-391.

Katsuya, K., and Green, C.J., 1967, Reproductive potentials of races 15 B and 56 of wheat stem rust, *Can. J. Bot.*, 45: 1077-1091.

Kerber, E.R., and Green, G.J., 1980, Suppression of stem rust resistance in the hexaploid wheat c.v. Canthatch by chromosome 7DL, *Can. J. Bot.*, 58:1347-1350.

Kochman, J.K., and Brown, J.F., 1975, Host and environmental effect on post-penetration development of *Puccinia graminis avenae* and *P. coronata avenae, Ann. Appl. Biol.*, 81:33-41.

Kuč, J., and Richmond, S., 1977, Aspects of protection of cucumber against *Colletotrichum lagenarium* by *C. lagenarium*, *Phytopathology*, 67:533-536.

Luke, H.H., Barnett, R.D., and Chapman, W.H., 1975, Types of horizontal resistance of oats to crown rust, *Pl. Dis. Rep.*, 59:332-334.

MacNeal, F.H., Konzak, C.F., Smith, E.P., Tate, W.S., and Russell, T.S., 1971, A uniform system for recording and processing cereal research data, *U.S. Dep. Agric. Res. Serv.*, Publ. 34-121, 42 p.

Main, C.E., and Gallegly, M.E., 1964, The disease cycle in relation to multigenic resistance of potato to late blight, *Am. Potato Journal*, 41:387-400.

Malcolmson, J.F., 1969, Factors involved in resistance to blight *(Phytophthora infestans* (Mont) de Bary) in potatoes and assessment of resistance using detached leaves, *Ann. Appl. Biol.*, 64:461-468.

Manibhushanrao, K.M., and Day, P.R., 1972, Low night temperature and blast disease development on rice, *Phytopathology*, 62:1005-1007.

Martin, T.J., and Ellingboe, A.H., 1976, Differences between compatible parasite/host genotypes involving the Pm4 locus of wheat and the corresponding genes in *Erysiphe graminis* f. sp. *tritici, Phytopathology*, 66:1435-1438.

McIntyre, J.L., Dodds, A.J., and Hare, D.J., 1981, Effects of localized infections of *Nicotiana tabacum* by Tobacco Mosaic Virus on systemic resistance against diverse pathogens and an insect, *Phytopathology*, 71:297-301.

Milus, E.A., and Line, R.F., 1980, Characterization of resistance to leaf rust in Pacific Northwest wheats, *Phytopathology*, 70:167-172.

Moraes, W.B.C., Martins, E.M.F., Musumeci, M.R., and Beretta, M.J.G., 1976, Induced protection to *Hemileia vastatrix* in coffee plants, *Summa Phytopathologica*, 2:39-43.

Nazareno, N.R.X., and Roelfs, A.P., 1981, Adult plant resistance of Thatcher wheat to stem rust, *Phytopathology*, 71:181-185.

Nelson, R.R., MacKenzie, D.R., and Scheifele, G.L., 1970, Interaction of genes for pathogenicity and virulence in *Trichometasphaeria turcica* with different number of genes for vertical resistance in *Zea mays, Phytopathology*, 60:1250-1254.

Nelson, R.R., 1978, Genetics of horizontal resistance to plant diseases, *Ann. Rev. Phytopath.*, 16:359-378.

Ou, S.H., Nuque, F.L., and Bandong, J.M., 1975, Relation between qualitative and quantitative resistance to rice blast, *Phytopathology*, 65:1315-1316.

Ou, S.H., 1980, Pathogen variability and host resistance in rice blast disease, *Ann. Rev. Phytopath.*, 18:167-187.

Parlevliet, J.E., 1975, Partial resistance of barley to leaf rust, *Puccinia hordei* I. Effect of cultivars and development stage on latent period, *Euphytica*, 24:21-27.

Parlevliet, J.E., 1976a, Evaluation of the concept of horizontal resistance in the barley/*Puccinia hordei* host-pathogen relationship, *Phytopathology*, 66:494-497.

Parlevliet, J.E., 1976b, The genetics of seedling resistance to leaf rust *Puccinia hordei* Otth. in some spring barley cultivars, *Euphytica* 25:249-254.

Parlevliet, J.E., 1978, Further evidence of polygenic inheritance of partial resistance in barley to leaf rust, *Puccinia hordei*, *Euphytica*, 27:369-379.

Parlevliet, J.E., 1979, Components of resistance that reduce the rate of epidemic development, *Ann. Rev. Phytopath.*, 17:203-222.

Parlevliet, J.E., 1981, Stabilizing selection in crop pathosystems: an empty concept or a reality?, *Euphytica*, 30:259-269.

Parlevliet, J.E., and Kuiper, H.J., 1977a, Partial resistance of barley to leaf rust, *Puccinia hordei* IV, Effect of cultivar and development stage on infection frequency, *Euphytica*, 26:249-255.

Parlevliet, J.E., and Kuiper, H.J., 1977b, Resistance of some barley cultivars to leaf rust, *Puccinia hordei*; polygenic partial resistance hidden by monogenic hypersensitivity, *Neth. J. Pl. Path.*, 83:85–89.

Parlevliet, J.E., and van Ommeren, A., 1975, Partial resistance of barley to leaf rust, *Puccinia hordei*. II, Relationship between field trials, microplot tests and latent period, *Euphytica*, 24:293–303.

Parlevliet, J.E., and Zadoks, J.C., 1977, The integrated concept of disease resistance; a new view including horizontal and vertical resistance in plants, *Euphytica*, 26:5–21.

Partiot, M., N'Guessan Kona, Zoumboi, A.M., 1977, La prémunition du cacaoyer contre le *Phytophthora* sp. Induction du phénomène et premier ensai d'utilisation comme moyen de lutte, *Cafe Cacao The*, 11:29–40.

Pope, W.K., 1968, Interaction of minor genes for resistance to stripe rust in wheat. *Proc. 3rd Int. Wheat Genetics Symp.*, Canberra, 251–257.

Randall, H., and Helton, A.W., 1977, Effect of inoculation date on induction of resistance to *Cytospora* in Italian Prune Trees by *Cytospora cyneta*, *Phytopathology*, 66:206–207.

Rees, R.G., Thompson, J.P., and Mayer, R.J., 1979, Slow rusting and tolerance to Rusts in Wheat. I. The progress and effects of epidemics of *Puccinia graminis tritici* in selected wheat cultivars, *Austr. J. Agric. Res.*, 30:403–419.

Robinson, R.A., 1976, "Plant Pathosystems", Springer, Berlin, Heidelberg, New York.

Robinson, R.A., 1979, Permanent and impermanent resistance to crop parasites. A re-examination of the pathosystems concept with special reference to rice blast, *Z. planzenzuecht.*, 83:1–39.

Roelfs, A.P., and McVey, D.V., 1979, Low infection types produced by *Puccinia graminis* f. sp *tritici* and wheat lines with designated genes for resistance, *Phytopathology*, 69:722–729.

Rowan, S.J., and Steinbeck, K., 1977, Seedling age and fertilization effect on susceptibility of loblolly pine to fusiform rust, *Phytopathology*, 67:242–246.

Rowell, J.B., and McVey, D.V., 1979, A method for field evaluation of wheat for low receptivity to infection by *Puccinia graminis* f. sp. *tritici*, *Phytopathology*, 69:405–409.

Samborski, D.J., 1963, A mutation in *Puccinia recondita* Rob. ex. Dosm. f. sp *tritici* to virulence on Transfer, Chinese Spring x *Aegilops umbellata* Zhuk, *Can. J. Bot.*, 41:475–479.

Schumann, G., and Thurston, H.D., 1977, Light intensity as a factor in field evaluations of general resistance of potatoes to *Phytophthora infestans*, *Phytopathology*, 67:1400–1402.

Schwarzbach, E., 1979, Response to selection for virulence against the ml–0 based mildew resistance in barley, not fitting the gene-for-gene hypothesis, *Barley Genetics Newsletter*, 9:85-88.

Shaner, G., and Dana Bess, F., 1978, Equations for integrating components of slow leaf rusting resistance in wheat, *Phytopathology*, 68:1464-1469.

Sharma, J.K., Heather, W.A., and Winer, P., 1980, Effect of leaf maturity and shoot age of clones of *Populus* species on susceptibility to *Melampsora larici-populina*, *Phytopathology*, 70:548-555.

Sharp, E.L., Sally, B.K., and Taylor, G.A., 1976, Incorporation of additive genes for stripe rust-resistance in winter wheat, *Phytopathology*, 66:794-797.

Simons, M.D., 1972, Polygenic resistance to plant diseases and its use in breeding resistant cultivars, *J. Environ. Qual.*, 1:232-239.

Simons, M.D., 1975, Heritability of field resistance to the oat crown rust fungus, *Phytopathology*, 65:324-328.

Skovmand, B., Roelfs, A.P., and Wilcoxson, R.D., 1978a, The relationship between slow rusting and some genes specific for stem rust resistance in wheat, *Phytopathology*, 68:491-499.

Skovmand, B., Wilcoxson, R.D., Shearer, B.L., and Stucker, R.R., 1978b, Inheritance of slow rusting to stem rust in wheat, *Euphytica*, 27:95-107.

Statler, G.D., and Jones, D.A., 1981, Inheritance of virulence and uredial color and size in *Puccinia recondita tritici*, *Phytopathology*, 71:652-655.

Stubbs, R.W., 1967, Influence of light intensity on the reaction of wheat and barley seedlings to *Puccinia striiformis*, *Phytopathology*, 57:615-619.

Stubbs, R.W., 1980, Environmental resistance of wheat to yellow rust (*Puccinia striiformis* Westend. f. sp.*tritici*). Fifth European and Mediterranean Cereal Rust Conference, Bari and Rome, Italy, Proceedings: 77-81.

Thurston, H.D., 1971, Relationship of general resistance: late blight of potato, *Phytopathology*, 61:620-626.

Tjamos, E.C., 1979, Induction of resistance to Verticillium wilt in cucumber (*Cucumis sativus*), *Physiol. Plant Pathol.*, 15:223-227.

Toxopeus, H.J., 1959, Notes on the inheritance of field resistance of the foliage of *Solanum tuberosum* to *Phytophthora infestans*, *Euphytica*, 8:117-124.

Umaerus, V., 1970, Studies on field resistance to *Phytophthora infestans* 5. Mechanisms of resistance and application to potato breeding, *Z. Planzenzuecht.*, 63:1-23.

Umaerus, V., 1973, Background paper for the late blight planning conference of CIP *in*:"Late Blight Strategy, Report of the late blight project planning conference held at CIMMYT, El Batan, Edo. de México, México, August 22-27", 43 p.

Vanderplank, J.E., 1968, "Disease resistance in plants", Academic Press, New York.

Vanderplank, J.E., 1978, "Genetic and molecular basis of plant
    pathogenesis", Springer, Berlin, Heidelberg, New York.
Van der Wal, A.F., Smeitink, H., and Maan, G.C., 1974, An
    ecophysiological approach to crop losses exemplified in the
    system wheat, leaf rust and glume blotch III. Effects of
    soil-water potential on development, growth, transpiration,
    symptoms and spore production of leaf rust infected wheat,
    Neth. J. Pl. Path. , 81:1-13.
Victoria, J.I., and Thurston, H.D., 1974, Light intensity effects
    on lesion size caused by *Phytophthora infestans* on potato
    leaves, *Phytopathology* , 64:753-754.
Watson, I.A., 1970, Changes in virulence and population shifts
    in plant pathogens, *Ann. Rev. Phytopath.*, 8:209-230.
Watson, I.H., and Luig, N.H., 1968, Progressive increase of
    virulence in *Puccinia graminis* f. sp. *tritici, Phytopathology* ,
    58:70-73.
Wilcoxson, R.D., Skovmand, B., and Atif, A.H., 1975, Evaluation
    of wheat cultivars for ability to retard development of stem
    rust, *Ann. Appl. Biol.* , 80:275-281.
Williams, E.B., and Kuč, J., 1969, Resistance in *Malus* to
    Venturia inaequalis, Ann. Rev. Phytopath., 7:223-246.
Wolfe, M.S., and Schwarzbach, E., 1978, Patterns of race
    changes in powdery mildews, *Ann. Rev. Phytopath.* , 16:
    159-180.
Zimmer, D.E., and Schafer, J.F., 1961, Relation of tempera-
    ture to reaction type of *Puccinia coronata* on certain oat
    varieties, *Phytopathology* , 51:202-203.

DISCUSSION

KRANZ: I was greatly pleased with your critical review of infection
types. In fact, from the epidemiological point of view, and for the
unification of terms, we do not need this term. Let us use "disease
efficiency" instead, i.e. number of sporulating lesions per unit
inoculum per unit host surface. This may be supplemented by the
parameter "sporulation intensity per lesion". All lower infection
types are epidemiologically irrelevant (though they are valuable
clues to possible resistance mechanisms). Hence, a resistant
cultivar will have low infection types and low disease efficiency
(approximately 0), but the latter is quantitative.

ESKES: Yes, thank you for this remark.

SHARP: I believe infection types can be a valid criterion to evaluate resistance, even that which may be durable. For example, in
stabilizing resistance to stripe rust for an infection type (2), it was
necessary to self, inoculate and select plants over several generations to stabilize plants for this infection type. This indicates
that a certain background must be developed before a (2) infection

type becomes meaningful. The background genes are very important. To use infection types efficiently as a criterion for disease
resistance, it is necessary to determine what is the genotypic
basis for their phenotypic expression. If studies show this to be
determined by polygenes, the resistance has a higher probability
of being durable.

ESKES: I agree fully with you that the genotypic basis of resistance may be more important in determining durability than its
phenotypic expression. However, I have one doubt. If the background genes are only backing-up a major gene for hypersensitive
resistance, then the resistance may be easily overcome by the
pathogen, although its inheritance may be polygenic. Thus, theoretically, this type of polygenic resistance may not be durable.
I expect polygenic resistance to be durable, when each polygene
has a small effect on the resistance by itself. This type of resistance should be additive.

RESISTANCE OF CROPS TO NEMATODES

F. Lamberti and C.E. Taylor[*]

Istituto di Nematologia Agraria del C.N.R.
70126 Bari, Italy
* Scottish Crop Research Institute
Invergowrie, Dundee; UK

INTRODUCTION

Resistance to plant parasitic nematodes has been described as a set of characteristics of the host plant which act more or less to the detriment of the parasite (Rohde, 1965). Such resistance is usually recognized as either resistance to the nematode and its development (reproduction), or resistance to the disease caused by the nematode (damage). Plant nematologists tend to evaluate resistance of the crop host on the basis of the reproductive potential of the parasite on the host, rather than on the capability of the host to overcome the attack or to withstand the injury of the parasite (Rohde, 1972). In most situations this is adequate because of the high positive correlation between reproduction of the nematode and crop damage. However, in some instances the actual penetration of the nematodes into the roots may be damaging to the plant, although reproduction may be inhibited, and thus a level of tolerance, as well as resistance, to the nematode is required if healthy crops and adequate yields are to be secured.

Nematodes have adapted to life as plant parasites by the gradual accumulation of appropriate genes; at the same time, plants have become adapted to nematodes. In the wild ecosystem most plant species may support low to moderate populations of nematodes with little

adverse effect on their growth. High populations of nematodes in
agro-ecosystems, with consequent severe damage to the growth of the
plants, results from a series of interactions between nematodes and
plants - starting with the initial attraction of nematodes to host
roots, continuing through penetration of the host tissues, feeding,
and the establishment of a specific interaction resulting in some
modification of the host tissues, such as the production of 'giant
cells', which allows the development and reproduction of the nematode.
Each stage in the series of interactions is a means of establishing
resistance to the nematode parasite (Peacock, 1959; Rohde, 1965)
through such mechanisms as lack or inhibition of a nematode hatching
factor; inhibition of nematode penetration; inhibition of development
or kill of nematodes upon entry of the roots; inhibition of enzymes
or other constituents of the nematode's secretions which cause 'giant
cell' formation; alteration of the sex ratio of the nematode.

Whilst there is a relative abundance of information on the suitability of crops as hosts of various nematode species and on their
reaction to the attacks of the pathogens (Rohde, 1972), little is
known about the genetics of resistance or susceptibility. Those
studies that have been made, and particularly where breeding programs
have been established for resistance to specific nematodes, are concerned mainly with the more widely grown staple food crops. A brief
account is given in this chapter of some of the better known examples.

POTATO

Resistance to the potato cyst nematode, *Globodera rostochiensis*,
was looked for among the wild and primitive potatoes of the Commonwealth Potato Collection by Ellenby (1952). Resistance was found in
the diploid *Solanum tuberosum* ssp. *andigena* and *S. vernei* and using
this material breeding programs were established in the Netherlands
(Toxopeus and Huijsman, 1952, 1953) and in Britain (Howard, 1953).
The Dutch workers suggested that resistance was controlled by a
single dominant gene.

Dunnett (1957) and Jones (1957, 1958) found populations of *G.
rostochiensis* capable of producing cysts on potato plants ex *andigena*.
Further work by Dunnett (1961) proved the presence in *Solanum
multidissectum* of a dominant gene preventing reproduction of the
nematode population which had formed many cysts on the ex *andigena*
plants, but not that of populations against which the ex *andigena*
plants were resistant. Jones and Parrott (1965) advanced the hypothesis that the dominant genes for resistance to *G. rostochiensis*
pathotypes in *S. tuberosum* ssp. *andigena* and *S. multidissectum* are
matched by corresponding recessive genes in the nematode, and indicated the possibility of obtaining potato cultivars resistant to
both pathotypes (A and B) of the nematode from breeding *andigena*
and *multidissectum*.

Further studies on the systematics and biology of the potato cyst nematode have shown that the pathotype B (not producing cysts on *S. multidissectum*) that Jones and Parrott (1965) were dealing with in reality is a different species (Howard, 1972; Parrott, 1972; Stone, 1972), *Globodera pallida* (Stone, 1973; Mulvey and Stone, 1976). Currently, five pathotypes of *G. rostochiensis* and three of *G. pallida* are known in Europe. They are classified according to an internationally recognised scheme as Ro1..... Ro*n* and Pa1.... Pa*n* (Kort et al., 1977).

Solanum vernei has been used to provide resistance to pathotypes Ro1, Ro2, Ro3 and Ro4 (Huijsman, 1970; Evans and Stone, 1977) and to pathotypes Pa1, Pa2 and Pa3 (Howard and Cotten, 1978). Resistance in *S. vernei* was thought to be polygenic (Evans and Stone, 1977; Kort et al., 1977), but the mechanism of inheritance, as studied by Huijsman (1970, 1974), indicates that it is governed by major genes and influenced by modifying genes. This conclusion was reached because cysts were found on the roots of potato clones, grown in the Netherlands, which were believed to have polygenic resistance.

Less is known about resistance to root-knot nematodes, *Meloidogyne* spp., but resistance has been reported in *S. chacoense* and other *Solanum* spp. (Brucher, 1967a, b), and in potato varieties in India (Singh et al., 1974), and in Peru (Franco, 1972). *S. vernei* shows resistance to root-knot nematodes (Brucher, 1967a; Nirula et al., 1967; Fuller and Howard, 1974) so that it would seem possible that this resistance could be introgressed into commercial potato varieties, as was done with resistance to the potato cyst nematodes.

Finally, resistance to the reniform nematode, *Rotylenchulus reniformis* has been found in commercial cultivars of potato (Rebois and Webb, 1979). There is no information available on the nature of such resistance, but it appears to be independent of the resistance to some pathotypes of *G. rostochiensis*, although clones and lines contain simultaneously genes for resistance to both the reniform nematode and to some pathotypes of the cyst nematodes (Rebois and Webb, 1979).

SUGARBEET

The release of commercial cultivars of sugarbeet resistant to the sugarbeet cyst nematode, *Heterodera schachtii*, is still far from being realised despite the considerable amount of work carried out in Europe and in the United States of America.

Heijbroek (1977) found several resistant plants among populations of *Beta vulgaris* and *B. maritima*, but failed to obtain worthwhile rates of transmission to commercial clones; he obtained a slight increase by selfing or sibmating, but the resistance was lost at the second backcross to commercial varieties of sugarbeet.

He concluded that it was not worthwhile pursuing this line of investigation because the resistance was multifactorial and recessive.

Savitsky (1975) discovered resistance to the sugarbeet cyst nematode in diploid wild beets, *Beta patellaris*, *B. procumbens* and *B.
webbiana*. Unfortunately hybrids with the tetraploid *B. vulgaris* were
difficult to grow because they failed to develop root systems and were
male-sterile, although grafting the F_1 seedlings onto sugarbeet and
then pollinating them with sugarbeet overcame this lack of viability
to some extent (Savitsky, 1975).

In experimental crosses of sugarbeet with *B. procumbens*,
Savitsky (1975) obtained two diploid plants with nematode resistance
and these contained a segment of the chromosome from *B. procumbens*
with the gene responsible for resistance. At this point there seemed
to be a clear way ahead for breeding for resistance, which was reported as dominant (Heijbroek, 1981). However, the rate of transmission of the resistance gene is very low because of abnormal meiosis,
including detachment and loss of the translocated chromosome segment
from *B. procumbens* (Savitsky, 1978; Yu, 1978). Yu (1978) suggests
that resistance is governed by three complementary dominant genes
and that the rate of transmission can be increased by phasing the
transfer of resistance genes from *B. procumbens* to *B. vulgaris* (into
different chromosomes or chromosome arms or on different segments
that do not contain all the three genes at the same time.

Although difficulties remain for the plant breeders, and commercial cultivars resistant to the cyst nematode must still remain
as a long term objective, it is encouraging to note that among the
large number of different populations of *H. schachtii* tested against
the resistance found in wild beets, there is no evidence of resistance breaking pathotypes (Heijbroek, 1981).

CEREALS

Recent reviews (Cook, 1974; Howard and Cotten, 1978) provide a
comprehensive picture of the situation regarding the resistance of
cereals to plant parasitic nematodes. The cereal cyst nematode,
Heterodera avenae, is recognised as the most serious nematode pest
of various cereals. Resistance to it has been found in barley,
oats and wheat.

In barley, multiple or single resistance to the four European
pathotypes and to the Australian pathotype of *H. avenae* seems to be
governed by single dominant genes (Brown and Meagher, 1970; Cook,
1974; Andersen, 1980). Closely linked genes were identified at a
minimum of three loci (Cook, 1974). In some varieties, which exhibited only partial resistance to pathotypes 1 and 2, several
genes may be involved (Cook, 1974).

In oats, resistance to the European pathotypes 1 and 2 observed
in selections of *Avena sativa* is attributed to a dominant gene
(Andersen, 1980). Conversely, resistance found in *A. sterilis* against
all the European and the Australian populations of *H. avenae* (Brown
and Meagher, 1970; Cook, 1974) is probably controlled by two dominant genes (Howard and Cotten, 1978). In the spring wheat cultivar
Loros resistance to the nematode is controlled by a single dominant
gene (Howard and Cotten, 1978).

Several other nematodes are reported to cause damage to cereals
(Cook, 1974) and for many of them resistant cultivars or lines have
been found and selected (Cook, 1974). However, information on the
inheritance of resistance is scanty or lacking.

SOYBEAN

Soybean yields are decreased considerably due to the attacks
of *Heterodera glycines*, *Meloidogyne* species and *Rotylenchulus
reniformis* (Hartwig, 1976). Resistance to the soybean cyst nematode,
H. glycines has been worked on for many years and numerous cultivars
resistant to the nematode have been produced in the USA and Japan.

It is thought that resistance to the cyst nematode is controlled
by three independent recessive genes and one dominant gene
(Triantaphyllou, 1975). This conclusion is based on studies of the
genetics of the races of *H. glycines* (Triantaphyllou, 1975). There
are four known races of the nematode which are essentially field
populations differing from each other in the frequencies of three
groups of genes for parasitism – acting quantitatively and controlling the ability of the nematode to reproduce on resistant cultivars
of soybean. Three of the races have one to three genes for parasitism,
and race 3 has none of these genes and does not reproduce on the
resistant cultivars (Triantaphyllou, 1975).

Further studies carried out in the soybean region of the United
States have led to the identification of sources of resistance to
the root-knot nematode, *M. incognita* and *M. arenaria*, and against
the reniform nematode, *R. reniformis*. Resistance to the cyst and
to the reniform nematodes appears to be governed by the same gene or
or by separate but closely linked genes (Rebois et al., 1970;
Birchfield et al., 1971).

Presently soybean cultivars are available which incorporate
multiple resistances to *H. glycines*, *M. incognita*, *M. arenaria*,
R. reniformis, phytophthora root rot and foliar bacterial pustule
(Hartwig, 1976).

TOMATO

Much has been written about the sources of resistance to

Meloidogyne incognita, M. javanica and *M. arenaria* originally found
in lines of *Lycopersicon peruvianum* (Hare, 1965; Singh et al., 1974;
Fassuliotis, 1979). Studies on the inheritance of resistance have
given contradictory results, but the majority of investigators have
concluded that it is controlled by a single dominant gene (Singh et
al., 1974). Further investigations carried out by Sidhu and Webster
(1973) have demonstrated that some cultivars each possess a single
gene for resistance to *M. incognita,* and this gene is dominant for
the cultivar Nematex (LmiRl) and for the cultivar Small Fry (LMiR2)
and recessive for the cultivar Cold Set (LMiR3). According to
Netscher and Taylor (1979) the two dominant genes might be combined
in lines of the cultivar Rossol.

There are tomato cultivars which incorporate multiple resistances to various diseases (Malo, 1964; Sidhu and Webster, 1974).
This is a highly desirable attribute where nematodes and fungal soil
pathogens, such as *Fusarium* or *Verticillium* occur together because,
as demonstrated by Sidhu and Webster (1974), the predisposition of
the plant to *Fusarium oxysporum* f.sp. *lycopersici* due to infestation by root-knot nematodes is governed genetically.

OTHER CROPS

Monogenic dominant resistance has been reported for several
other crops to various nematodes (Malo, 1964; Hare, 1965), e.g.
alfalfa to *M. hapla* and to *Ditylenchus dipsaci*, pepper to *M. incognita*
and *M. javanica*, tobacco to *M. incognita*, lima beans to *M. incognita,*
Vitis spp. to *M. incognita* and probably trifoliate orange to
Tylenchulus semipenetrans. Conversely, resistance of cotton,
Gossypium barbadense, and 'common' beans to *M. incognita* is thought
to be polygenic recessive (Malo, 1964; Hare, 1965).

CONCLUDING REMARKS

Most cultivars of crops resistant to plant parasitic nematodes
incorporate a monogenic type of resistance – vertical resistance
according to Van der Plank (1968). This means that only one or a
few pathotypes of the parasite are controlled and that before introducing such a cultivar in a new area its resistance should be
tested with the local populations of the pathogen (Fassuliotis,
1979).

A more stable resistance controlled by polygenic combinations –
the horizontal resistance of Van der Plank (1968) – would be desirable against nematode species having high genetic variability as
expressed by pathotypes or physiological races. However, because
of the restricted movements of nematodes in soil and their slow
spread in different areas, breaking of resistance by new pathotypes
is likely to be a slow process compared with the spread of disease
involving air-borne propagules.

New pathotypes or virulent populations may originate by mutation or introduction. In either case the establishment of a mutant
or an immigrant in the new environment will take some time. Jones
(1972) has calculated hypothetical periods for the establishment of
a mutant or an immigrant of *Globodera rostochiensis* able to overcome
resistance in potato controlled by a single gene. He assumed that
with an initial frequency of 1 in 10,000, the new mutant will overcome resistance in 9 years or 40 years in monoculture depending on
whether its ability to overcome resistance is controlled by a dominant or a recessive gene respectively. The same mutant or immigrant
with the dominant gene, at an initial frequency of 1 in 100,000, will
reach a 50 percent frequency in 7 years and 90 percent frequency in
322 years when a resistant variety is grown continuously (Jones,
1981). The mutant or immigrant having a recessive gene to overcome
resistance will reach a 50 percent frequency in 50 years under the
same conditions.

In selecting for resistance to plant nematodes it should be
noted that within certain genera, such as *Globodera* (Jones and
Parrott, 1965), *Heterodera* (Viglierchio, 1960), and *Meloidogyne*
(Fassuliotis, 1979) the unfavourable host reaction will allow matu-
ration of a few recessive homozygous females but the population will
tend towards maleness (Jones and Parrott, 1965). In these cases
recombination may occur and the period of segregation will be extended. The effect of the environmental conditions also may often
modify the reaction of plants to pathogens (Rohde, 1972).

Breeding for nematode resistance has the objective of impairing
the establishment of genetic compatibility between host crop and
parasite, and is achievable by the effective employment of R-genes
in various breeding and complementary agronomic strategies (Marshall
and Pryor, 1978). The monogenic, vertical resistance which is the
main type available against nematode pathogens must be managed pru-
dently, to minimize the build up of resistance breaking pathotypes
and to provide the 'durable resistance' that is the essential requirement for crop production in those countries where nematodes are
of economic importance.

REFERENCES

Andersen, S., 1980, The breeding and use of varieties resistant to
    Heterodera avenae, *EPPO Bull.*, 10:303-310.
Birchfield, W., Williams, C., Hartwig, E.E. and Brister, L.R., 1971,
    Reniform nematode resistance in soybeans, *Plant Dis. Reptr*,
    55:1043-1045.
Brown, R.H. and Meagher, J.W., 1970, Resistance in cereals to the
    cyst nematode (*Heterodera avenae*) in Victoria, *Austr. J. Exp.
    Agr. Anim. Husb.*, 10:360-365.

Brucher, H., 1967a, Genetic resistance against nematodes in Argentine *Solanum* spp., *Phytopathology*, 57:7.

Brucher, H., 1967b, Root knot-eelworm resistance in some South American tuber-forming *Solanum* species, *Am. Pot. J.*, 44:370-375.

Cook, R., 1974, Nature and inheritance of nematode resistance in cereals, *J. Nematol.*, 6:165-174.

Dunnett, J.M., 1957, Variation in pathogenicity of the potato root eelworm (*Heterodera rostochiensis* Woll.) and the significance in potato breeding, *Euphytica*, 6:77-89.

Dunnett, J.M., 1961, Inheritance of resistance to potato root eelworm in breeding lines stemming from *Solanum multidissectum* Hawkes, *Ann. rep. Scott. Pl. Breed. Stn.*, pp. 39-46.

Ellenby, C., 1952, Resistance to the potato root eelworm, *Nature, Lond.*, 170:1016.

Evans, K. and Stone, A.R., 1977, A review of the distribution and biology of the potato-cyst nematodes *Globodera rostochiensis* and *G. pallida*, *PANS*, 23:178-189.

Fassuliotis, G., 1979, Plant breeding for root-knot nematode resistance, In: 'Root-knot Nematodes (*Meloidogyne* Species), Systematics, Biology and Control', F. Lamberti and C.E. Taylor eds, Academic Press, London, New York.

Franco, J., 1972, Comportamiento de variedades de papa en la costa central del Peru al ataque del nematode del nudo de la raiz, *Meloidogyne incognita* (Kofoid y White) Chitwood, *Invest. Agropecuarias*, 3:25-39.

Fuller, J.M. and Howard, H.W., 1974, Breeding for resistance to the white potato cyst-nematode, *Heterodera pallida*, *Ann. appl. Biol.*, 77:121-128.

Hare, W.W., 1965, The inheritance of resistance to nematodes, *Phytopathology*, 55:1162-1167.

Hartwig, E.E., 1976, Breeding soybeans for resistance to nematodes and other diseases, *Proc. Second Cyst Nematode Workshop, Jackson, Tenn., Aug. 10-12, 1976*, pp. 45-50.

Heijbroek, W., 1977, Partial resistance of sugarbeet to beet cyst eelworm (*Heterodera schachtii* Schm.), *Euphytica*, 26:257-262.

Heijbroek, W., 1981, Nematodes, *Proc. IX Int. Congress Plant Protec., Washington DC, Aug. 5-11, 1979*, Vol. II, pp. 511-513.

Howard, H.W., 1953, Crops and plant breeding, *F.R. Agric. Soc.*, 114:90-106.

Howard, H.W., 1972, Pathotypes of potato cyst nematode, *Ann. appl. Biol.*, 71:263-266.

Howard, H.W., and Cotten, J., 1978, Nematode-resistant crop plants, In: 'Plant Nematology', J.F. Southey, ed., Ministry of Agriculture, Fisheries and Food, London, pp. 313-325.

Huijsman, C.A., 1970, Present state of breeding work for resistance to *Heterodera rostochiensis* (Wollenweber) in the Netherlands, *Proc. Int. Nematol. Symposium, Warsaw, Poland, Aug. 21-24, 1976*, pp. 255-258.

Huijsman, C.A., 1974, Host-plants for *Heterodera rostochiensis* Woll.
     and the breeding for resistance, *EPPO Bull.*, 4:501-509.
Jones, F.G.W., 1957, Resistance breeding biotypes of potato root
     eelworm (*Heterodera rostochiensis* Woll.), *Nematologica*, 2:
     185-192.
Jones, F.G.W., 1958, Resistance-breaking populations of potato root
     eelworm, *Plant Path.*, 9:24-25.
Jones, F.G.W., 1972, Pathotypes in perspective, *Ann. appl. Biol.*,
     71:296-300.
Jones, F.G.W., 1981, Management of potato nematodes, *Proc. IX Int.
     Congress Plant Protec., Washington DC, Aug. 5-11, 1979*, Vol. II,
     pp. 480-484.
Jones, F.G.W., and Parrott, D.M., 1965, The genetic relationships of
     pathotypes of *Heterodera rostochiensis* Woll. which reproduce on
     hybrid potatoes with genes for resistance, *Ann. appl. Biol.*,
     56:27-36.
Kort, J., Ross, H., Rumpenhorst, H.J. and Stone, A.R., 1977, An
     international scheme for identifying and classifying pathotypes
     of potato cyst-nematodes *Globodera rostochiensis* and *G. pallida*,
     Nematologica, 23:333-339.
Malo, S.E., 1964, A review of plant breeding for nematode resistance,
     Soil Crop Sci. Soc. Fl., 24:354-365.
Marshall, D.R., and Pryor, A.J., 1978, Multiline varieties and
     disease control. I. The 'dirty crop' approach with each com-
     ponent carrying a unique single resistance gene, *Theor. Appl.
     Genet.*, 51:177-184.
Mulvey, R.H. and Stone, A.R., 1976, Description of *Punctodera
     matadorensis* n. gen., n. sp. (Nematoda: Heteroderidae) from
     Saskatchewan with lists of species and generic diagnosis of
     Globodera (n. rank), *Heterodera* and *Sarisodera*, *Can. J. Zool.*,
     54:772-786.
Netscher, C. and Taylor, D.P., 1979, Physiological variation with
     the genus *Meloidogyne* and its implications on integrated control,
     In: 'Root-knot Nematodes (*Meloidogyne* Species), Systematics,
     Biology and Control, F. Lamberti and C.E. Taylor, eds, Academic
     Press, London, New York.
Nirula, K.K., Nayar, N.M., Bassi, K.K. and Singh, G., 1967, Reaction
     of tuber-bearing *Solanum* species to root-knot nematode,
     Meloidogyne incognita, *Am. Pot. J.*, 44:66-69.
Parrott, D.M., 1972, Mating of *Heterodera rostochiensis* pathotypes,
     Ann. appl. Biol., 71:271-273.
Peacock, F.C., 1959, The development of a technique for studying
     the host/parasite relationship of the root-knot nematode
     Meloidogyne incognita under controlled conditions, *Nematologica*,
     4:43-45.
Rebois, R.V., Epps, J.M. and Hartwig, E.E., 1970, Correlation of
     resistance in soybeans to *Heterodera glycines* and *Rotylenchulus
     reniformis*, *Phytopathology*, 60:695-700.
Rebois, R.V. and Webb, R.E., 1979, Reniform nematode resistance in
     potato clones, *Am. Pot. J.*, 56:313-319.

Rohde, R.A., 1965, The nature of resistance in plants to nematodes, *Phytopathology*, 55:1159-1162.

Rohde, R.A., 1972, Expression of resistance in plants to nematodes, *Ann. Rev. Phytopathol.*, 10:233-252.

Savitsky, H., 1973, Meiosis in hybrids between *Beta vulgaris* L. and *Beta procumbens* Chr. Sm. and transmission of sugarbeet nematode resistance, *Genetics*, 74:241.

Savitsky, H., 1975, Hybridization between *Beta vulgaris* and *B. procumbens* and transmission of nematode (*Heterodera schachtii*) resistance to sugarbeet, *Can. J. Genet. Cytol.*, 17:197-209.

Savitsky, H., 1978, Nematode (*Heterodera schachtii*) resistance and meiosis in diploid plants from interspecific *Beta vulgaris* x *B. procumbens* hybrids, *Can. J. Genet. Cytol.*, 20:177-186.

Sidhu, G. and Webster, J.M., 1973, Genetic control of resistance in tomato. I. Identification of genes for host resistance to *Meloidogyne incognita*, *Nematologica*, 19:546-550.

Sidhu, G. and Webster, J.M., 1974, Genetics of resistance in tomato to root-knot nematode - wilt fungus complex, *J. Heredity*, 65:153-156.

Singh, B., Bhatti, D.S. and Singh, K., 1974, Resistance to root-knot nematodes (*Meloidogyne* spp.) in vegetable crops, *PANS*, 20:58-67.

Stone, A.R., 1972, The round-cyst species of *Heterodera* as a group, *Ann. appl. Biol.*, 71:280-283.

Stone, A.R., 1973, *Heterodera pallida* n. sp. (Nematoda: Heteroderidae), a second species of potato cyst nematode, *Nematologica*, 18:591-606.

Toxopeus, H.J. and Huijsman, C.A., 1952, Genotypical background of resistance to *Heterodera rostochiensis* in *Solanum tuberosum* var. *andigena*, *Nature, Lond.*, 170:1016-1017.

Toxopeus, H.J. and Huijsman, C.A., 1953, Breeding for resistance to potato root eelworm, *Euphytica*, 2:180-186.

Triantaphyllou, A.C., 1975, Genetic structures of races of *Heterodera glycines* and inheritance of ability to reproduce on resistant soybeans, *J. Nematol.*, 7:356-364.

Van der Plank, J.E., 1968, 'Disease Resistance in Plants', Academic Press, New York.

Viglierchio, D.R., 1960, Resistance in *Beta* species to sugarbeet nematode, *Heterodera schachtii*, *Exper. Parasitol.*, 10:389-395.

Yu, M.H., 1978, Meiotic behaviour of a disomic nematode-resistant sugarbeet, *Crop Science*, 18:615-618.

DURABLE RESISTANCE OF BARLEY CULTIVARS TO THE NEMATODE

Heterodera avenae

Francoise Person-Dedryver

I.N.R.A., Laboratoire de Zoologie
Domaine de la Motte-au-Vicomte
Le Rheu, France

INTRODUCTION

The resistance of barley cultivars to *Heterodera avenae* is often controlled by one dominant gene. The durable character of some of these resistant barley cultivars against the northern and southern races Fr 4 and Fr 1 of *H. avenae* in France is discussed in this paper.

STUDY OF THE RESISTANCE OF BARLEY TO FRENCH RACES Fr 1 AND Fr 4 OF *H. avenae*

Although adult males are able to develop on barley cultivars resistant to *H. avenae*, the larvae of this nematode are unable to develop into the adult gravid female, *Heterodera avenae* cannot therefore reproduce on these resistant cultivars. The barley cultivars, Siri and P 31-322-1 are resistant to the northern race Fr 4 and Ortolan is resistant to the southern race Fr 1. The mode of inheritance of the resistance of these cultivars is shown in Table 1.

Andersen (1970) showed that the resistance in barley cultivars, Siri and Ortolan, depends upon one dominant gene in each cultivar, respectively Ha_2 and Ha_1. Similar results were obtained with resistant barley Siri to the French race Fr 4 (Person-Dedryver, 1981). Both genes, Ha_1 and Ha_2 were responsible for the resistance of the barley cultivar P 31-322-1 to *H. avenae* (Andersen, 1975), but only one dominant gene (probably Ha_2) induced resistance of this cultivar to the French race Fr 4 (Person-Dedryver, 1981).

Table 1. Mode of inheritance of the resistance of barley cultivars
          to French races of *Heterodera avenae* Fr 1 and Fr 4.

Barley cultivars	Genes resistant to *H. avenae* (Andersen, 1970, 1975)	Inheritance of the resistance to French races of *H. avenae*	
		Fr 1	Fr 4
Siri	Ha_2 dominant	susceptible	1 dominant gene
Ortolan	Ha_1 dominant	not tested	susceptible
P 31-322-1	Ha_1 and Ha_2 dominant	susceptible	1 dominant gene

STUDY OF THE VIRULENCE OF THE FRENCH RACES Fr 1 AND Fr 4 OF
H. avenae TO THE CULTIVARS OF BARLEY

     Crosses between the French races Fr 1 and Fr 4 provide a means
to study the genetics of virulence to barley cultivars. The cre-
ation of a new and more virulent race from the cross of these races
seems probable and because both races are sometimes mixed in the
same field, the results of this cross were relevant to durability
of resistance in barley.

     The larvae of the F_1 hybrid from crosses between Fr 1 and Fr 4
races developed into males or females on cultivars susceptible to
both races (barley Aramir). The development of other F_1 hybrid
larvae was studied on selected barley cultivars on which only one
of the two parents could develop (barley Siri, P 31-322-1 and
Ortolan). No F_1 females were obtained as the genes for virulence
in the pathogen to those resistant barley cultivars are recessive
(Person & Rivoal, 1979).

     A few F_2 nematodes resulting from the crosses between Fr 1 and
Fr 4 were obtained. The culture of cereals in Petri dishes under
standardised conditions permited the development of a few infective
larvae of *H. avenae* (Person & Doussinault, 1978). The F_2 larvae
were put on a Petri dish culture of a hybrid plant resulting from
the crosses between different barley cultivars resistant to each
of the races (Ortolan x P 31-322-1 and Siri x Ortolan). The
resulting development is shown in Table 2.

     A few F_2 larvae developed into males on hybrid plants resistant
to both races. A few small F_2 females containing ten or twenty eggs
developed only on plants from the cross: Ortolan x P31-322-1.
The F_2 larvae resulting from this cross were no more virulent than
the parents. Similar results were obtained with the F_3 larvae from

Table 2.   Developement of F$_2$ hybrid larvae of *H. avenae* from crosses
between Fr 1 and Fr 4 races on different barley hybrids.

Crosses between barley cultivars	Number of inoculated larvae	Number of gravid females	Number of small females with a few eggs
Ortolan x P 31-322-1	180	0	13
Siri x Ortolan	110	0	0

crosses between Fr 1 - Fr 4 and the larvae were also no more
virulent than the parents.

DISCUSSION

It can be concluded that the resistance of the barley cultivars
Siri, Ortolan and P 31-322-1 to the races Fr 1 and Fr 4 of *H. avenae*
is durable, although it is only regulated by one dominant gene.

These results have also been confirmed under field conditions
where a few females of *H. avenae* were sometimes present on the roots
of a resistant barley. The larvae resulting from these females were
put on the same resistant cultivar, but a new race more virulent
than Fr 1 and Fr 4 was never noticed. In Sweden populations of
H. avenae races, two similar to the Fr 4 race, are entirely controlled
by the use of resistant barley cultivars such as Prisca and Ansgar,
which have the Ha$_2$ gene for resistance (Anderson, S. pers. comm.).
The monoculture of a resistant barley cultivar in the same field for
several years did not induce a new race of *H. avenae*.

The author thanks Mr. G. Doussinault for making the crosses
between the barley cultivars.

REFERENCES

Andersen, S., and Andersen, K., 1970, Sources of genes which promote
    resistance to races of *Heterodera avenae* Woll., *EPPO Publ. Ser.
    A* No. 54:29-36.
Andersen, S., 1975, International test sortiment for *Heterodera
    avenae* a.o. species: 1975, 3 pp.
Person, F., and Doussinault, G., 1978, Influence de la température,
    de l'agressivité et de la virulence des races d'*Heterodera
    avenae* Woll. dans la mise au point d'un test de résistance de
    céréales en conditions contrôlé: utilisation en sélection.
    Ann. Amélior. Pl., 28:513-527.
Person, F. and Rivoal, R., 1979, Hybridation entre les races Fr 1 et
    Fr 4 d'*Heterodera avenae* Wollenweber en France et étude du

comportement d'agressivité des descendants F_1. *Revue Nématol.*,
2:177-183.
Person-Dedryver, F., 1982, Les Variétés d'orge facteurs de
discrimination des races d'*Heterodera avenae* Woll., *EPPO Bull.*
12: (in press).

RESISTANCE TO INSECTS PROMOTES THE STABILITY OF INTEGRATED PEST

CONTROL

O.M.B. de Ponti

Institute for Horticultural Plant Breeding (IVT)
P.O. Box 16, Wageningen, The Netherlands

SUMMARY

Although a distinct terminology suggests the opposite, essential differences between resistance to pests and resistance to diseases hardly exist. The same phenomena have been identified, but they are of different importance. Race specific, complete resistance, for example, also occurs in resistance to pests, but race-nonspecific, partial resistance is by far the most frequent type of resistance. This partiality might be one of the reasons, why resistance to pests was distinguished relatively late as a valuable method of crop protection. Moreover this partiality emphasized the limited control potential of resistant cultivars alone. Therefore resistance was incorporated in Integrated Pest Control (IPC) or Pest Magagement (PM) systems as soon as these new crop protection strategies were developed.

Because the population development of insects and mites is slowed down on resistant cultivars, the effectiveness and profitability of many other IPC tactics are markedly improved. A totally different contribution of plant breeding to IPC is the development of cultivars on which the effectiveness of parasites and predators is enhanced by differences in morphology and general plant architecture. During the so-called pesticide era, when insect control was based almost exclusively on chemical control, most cultivars were bred under a constant pesticide protection. This might have led to a gradual erosion of resistance genes from breeding populations. The present interest in insect resistant cultivars will hopefully change this dangerous trend.

INTRODUCTION

In the previous chapters the newest insights into resistance
have been discussed in detail, all tuned to the durability of
resistance, the leading theme at this conference. Although insects
have been mentioned only sporadically, there is little sense in
repeating all the foregoing with examples taken from resistance to
insects and other animal parasites, the more so as it will be clear
from the following that resistance phenomena are universal, irrespective of the kind of parasite concerned.

Furthermore resistance to insects has recently been thoroughly
reviewed. Russell (1978), Panda (1979), Harris (1980) and Maxwell &
Jennings (1980) covered resistance to insects as broadly as this
conference does. Therefore this paper will concentrate on the role
resistance can play in promoting the stability of the entire system
of control measures. This is far more important than the durability
of one aspect, resistance, alone. In addition, the influence of
modern practices in plant breeding and cultivar registration on the
maintenance of resistance will be discussed.

RESISTANCE: A UNIVERSAL PHENOMENON

The classic book of Painter (1951) 'Insect Resistance in Crop
Plants' demonstrates that people have long been aware of the similarities in resistance to plant (fungi, bacteria, viruses) and
animal (insects, mites, nematodes) parasites. This is also illustrated by a comparison of the population dynamic components of
pathogen and arthropod host plant interactions (Table 1). Although
using a different terminology, Painter defined the three basic
defence mechanisms of a host plant, which are generally distinguished (Parlevliet, 1980) as nonpreference (= avoidance),

Table 1. A comparison of the population dynamic components
         of pathogen-host plant and arthropod-host plant
         interactions (Carter and Dixon, 1981).

Pathogen-host plant	Arthropod-host plant
Infection ratio	Immigration rate
	Survival rate
Latent period	Larval period
Lesion growth	Population growth
Sporulation rate	Reproductive rate
Infectious period	Longevity of adults
Spore dispersal	Emigration

antibiosis (= resistance) and *tolerance* (Table 2). The importance of non-preference as a defence mechanism rather specific to animal parasites points to the only essential difference between resistance to plant and animal parasites. The combination of sense-organs and mobility of the animal parasites provides an extra dimension to this defence mechanism, influencing the behaviour of the parasite. Because both nonpreference and antibiosis reduce the population development of the parasite on a particular crop, De Ponti (1977) advocated considering them not as distinct phenomena, but as distinct mechanisms of resistance. Then *nonpreference* refers to plant characters causing a reduced acceptability for shelter, food or oviposition by an insect, whereas *antibiosis* refers to antibiotic effects of the plant on the development and reproduction of an insect. Because of the ambiguity of the term nonpreference Van Marrewijk & De Ponti (1975) proposed the term *nonacceptance* for this phenomenon, to emphasize its activity in choice and no-choice situations. Later Kogan and Ortman (1978) suggested replacing the term nonpreference by *antixenosis*. The latter term is better in tune with the term antibiosis and better indicates that a plant character is concerned. Although it will be difficult to replace the original terms of Painter, which have been generally accepted, the terms resistance and tolerance should be used for all kinds of parasites to emphasize their universality. The other terms have only a descriptive value.

Resistance and tolerance often occur simultaneously. In a routine selection program it will be difficult, but on the other hand seldom necessary, to distinguish these defence mechanisms. With insects and other animal parasites it is easier to determine the population development of the parasite separately from the direct and indirect symptoms of the plant. Therefore animal parasites seem to be better suited for studying the tolerance phenomenon The supposed advantages (no biotype selection) and disadvantages (no reduction of the parasite population) of tolerance are of course the same as with plant parasites.

Table 2. Comparison of terms for resistance to diseases and pests.

DISEASES (cf Parlevliet, 1980)	PESTS (cf Painter, 1951)	GENERAL (cf De Ponti, 1977)
Avoidance Resistance Tolerance	Nonpreference Antibiosis Tolerance	Resistance Tolerance

In screening 800 cucumber cultivars for resistance to the two-spotted spider mite, *Tetranychus urticae,* De Ponti (1978) scored acceptance, oviposition and damage index. Only nine cultivars were significantly better than the susceptible control for the three parameters, PI 220860 being the most resistant cultivar (Table 3). Two other cultivars showed a very interesting pattern of results: PI 181756 causes a low acceptance and oviposition of the spider mite, but is seriously damaged, and PI 279466 shows the opposite: acceptance and oviposition of the spider mite are high, but the plants are only slightly damaged. Because acceptance and oviposition are resistance parameters and the damage index reflects the result of resistance and tolerance, based on these figures, PI 181756 can be classified as resistant but sensitive and PI 279466 as susceptible but tolerant.

Since Painter's book, knowledge of resistance to all kinds of phytophagous parasites, pathogens and arthropods has increased enormously. It has become evident that no matter which aspect of resistance was investigated (expression, inheritance, parasite differentiation, mechanisms, epidemiology), resistance is a universal phenomenon.

DURABILITY OF RESISTANCE TO INSECTS

The occurrence of 'resistance breaking' races or biotypes is not restricted to pathogens. A classic and closely studied example is the Hessian fly, *Mayetiola destructor,* a major pest of wheat in the USA. Eight biotypes have been identified. A more recent example is the brown planthopper, *Nilaparvata lugens,* one of the most

Table 3. Average acceptance, oviposition and damage index of four cucumber cultivars tested for resistance to the twospotted spider mite.

CULTIVAR	ACCEPTANCE (%)	OVIPOSITION (3 days)	DAMAGE INDEX (0-5)
Susc. control	76	27	4.3
PI 220860	24[xx]	12[xx]	1.2[xx]
PI 181756	46[xx]	15[xx]	4.1
PI 279466	64	24	1.5[x]

x, xx   Significantly different from the susceptible control for $P \leqslant 0.05$ and $P \leqslant 0.01$

serious pests of rice. Four biotypes have been identified which have caused dramatic losses in many developing countries. Table 4 summarizes the presently known insect-host plant relations, in which biotype differentiation has been reported. Because entomologists speak often only of biotypes, where phytopathologists usually distinguish races and isolates (by virulence and aggressiveness), it is unclear whether in all these cases true breakdown of resistance occurred. Anyhow differential interaction in terms of a gene-for-gene relationship has not always been demonstrated.

In view of the many reports on resistance to insects the occurrence of resistance breaking biotypes with animal parasites seems far less frequent than with pathogens. This difference will be due to a variety of factors which act collectively rather than individually. Three will be discussed.
1) Resistance to insects is less operational than resistance to pathogens: the number of insect resistant cultivars that is grown over vast areas is still relatively small. Whenever insect resistant cultivars are widely applied the sequence of resistance and breakage events can be similar to that with pathogens (e.g. Hessian fly resistant wheat cultivars and brown plant hopper resistant rice cultivars).
2) Resistance to insects is mostly of a complex nature and polygenically inherited. The biotype-specific insect resistances (Table 4) are often monogenically or oligogenically regulated. The underlying mechanisms are perhaps simple, but little is yet known about them.

Table 4. Cases of resistance to insects where biotype differentiation has been reported. Sometimes the number of biotypes is not known (?).

CROP	PEST	INHERITANCE	BIOTYPES
Wheat	Hessian fly	oligogenic	8
	Greenbugs	monogenic	3
Barley	Greenbugs	oligogenic	?
Rice	Planthopper	monogenic	4
	Leafhopper	monogenic	2
	Gall midge	oligogenic	2
Alfalfa	Spotted aphid	polygenic	7
	Pea aphid	oligogenic	5
Brassica	Cabbage aphid	polygenic	?
Lettuce	Leaf aphid	polygenic	2
Apple	Woolly aphid	monogenic	2
Raspberry	Rubus aphid	monogenic	4

3) Most crucial perhaps, is that resistance to insects is often
partial, so that selection for new biotypes is markedly retarded.
The complex nature of the resistance, its polygenic inheritance
and its partiality are probably often interrelated. The partiality
of resistance to insects has been very important in the application
of insect resistant cultivars in plant protection. Therefore it
deserves further attention.

Because partial resistance often does not give complete control,
its value was for long underestimated by many phytopathologists,
entomologists and plant breeders. For resistance to pathogens, large
reservoirs of genes causing complete resistance were available.
Despite the occurrence of new races the control power of complete
resistance was convincing and resistance soon became the corner stone
of disease control. The same is true for resistance to some insects
(e.g. Hessian fly) where almost complete control was also achieved.
Partial resistance to pathogens has received the attention it deserves
only after the experience of the permanent resistance-breakage 'game'
between pathogens and phytopathologists/plant breeders. Only then were
the control potential and advantages of partial resistance appreciated.

With insect resistance the situation is different. Being known,
but hardly applied before the 1940's, when synthetic pesticides were
thought to become the panacea for all pest problems, partial resistance and other natural, biological and cultural control measures
were badly neglected. Only a minority of entomologists remained
alert. Soon after the large scale introduction of these new pesticides
it became evident that in addition to health and environmental drawbacks the durability of chemical control was also limited: pesticide
resistant biotypes developed, sometimes dramatically fast.

Here we meet an interesting parallel between two seemingly very
potent control measures, chemical control and complete resistance.
Both cause more or less acute death of pathogens and insects, but
are often non-durable due to loss of their effect by development of
resistant biotypes.

The broad biocidal action of the synthetic pesticides caused a
dramatic decline in the populations of parasitizing and predating
insects, and other fauna. This loss of natural control elements
increased the pest potential, so that the durability of the entire
natural control system was endangered. Not only did secondary pests
evolve which were previously entirely controlled by these parasites
and predators, but also the key-pest became more serious than before,
particularly when the pesticide used against it became ineffective.
This is a second parallel with complete resistance, where the selection for major genes regulating race-specific resistance can be
attended by a gradual erosion of minor genes regulating race-non-
specific resistances, the so-called 'Vertifolia-effect' (Van der
Plank, 1963).

The drawbacks of a relatively large dependence on one control measure, be it pesticides or resistance, stimulated entomologists to search for alternatives in the 1950's. These were sought in an approach, which is now called integrated pest control or pest management, whose main objective is safeguarding the naturally occurring biological control elements as a basic control strategy (Brader, 1980). Partial resistance is thought to be one of the most essential elements of this strategy. Here phytopathologists and entomologists meet again, both accepting that as an alternative to complete resistance, partial resistance deserves more attention for implementation in an integrated control system with additional control elements. Wolfe (these proceedings) already pointed to the combination of partial resistance and limited fungicide use. Disease and pest monitoring systems as developed by Zadoks et al. (1982) will favour this approach.

HOST PLANT RESISTANCE PROMOTES THE STABILITY OF INTEGRATED PEST CONTROL

Integrated control aims at keeping pest populations below damaging levels. It is most effective on pest populations which develop slowly and with limited fluctuations. On susceptible cultivars, pest populations can develop explosively which affects the stability and so the reliability of integrated control systems. Host plant resistance seems to be one of the best methods to keep pest populations at controllable levels.

Both chemical and many integrated control measures are geared mostly towards increasing the *death rate* of an insect population. Host plant resistance however, aims at decreasing the *birth rate* thus preventing an insect population from reaching damaging levels. This decreasing effect on pest development will continue during a number of pest generations until a new equilibrium at a lower level is obtained. Using simulation models of pest population development Knipling (1979) demonstrated that a reduction in the reproduction rate of a pest with two generations per year by 25 and 40% stabilizes the peak population in respectively 4 and 9 years at respectively two thirds and one third of the original level.

When resistance alone does not give sufficient control, the characteristically slow development of insect populations on resistant cultivars will markedly increase the efficacy and consequently the profitability of other control measures. This high compatibility with other measures of control is one of the greatest advantages of resistance as a basic component of integrated control. The effectivity of parasites and predators, either natural or released, is greatly improved, even on varieties with only a low degree of resistance (Van Emden, 1966). This is also clearly demonstrated by Lowe (1975a) for resistance in sugar beet to the leaf aphids

Myzus persicae and *Aphis fabae*. Besides resistance a higher effectivity of coccinellids and other natural enemies at lower prey
densities contributed to the fewer aphids on the resistant sugar
beet. This synergism of resistance and natural enemies indicates
that hitherto inefficient beneficial insects might become valuable
for biological control on resistant cultivars. Lowe (1975b) also
demonstrated a lower incidence of virus yellows on these aphid
resistant sugar beets. Moreover the combination of host plant
resistance with chemical control resulted in a much lower incidence
than with either measure alone. Addition to this breeding stock of
genes governing tolerance to the viruses will further decrease the
disease potential. This is a good example of the effects of integrating a multitude of control measures, which are insufficiently
effective if used alone.

The use of male sterile and otherwise manipulated insects will
also be favoured by the application of resistant cultivars, because
smaller quantities are needed. Thus the profitability of these
rather expensive strategies will increase, promoting their applica-
bility.

If resistance alone or combined with other nonchemical measures,
does not provide satisfactory control, insecticides should be added
to the system of control measures. Resistant cultivars will however,
markedly reduce the amount of insecticides needed. Studying the
resistance of cucumber to the twospotted spider mite (*Tetranychus
urticae*) De Ponti (1979) observed that on some partially resistant
cultivars it took longer for the spider mite population to reach
the economic injury level than on a susceptible cultivar. On the
most resistant breeding lines, derived from crosses between these
partially resistant cultivars, it took two to three times as long.
The number of acaricide applications required will be inversely
proportional to the length of this period. Kea et al. (1978)
demonstrated that resistant cultivars sometimes need lower insecticide doses to achieve satisfactory control. They observed that
some lepidopterous pests of soybean were more sensitive to several
insecticides when fed with foliage of a resistant cultivar than
when fed with foliage of a susceptible cultivar. As a consequence
the resistant cultivar in a field test needed lower insecticide
doses. Of particular interest was the biological insecticide
Bacillus thuringiensis, which was highly effective on the resistant
cultivar, but not at all on the susceptible cultivar. Stubbs and
De Bruin (1970) had similar experiences with the fungicide oxycarboxin against yellow rust on wheat cultivars with different levels
of resistance.

Within the theme of this conference it is inevitable to
mention a fascinating way of controlling insects which have developed new virulent biotypes. The method was suggested by Hatchet and

Gallun (1967) and tested by Foster (1976) with Hessian flies in wheat. Because avirulence of the Hessian fly biotype GP is dominant over virulence, it appeared possible to reduce a population of virulent Hessian flies markedly by a mass release of avirulent Hessian flies of the GP-biotype. The females could not survive on the GP-resistant cultivar 'Monon', whereas the males mated freely with virulent females of the wild population causing avirulent offspring, that died in the early larval phases, when feeding on 'Monon'. In this way all the virulent biotypes were eradicated.

OTHER CONTRIBUTIONS OF PLANT BREEDING TO INTEGRATED PEST CONTROL

Although genetic resistance is the most obvious contribution, plant breeding can also promote integrated control through other heritable plant characters. Cultivars may, for example, differ in morphological characters which influence either the target insect or their parasites and predators. Two examples will illustrate this.

In our own work on resistance to the glasshouse whitefly (*Trialeurodes vaporariorum*) we crossed common tomato cultivars with the resistant related species *Lycopersicon hirsutum glabratum* Besides partially resistant plants with an indeterminate growth habit the F_2 produced some determinate sideshootless offspring, which are highly resistant. This was confirmed in later generations. When however, a sideshoot was formed by accident on these plants, it appeared susceptible. Therefore these genotypes are potentially susceptible, but resistance is due to a premature cessation of growth and early senescense of the leaves. The shorter growing period also provides less opportunity for a massive pest development. The determinate sideshootless character is regulated by only a few genes and can easily be incorporated in new cultivars. Unfortunately these traits are attended by bad setting of fruits.

In the Netherlands the glasshouse whitefly, is biologically controlled on 600 ha of glasshouse tomatoes by repeated releases of the parasitizing wasp *Encarsia formosa*. Biological control of the whitefly on glasshouse cucumber is insufficiently effective and therefore not commercialized. Van Lenteren et al. (1980) ascribed this to the relatively large hairs on the cucumber leaf, especially on its veins, which reduce the mobility and consequently the parasitizing activity of the wasp. Therefore De Ponti (1980) introduced a spontaneous glabrous mutant from the USSR, on which the wasp appeared to move about 3.5 times faster than on normal hairy leaves (Hulspas-Jordaan & Van Lenteren, 1978). A glasshouse experiment showed that the parasitization efficiency was about 20% higher on that mutant.

The   general plant architecture can also have a significant
effect on insect development and control. Open plant structures can
reduce an insect population in several ways: (1) the target insects
meet less suitable microenvironmental conditions, (2) they are less
protected from natural enemies, and (3) they are easier hit by
insecticides. In field experiments screening hunderds of barley
accessions for resistance to the leaf aphid complex, Van Marrewijk
& Dieleman (1980) found less aphids on two-row than on four-row
and six-row barley. Although resistance was not ruled out, they
were inclined to explain this by the greater accessibility of two-
row barley to predaceous insects, especially coccinellids. These
rather peculiar elements of integrated control through plant
breeding are supposed to be very durable, but can only be traced
by careful study of the entire biocenosis of a crop. Evidently
this requires a very close cooperation between entomologists and
plant breeders.

RECENT TRENDS IN PLANT BREEDING AND CULTIVAR REGISTRATION WHICH
MIGHT THREATEN RESISTANCE BREEDING

Since the introduction of synthetic pesticides it became
common practice to apply insecticides routinely in selection fields.
Because in this way all plants were protected from attack, possible
differences in resistance were masked, so that natural and breeder's
selection for resistance no longer occurred. This unfavourable
trend, which has been called 'breeding under the pesticide umbrella'
(De Ponti, 1981), may lead to a gradual erosion of resistance genes
from breeding populations and should be stopped as soon as possible
by drastically limiting the use of insecticides in selection fields.
Better and more rapid results can be expected from a deliberate
selection for resistance by using the pesticide umbrella in its
correct metaphoric sense: protecting the selection fields from the
pesticides used on surrounding fields and thus enhancing natural
selection (De Ponti, 1981). Some surprising results in tracing or
enhancing the resistance to insects in cultivars commercially grown
in Europe will illustrate the potential of this approach. In testing
a number of wheat cultivars currently used in Britain, for resistance
to the cereal leaf aphid, *Sitobion avenae*, Lowe (1980) found significant differences in repeated glasshouse and field tests. Eenink and
Dieleman (1977) and Bintcliffe (1981) had the same experience in
testing commercially available lettuce and potato cultivars for
resistance to the leaf aphid *Myzus persicae*. During breeding of
these wheat, lettuce and potato cultivars no attention was paid to
the resistance against these insects, so that it is remarkable that
some of them proved to be resistant and others susceptible. This
suggests a wide genetic variation for these characters in the current
breeding populations, which should be better exploited by deliberate
selection. In outbreeding crops it might also be possible to trace

differences in resistance *within* open pollinated varieties, in addition to differences *between* varieties. After some carrot cultivars had been found, which were less attacked by the carrot fly (*Psila rosae*), in repeated field tests, De Ponti & Freriks (1980) succeeded in increasing the resistance level of the varieties 'Nantes', 'Pioneer', 'Vertou', and 'Signal' by two generations of line selection. On average the resistance of the selected I_2 lines was about 25 precent higher than of the parental varieties.

Although fewer fungicides than insecticides have sofar been used in selection fields, because of the great interest plant breeders attach to developing disease resistant cultivars, there are some signs that a change for the worse may occur. Recently the cultivar registration authority of the Federal Republic of Germany, das Bundessortenamt, decided that in all cereal performance trials fungicides should be used more or less routinely according to the common agricultural practice in some regions of Germany (Schlumbohm, 1980). This approach of spraying all cultivars, even the most resistant ones, with fungicides, is based on an underestimation of the agronomic, economic and environmental value of resistance.This is contradictory to the trend to stimulate nonchemical plant protection, a policy which is rapidly gaining ground in governmental institutions. If the above decision is maintained and, still worse, taken over by cultivar registration authorities in other countries, plant breeders might be discouraged in their efforts to breed for resistance.

Let us hope that plant breeders continue breeding for resistance, not only to pathogens but also to insects and that these activities will be stimulated by governmental institutions and policy makers to provide the hungry world with sound and durable plant protection strategies.

REFERENCES

Bintcliffe, E.J.B., 1981, Resistance to the aphid *Myzus persicae* (Sulz.) in potato cultivars, *SROP/WPRS Bull.*, 1981/IV/1: 29-34.
Brader, L., 1980, Advances in applied entomology, *Ann. appl. Biol.*, 94: 349-365.
Carter, N. & Dixon, A.F.G., 1981, The use of insect population simulation models in breeding for resistance, *SROP/WPRS Bull.*, 1981/IV/1: 21-24.
Eenink, A.H. & Dieleman, F.L., 1977, Screening *Lactuca* for resistance to *Myzus persicae*, *Neth.J.Pl.Path.*, 83: 139-152.
Emden, H.F. van, 1966, Plant insect relationships and pest control, *World Rev.Pest.Control*, 5: 115-123.

Foster, J.E., 1976, Current status of genetic control of Hessian
    fly populations with the dominant great plains race, *Proc.*
    XV Int.Congr.Entomol., 157-163.
Harris, M.K., 1980, Biology and breeding for resistance to
    arthropods and pathogens of cultivated plants, Texas A&M
    University: 605 pp.
Hatchett, J.H. & Gallun, R.L., 1967, Genetic control of the Hessian
    fly, *Proc.N.Cent.Br.Entomol.Soc.Am.*, 22: 100-101.
Hulspas-Jordaan, P.M. & Van Lenteren, J.C., 1978, The relationship
    between host-plant leaf structure and parasitization efficiency
    of the parasitic wasp *Encarsia formosa* Gahan (Hymenoptera:
    Aphelinidae), *Med.Fac.Landbouw.Rijksuniv.Gent*, 43/2: 431-440.
Kea, W.C., Turnipseed, S.G. & Carner, G.R., 1978, Influence of
    resistant soybeans on the susceptibility of lepidopterous pests
    to insecticides, *J.econ.entomol.*, 71: 58-60.
Knipling, E.F., 1979, The basic principles of insect population
    suppression and management, *USDA Agric.Handbook* 512: 659 pp.
Kogan, M. & Ortman, E.E., 1978, Antixenosis - a new term proposed to
    replace Painter's "Nonpreference" modality of resistance,
    ESA Bull., 24: 175-176.
Lenteren, J.C. van, Ramakers, P.M.J. & Woets, J., 1980, Integrated
    control of vegetable pests in greenhouses, *In:* Integrated
    Control of Insect Pests in the Netherlands, A.K. Minks &
    P. Gruys, eds, PUDOC, Wageningen: 109-118.
Lowe, H.J.B., 1975a, Infestation of aphid resistant and susceptible
    sugar beet *Myzus persicae* in the field, *Z.angew.Ent.*, 79:
    376-383.
Lowe. H.J.B., 1975b, Crop resistance to pests as a component of
    integrated control systems, *Proc. 8th Brit.Insect. & Fung.Conf.*:
    87-92.
Lowe, H.J.B. 1980, Resistance to aphids in immature wheat and
    barley, *Ann.appl.Biol.*, 95: 129-135.
Marrewijk, G.A.M. van & Dieleman, F.L., 1980, Resistance to aphids
    in barley and wheat, *In:* Integrated Control of Insect Pests
    in the Netherlands, A.K. Minks & P. Gruys, eds, PUDOC,
    Wageningen: 165-167.
Marrewijk, G.A.M. van & Ponti, O.M.B. de, 1975, Possibilities and
    limitations of breeding for pest resistance, *Med. Fac. Landbouw.*
    Rijksuniv. Gent, 40: 229-247.
Maxwell, F.G. & Jennings, P.R., 1980, Breeding plants resistant to
    insects, Wiley & Sons, New York: 683 pp.
Painter, R.H., 1951, Insect resistance in crop plants, MacMillan,
    New York: 520 pp.
Panda, N., 1979, Principles of host-plant resistance to insect pests,
    Allenheld/Universe, Montclair & New York: 386 pp.
Parlevliet, J.E., 1980, Disease resistance in plants and its
    consequences for plant breeding, *In:* Plant breeding II,
    K.J. Frey, ed., Iowa St. Univ. Press, Ames: 309-364.

Ponti, O.M.B. de, 1977, Resistance in *Cucumis sativus* L. to *Tetranychus urticae* Koch. 1. The role of plant breeding in integrated control, *Euphytica*, 26: 633-640.

Ponti, O.M.B. de, 1978, Ibid. 3. Search for sources of resistance, *Euphytica*, 27: 167-176.

Ponti, O.M.B. de, 1979, Ibid. 5. Raising the resistance level by exploitation of transgression, *Euphytica*, 28: 569-577.

Ponti, O.M.B. de, 1980, Breeding glabrous cucumber (*Cucumis sativus*) varieties to improve the biological control of the greenhouse whitefly (*Trialeurodes vaporariorum*), In: Integrated Control of Insect Pests in the Netherlands, A.K. Minks & P. Gruys, eds, PUDOC, Wageningen: 197-199.

Ponti, O.M.B. de, 1981, Conserving the natural resistance to insects by a proper use of the pesticide umbrella, *Proc. IX Eucarpia Congress*, Leningrad, in press.

Ponti, O.M.B. de & Freriks, J.C., 1980, Breeding carrot (*Daucus carota*) for resistance to the carrot fly (*Psila rosae*), In: Integrated Control of Insect Pests in the Netherlands, A.K. Minks & P. Gruys, eds, PUDOC, Wageningen: 169-172.

Russell, G.E., 1978, Plant breeding for pest and disease resistance, Butterworths, London-Boston: 485 pp.

Schlumbohm, F.W., 1980, Prüfung der Anfälligkeit für Krankheiten und deren Beurteilung im Rahmen der Wertprüfung in der Bundesrepublik Deutschland, *Bericht Arbeitstagung 1980*, Gumpenstein: 85-93.

Stubbs, R.W. & Bruin, T. de, 1970, Bestrijding van gele roest in wintertarwe met het systemische fungicide oxycarboxin "Plantvax", Gewasbescherming, 1: 99-103.

Van der Plank, J.E., 1963, Plant diseases: Epidemics and control, Academic Press, New York & London: 349 pp.

Zadoks, J.C., Rijsdijk, F.H. & Rabbinge, R., 1982, Systems approach to supervised control of pests and diseases of wheat in the Netherlands. *Proc. IIASA Conference*, Laxemburg, Austria 1979: in press.

DISCUSSION

PARLEVLIET: Have you studied wild carrot populations for resistance to carrot root fly?

DE PONTI: Yes, in 1979 we tested a large number of wild carrot accessions originating from the Netherlands, botanic gardens in Europe and an expedition to Israel.   In general, these wild accessions were very susceptible and the few exceptions did not exceed the level of resistance we had developed in some lines of common cultivars.   Wild carrots are mostly annual, they flower and set seeds despite attack by the carrot fly. Therefore natural selection towards higher resistance is rather weak. On the contrary, heavily attacked carrots of the cultivated type die before they flower in the second year.

JOHNSON: The determinate, side-shootless growth habit gives resistance to white flies in tomatoes. Can this be used by the plant breeder?

DE PONTI: At present not for glasshouse tomato cultivars bred in the Netherlands, but it may be useful for field grown canning tomatoes grown in vast areas in the Mediterranean countries and elsewhere.

JOHNSON: Did the data showing a combined effect of moderate resistance and fungicides indicate a synergism or just an additive affect of the two components?

DE PONTI: There were several cases and both additive effects and synergism occurred.

JOHNSON: Does partial resistance give low selection pressure for the evolution of increased pathogenicity merely in proportion to its degree of incompleteness or is its apparent durability due to other factors? In resistance of wheat to yellow rust some intermediate levels of resistance are durable but others, equally high, are not.

DE PONTI: I believe that the incompleteness of partial resistance is only part of the cause of reduced selection pressure towards new biotypes. A shift in the population will last longer when new bio-types mate freely with the existing population. On the other hand I think that the complex nature which is common to partial resist-ances is perhaps more crucial for durability.

ZADOKS: You elevated the discussion to a higher level by pointing out the value of durability of an integrated control system based on high levels of insect resistance. Experiences with insects are parelleled by those with fungi. Partial resistance to fungi also makes fungicides more efficient. So there is hope for durability of comprehensive integrated control systems, effective against many pests and diseases, and based on high levels of resistance. How-ever, should a durable integrated control system be considered as a goal *per se*. In some countries we can rely upon an infra-structure for pesticide application and in others we cannot. Objectives have to be adjusted accordingly, but the prime responsibility stays with the plant breeders.

DE PONTI: In general I agree with you, but when you suggest that
the implementation of integrated control systems in developing
countries is perhaps too difficult, I disagree. Integrated control
is as old as agriculture, but has been upset by the extensive use
of synthetic pesticides. Local farming systems in many developing
countries often still retain many features of integrated control,
whereas in developed countries, extension services are often very
reluctant to adopt new approaches to regain integrated control.

INSECT RESISTANCE IN COTTON IN SUDAN

O.S. Bindra

Food and Agricultural Organisation of the United Nations
P.O. Box 126, Wad Medani
Sudan

In Sudan, cotton is the most important cash crop. About 80 percent of it is irrigated and grown in the Gezira and neighbouring areas. Here, the cotton whitefly, bollworms (mainly *Heliothis*), jassid and aphid are serious pests. The whitefly, since mid 1950s, and *Heliothis*, since mid 1960s, have attained notoriety. *Heliothis* damages flower buds, flowers and bolls and reduces yield. The whitefly debilitates plants and reduces both yield and quality. Further, whitefly and aphid excrete copious amounts of sugary fluid excreta (honeydew) which makes the cotton sticky and difficult to gin and spin. The sticky cotton has poorer fibre length, strength, maturity ratio, micronaire value, fibre weight per cm, yarn strength and appearance (Khalifa, unpub.). In 1968, a special Committee on Cotton Stickiness recommended two additional sprays for insect control at the end of the season to reduce stickiness (Abdel Rahman & Eveleens, 1979). Since then, pesticide use has increased greatly. However, whitefly and stickiness problems have continued unabated, and the average cotton yield dropped to an all-time low in 1980-81 despite an expenditure of about 65 million dollars on chemical control of cotton pests during the said season.

The serious situation led to the launching, on May 1, 1979, of the FAO/Sudan Government project on Integrated Pest Control (IPC) for cotton and rotational food crops in Sudan. Work on host-plant resistance (HPR) was started in August 1980, since varietal resistance is often quite compatible with other methods of pest control (Bindra, 1981).

In IPC of cotton, we can use all types of resistance, viz: (1) true resistance - (a) antibiosis, (b) non-acceptance; (2) pseudo-resistance - (a) klineducity or phenological escape, (b) induced resistance, (c) non-preference; and (3) tolerance. One or more of a number of morphological and biochemical factors may make a plant resistant to a certain pest. Plants having these factors may show increased resistance or susceptibility, or a neutral reaction to other pests, and have other advantages or disadvantages (Schuster, 1979). The effects of incorporating these factors into cultivars and of using klineducity and induced resistance must, therefore, be weighed carefully.

In Sudan, we need resistance in cotton to whitefly, *Bemisia tabaci* (Genn.), and *Heliothis armigera* (Hub.). Cotton cultivars resistant to these species and suitable for Sudan have not been developed so far. Narrow shape and glabrousness of leaves may prove useful against the whitefly. Cultivars with three-okra and super-okra leaf characters are recommended in Louisiana where the banded-wing whitefly, *Trialeurodes abutilonia* (Haldeman) is a pest. Narrow leaf may reduce the humidity to levels less suitable for the whitefly. Further, detection of pests by predators and parasites, spray penetration and deposition of fine droplets may be more effective on such cottons. The foliar glabrousness may lead to improved parasite efficiency (Hulspas-Jordaan & Van Lenteren, 1978).

Leaf glabrousness provides oviposition anti-xenosis against *Heliothis* moths and nectarilessness may reduce their fecundity. Glabrous nectariless cultivars have already been developed commercially in the USA. Further, some chemicals, high gossypol and related isoprenoids (hemigossypolene and 'heliocides'), cycloprenoid fatty acids, several flavenoids, e.g. quercetin, and polymeric condensed tannins, found in young leaves, flower buds, flowers and bolls on which neonate and young bollworms feed and an unknown factor in yellow pollen, are known to be antibiotic to larvae of *Heliothis* spp. and some other Lepidoptera (Hanny *et al.*, 1979; Waiss *et al.*, 1981). High gossypol imparts resistance to fleahoppers and *Spodoptera littoralis* (Boisd.) but increases whitefly population (Lukefahr, personal communication of 31 December 1980). Further, early-maturing determinate cottons may suffer lesser bollworm damage and require fewer sprays.

A large number of exotic cottons resistant to different insect pests, including narrow-leaf whitefly resistant cultivars together with many advanced lines from Louisiana, glabrous nectariless cultivars, and cultivars and advanced lines from the Multiple Adversity Resistance Programme at Texas A & M as well as early-maturing determinate cottons, have been procured. These are being bio-assayed along with the local material for resistance to the local pest species. Work done during 1980-81 showed that okra-leaf character should prove useful against the cotton whitefly.

REFERENCES

Abdel Rahman, A.A., and Eveleens, K.G., 1979, Cotton pest manage-
    ment in Sudan, 9th Session, FAO/UNEP IPC Panel, Wad Medani,
    Dec. 1979, Mimeogr., 32 pp.
Bindra, O.S., 1981, Development of resistant crop varieties for use
    in the integrated pest control for cotton and rotational food
    crops in Sudan, Mimeigr., 36 pp. (AGRIS No. 699795, FAO Micro-
    fiche No. 8111861-E.)
Hanny, B.V., Bailey, J.C., and Meredith, W.R. Jr., 1979, Yellow
    cotton pollen suppresses growth of larvae of tobacco budworm,
    Environ. Ent., 8:706-707.
Hulspas-Jordaan, P.M., and Van Lenteren, J.C., 1978, The relation-
    ship between host-plant leaf structure and parasitization
    efficiency of the parasitic wasp *Encarsia formosa* Gahan
    (Hymenoptera: Aphelinidae), *Med. Fac. Landbouw. Rijksuniv.
    Gent,* 43/2: 431-440.
Khalifa, H., (Unpublished), A review of cotton stickiness, Mimeogr.
    12 pp.
Schuster, M.F., 1979, Insect resistance in cotton. In: *Biology and
    Breeding for Resistance to Arthropods and Pathogens in Agri-
    cultural Plants,* M.K. Harris (ed.), Texas A & M University
    System, College Station, Texas.
Waiss, A.C. Jr., Chan, B.G., Elliger, C.A., and Binder, R.G., 1981,
    Biologically active cotton constituents and their significance
    in HPR, *1981 Proceedings, Beltwide Cot. Prod. Res. Conf.,* 61-
    62.

BREEDING COTTON CULTIVARS RESISTANT TO WHITEFLY

(*Bemesia tabaci* (Genn))

Hassan Khalifa and O.I. Gameel

Agricultural Research Corporation
Wad Medani
Sudan

INTRODUCTION

Cotton is the main cash crop in the Sudan. The three main categories: long and extra long staple, medium staple and short staple cotton are grown in an area of about one and a half million acres. All cotton cultivars are susceptible to whitefly infestation. Heavy infestation checks the vegetative growth of the cotton plant and results in excessive shedding of leaves and fruits. Final yield of seed cotton and fibre quality are eventually reduced (Mound, 1963; Gameel, 1969).

The cotton whitefly has a wide range of host plants which begin to dry out when cotton is planted in July–August. The whitefly then migrate to the cotton plants and start to breed rapidly during September, October and November. They have about 10-12 generations per growing season. The insects excrete a sugary substance known as honeydew which causes cotton stickiness. The problem of cotton stickiness is becoming a limiting factor in marketing Sudan cotton, because it causes a lot of trouble in the spinning and weaving processes.

The main control measure practised is spraying with chemical insecticides. The excessive use of insecticides besides being expensive, has harmful effects, e.g. hazards to humans, animals, natural enemies and the insect would develop resistance to these insecticides if used continuously for a long time. Therefore, a breeding project has been undertaken to synthesize cotton cultivars resistant to whitefly. The main objective of the project is to manipulate the morphological and physiological characteristics of the cotton plant in such a way as to reduce the whitefly population and allow for easy biological, chemical and cultural control.

MATERIALS AND METHODS

Breeding Lines

The characters okra leaf-shape, glabrous plant body and high gossypol content were transferred to the cultivar "Barac (67)B" (*Gossypium hirsutum* L.) from different genetic stocks. Plants homozygous for the genes conferring these characters were selected from the F_2 of the third backcross generation. A hairy line together with three standard cultivars, namely "Barakat" and "Huda" long staple cultivar (*G. barbadense* L.) and "Barac (67)B" a medium staple cultivar were chosen for comparison.

Field Experiments

The synthesized lines togehter with the standard cultivars were tested for whitefly infestation in a randomized block design with ten replications for three seasons, at the Gezira Research Station. The plot size was four half-rows, 80 cm apart, each with 10 holes spaced at 50 cm. Sowing dates were the last week of August for 1978/79 experiment, the middle of July for 1979/80 and 1980/81 experiments. Later the plants were thinned to three per hole. Irrigation, fertilization and other cultural practices were carried out according to the local standard practices. The experiments were carried out in isolation and no chemical insecticides were used.

Whitefly Counts

At weekly intervals and for about eight weeks, counts were taken for the whitefly adults from 10 randomly selected plants for each replication. Counts were actually taken from five leaves of each randomly selected plant, two leaves from the top, one from the middle and two from the bottom, i.e. 50 leaves per plot. The data were then transformed and analysed. Whitefly build-up and distribution were plotted in graphs.

Stickiness Grades

Seedcotton samples of the different lines were collected at random from each plot. The samples for each line were bulked and ginned separately. Forty grams of lint were taken from the bulk lint of each line and run through a miniature carding machine. The deposit of the sticky material and adhering fibres in the calendar roller were collected and weighed. Each sample was run twice, under atmospheric relative humidities of 55% and 75% and temperatures of 70 to 80°F. Then 10 samples each of 2 gm were taken from the bulk lint of each treatment and the total reducing and non-reducing sugars contaminating the lint were determined (Ali and Khalifa, 1980).

RESULTS AND DISCUSSION

In this study the results indicate that the synthesized line with okra leaf-shape, glabrous plant body and high gossypol content has significantly the lowest number of whitefly adults as compared to the other lines for the three successive season (Table 1). The hairy line, on the other hand, has the highest population of whitefly. These results are substantiated by the lowest sugar content on the lint collected from the okra leaf-shape line and the stickiness grade of "0" at 55% relative humidity. The cultivar "Barac (67)B" of *G. hirsutum* has a relatively higher whitefly population as compared to the two cultivars "Barakat" and "Huda" of *G. barbadense*.

The variation of whitefly populations on the okra leaf line compared to the three standard cultivars are shown in Figs. 1a,b, c for seasons 1978/79, 1979/80 and 1980/81 respectively. In all seasons the whitefly build-up is greater in the hairy line compared to others with the okra leaf-shape line showing the lowest, in the three successive seasons.

The okra leaf-shape has been chosen to create an unfavourable agro-ecosystem for the breeding of the insect. The cotton plants with the character have an open plant canopy which lacks the shadey, humid and warm conditions usually favoured by the whitefly. Another advantage of open plant canopy is that sprayed insecticides will easily reach the target insect especially towards the end of the season. In addition, the okra leaf-shape has less leaf surface area compared to the normal leaf, and hence may harbour less whitefly adults and nymphs.

The glabrous plant body, on the other hand, constitutes less favourable micro-conditions for egg protection and development of the juveniles stages which may be more exposed to natural enemies. This is in contrast to the hairy plant body line which showed the highest level of whitefly infestation compared to all other entries. It has been shown that hairy cotton cultivars harbour large numbers of whiteflies (Mound, 1965).

Gossypol is a poisonous alkaloid compound naturally found in most cotton plants in varying amounts. In high concentrations it may confer an antibiosis type of resistance against puncturing and sucking insects. It renders the plant unpalatable, and this may result in starvation and in some cases the eventual death of the feeding juvenile or adult stages.

Table 1. Average number of Whitefly Adults per 100 leaves and Grades of Cotton Stickiness in the Different Breeding Lines for Seasons 1978/79 - 1980/81 (Data transformed to $\sqrt{x}$).

Breeding Lines	1978/79			1979/80			1980/81		
	Mean No. of Whitefly Adults	Sticki-ness grades	Total sugars (mg/100 gm)	Mean No. of Whitefly Adults	Sticki-ness grades	Total sugars (mg/100 gm)	Mean No. of Whitefly Adults	Sticki-ness grades	Total sugars (mg/100 gm)
Okra leaf-shape	13.7[a]	0-1	684	12.6[a]	0-1	591	20.13[a]	0-1	434
Hairy Line	33.9[d]	1-2	908	25.8[c]	1-2	822	27.03[d]	1-1	697
Barac (67)B	29.9[c]	1-2	904	20.2[b]	1-1	789	25.73[c]	1-1	625
Barakat	27.3[b]	0-1	422	21.5[b]	1-1	672	24.20[b]	0-1	512
Huda	27.7[b]	0-1	393	21.5[b]	1-1	692	24.54[b]	0-1	452

Stickiness Grade:- (0) = Free; (1) = Light; (2) = Medium; (3) = Heavy; (4) = Very Heavy.

a, b, c and d; means having the same letter are not significantly different.

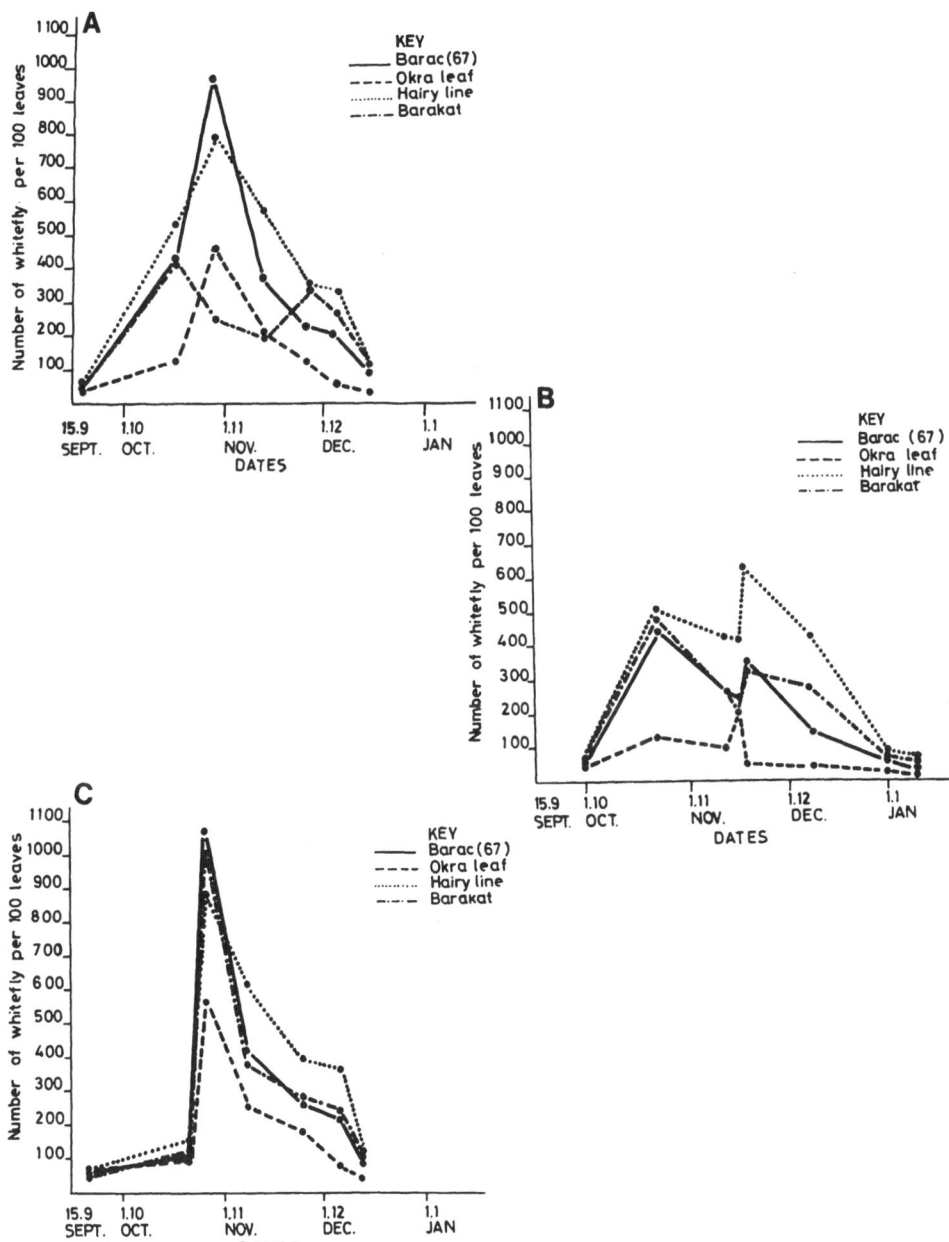

Fig. 1. (a) Variation of whitefly population on four cultivars
           of cotton, season 1978/79;
           (b) Variation of whitefly population on four cultivars
           of cotton, season 1979/80;
           (c) Variation of whitefly population on four cultivars
           of cotton, season 1980/81.

The different levels of whitefly infestation are reflected in the results of the physical and chemical tests of the lint. The lint samples of the long-staple cultivars showed similar grades of lint stickiness, grade O at 55% relative humidity and grade 1 at 75% relative humidity, with almost equivalent amounts of total sugars, whereas the medium-staple cultivars showed higher grades with higher amounts of contaminating sugars in the three seasons. This is because the medium-staple cultivars are more susceptible to whitefly infestation. When comparing the okra leaf-shape line with the hairy line and "Barac (67)B" it is evident that the okra leaf-shape line has lower amounts of contaminating sugars and consequently a lower stickiness grade (Table 1).

Successive backcrossing of the okra leaf-shape line to the standard medium-staple cultivars followed by selection to restore yield and fibre quality resulted in plants with the okra leaf-shape and good agronomic characters. These lines are now under propagation for further testing to determine yield and fibre characteristics.

REFERENCES

Ali, A.N., and Khalifa, H., 1980, Development of methods to measure cotton stickiness, *In*: *Proceedings of the Workshop on Cotton Stickiness, Agric. Res. Corp., Sudan.*

Gameel, O.I., 1969, The effects of whitefly on cotton, *In*: Cotton Growth in the Gezira Environment, M.A. Siddig and L.C. Hughes, eds.

Mound, L.A., 1963, *Emp. Cott. Gr. Rev., Prog. Rep. Exp. Sta., Sudan.*

Mound, L.A., 1965, Effect of leaf hair on cotton whitefly in Sudan Gezira, *Cott. Gr. Rev.,* 13:33-40.

PATHOSYSTEM MANAGEMENT

Raoul A. Robinson

Simon Fraser University
Burnaby, B.C.
Canada, V5A 1S6

INTRODUCTION

In my first paper, I talked about the wild plant pathosystem
and, because so few studies have been made of wild plant
pathosystems, we have to discuss them in terms of theoretical
models. You will recall that I spoke of analogies. There was the
analogy of islands; and migration and colonisation. Locks and keys
provided an analogy for the gene-for-gene relationship which
controls migration only and it does so on a basis of heterogeneity.
Computers provided an analogy for structure and behavior while
books provided an analogy for systems levels. These various
analogies permit the construction of theoretical models of wild
pathosystems. My final conclusion was that the gene-for-gene
relationship controls a behavioural strategy; that it can only do
this; and that it evolved by group selection.

Whether or not these models are an accurate reflection of
truth is anybody's guess. What matters is that they appear
reasonable and logical and, secondly, that they are vastly different
from what we know of the crop pathosystem. Let me put the problem
the other way round. Is it possible that the crop pathosystem is an
accurate reflection of what happens in a wild plant pathosystem? I
myself find it difficult to believe that an elaborate system of
locks and keys could have evolved only to be rendered useless by
uniformity. I am going to assume that uniformity is an artifact of
agriculture. We really must try to sort out what other features of
the crop pathosystem are artifacts also.

The crop pathosystem clearly differs from the wild
pathosystem in three respects. The crop pathosystem is not an

237

autonomous system; it has a major element of artificial control
imposed on it by man. Secondly, compared with the wild pathosystem,
the crop pathosystem is a notoriously unstable system. Thirdly, we
do not need theoretical models when discussing the crop pathosystem
because we are dealing with facts and history.

Let us briefly consider this history. At the turn of the
century, there was great optimism about breeding plants for
resistance to crop parasites. Mendel's laws had been recognised,
the role of chromosomes was understood and Biffen (1905) produced
his classic paper showing that the inheritance of resistance was
controlled by Mendelian genes. At last, it seemed, plant breeding
was on a scientific basis.

This 'scientific' basis for plant breeding led to what was
then an entirely new methodology. This methodology depended on
first finding a "good source" of resistance and then incorporating
it into a suitable cultivar by gene-transfer methods. Indeed, the
need for a good source of resistance became axiomatic and is now so
deeply entrenched that it has spawned other concerns, such as the
need for genetic conservation. I am now suggesting that this axiom
is wrong. It is entirely possible to breed for resistance without a
good source of resistance. However, the converse is not true; not
all sources of resistance are harmful, and some are very useful.

In one sense, the belief in the necessity for a good source
of resistance was a disaster. Indeed, it was a double disaster. If
no good source of resistance could be found, the breeding was not
even attempted; this has been particularly true of most breeding for
resistance to insect parasites. Secondly, if a good source was
found, preference was always given to a single, prominent,
resistance mechanism which was inherited by a single Mendelian gene.
Unfortunately, this kind of resistance usually proved to be
temporary; as we know, Vanderplank (1963) called it 'vertical'
resistance.

It is probably true to comment that, for the past three
quarters of a century, our breeding for resistance to crop parasites
has been so inappropriate that no breeding at all could be attempted
for about half of all the world's crop parasites; and that the
breeding for resistance to most of the other half was such that the
work had to be repeated endlessly.

In terms of the analogies in my last paper, we have used the
gene-for-gene relationship so incorrectly that we have employed only
one lock at a time, on a basis of crop uniformity. Every island has
the same lock and, initially at least, none of the parasite keys fit
this lock. Under these circumstances, the control of the parasite
appears to be perfect. All migrations fail and, as a result, there
is no colonisation either. Then, quite suddenly, all the keys fit;

there is now no control of migration because all allo-infection is matching infection; and there is no control of colonisation because the breeding was for vertical resistance only. The resulting damage is then so severe that the cultivar is abandoned; and the breeding must be repeated.

ARTIFACTS OF AGRICULTURE

Studies of plant pathosystems have been exclusively concerned with the crop pathosystem. As a result, we have no direct knowledge of wild plant pathosystems and, consequently, we tend to extrapolate from the crop to the wild pathosystem. In so far as major features of the crop pathosystem are artificats of agriculture, these extrapolations must be inaccurate. Equally, so long as we do not understand the wild pathosystem, we cannot employ the various components of the wild pathosystem to maximum advantage in the crop pathosystem. This is a "Catch 22" situation. The crop pathosystem is different from the wild pathosystem; we must understand the wild pathosystem in order to manage the crop pathosystem; but we only study the wild pathosystem in terms of the crop pathosystem.

Until such time as we have a large body of data concerning them, we really have only two ways in which to analyse wild pathosystems. The first is to construct theoretical models, as I did in my first paper. Given sound logic, and accurate evolutionary, ecological and other biological principles, there is no reason why these models should not be accurate representations. The second approach is to try to determine which components of the crop pathosystem are an accurate reflection of the wild pathosystem; or, alternatively, to identify those componenets which are the very opposite of accurate reflections; that is, the artifacts.

The first and most obvious artifact is uniformity; the pure line or clonal cultivar grown, perhaps, over millions of hectares. This is an artificial, spatial continuity which never happens in nature; it is the "all cars have identical locks" situation. More than any other factor, it seems to be responsible for the false idea that all resistance is temporary and is bound to break down sooner or later.

Possibly the most damaging artifact is the one revealed relatively recently by Vanderplank's (1963) very important discovery of the vertifolia effect, in which horizontal resistance declines during the process of breeding for vertical resistance. This artifact is damaging in the epidemiological sense because it leads to a very high susceptibility when vertical resistance 'breaks down'. It is even more damaging in the psychological sense because it is largely responsible for the not uncommon belief that horizontal resistance does not exist at all.

In my opinion, the artifact which has contributed most to misunderstanding the crop pathosystem has been the phenomenon of interplot interference, which is yet another of Vanderplank's (1963) discoveries. It can exaggerate differences in the level of disease by more than a hundred-fold. Because it involves allo-infection, which is either matching or non-matching infection, unmatched vertical resistance appears perfect and neither its impermanence nor a low level of horizontal resistance are apparent. Horizontal resistance, on the other hand, can look terrible because of interplot interference. Neither its permanence nor its value under conditions free from interference are apparant. The early workers cannot be blamed for having failed to comprehend such an utterly misleading situation.

Vanderplank (1968) also formulated the concept of 'strength' of vertical resistance. A strong vertical resistance has a matching vertical pathotype with a considerably reduced epidemiological competence; the pathotype then tends to disappear unless there is selection pressure for it. As a result, the pathotype is rare; the strong resistance endures for a longer period and, following its breakdown, it can be employed again sooner. With a weak resistance, on the other hand, the matching pathotype does not have a seriously reduced epidemiological competence; it is consequently common and it remains common; the resistance breaks down quickly and its prospects of being re-employed are slight. This concept of strength can also be applied to pesticide chemicals; DDT provides a weak protection against houseflies but a relatively strong protection against malarial mosquitoes.

The significance of this artifact is that, without it, vertical resistance would probably never have been employed in agriculture at all. If the sole evolutionary function of the vertical subsystem is to reduce the proportion of allo-infections which are matching infections, then any vertical resistance would normally breakdown in the course of a single epidemic. No breeder would employ resistance which failed in the course of one season; indeed, it is doubtful if such resistance, and the matching pathotype, would even be observed. As a consequence, only the strong vertical resistances would be employed. In other words, the abnormal resistances would be used; the normal ones would not even be noticed. It is probable, therefore, that all the vertical resistances we know about are artifacts in that they have an artificially high strength.

The importance of these various artifacts is that they are not recognised as such. As phenomena, they are taken for normality. It is then likely to be concluded, incorrectly, that all resistance is vertical resistance; that horizontal resistance does not exist; that all resistance is bound to fail eventually; and that breeding plants for resistance is, at best, of doubtful value. The sole purpose of this paper is to try to demonstrate that this very common attitude is wrong.

BREEDING FOR RESISTANCE WHICH IS PERMANENT, COMPLETE AND COMPREHENSIVE

I have already published (Robinson, 1979, 1980, 1981) details about breeding for horizontal resistance and, in this paper, I want only to mention three breeding programs by way of illustration.

BLUE MOLD OF TOBACCO

The first program involves the breeding of tobacco for resistance to blue mold (*Peronospora tabacina*) in Cuba. This disease took 95% of the 1979/80 Cuban tobacco crop. The 1980/81 crop was saved by spraying with a metalaxyl fungicide but this chemical provides a 'weak' protection against blue mold and its effectiveness is not expected to last for long. There are no other equally suitable fungicides and the conventional approach to breeding for resistance has some apparently insuperable problems. It is liable to produce temporary resistance which is of no more value than a temporary fungicide. Second, any attempt to cross the 'Criollo' tobacco, which is responsible for the famous Havana cigars, with a 'good source' of resistance leads to an unacceptable loss of quality. Lastly, a full breeding cycle normally requires 7-9 years.

This last point deserves elaboration. Tobacco is seed-propagated and the requirements of uniformity and quality necessitate its cultivation as a pure line. Breeding and selection, however, require variation. Homozygosity is thus essential for cultivation; and heterozygosity is essential for breeding. It is the process of converting heterozygous selections into homozygous pure lines which is the really time-consuming part of plant breeding.

It so happens that tobacco is the crop in which all the original research on pollen cell culture was undertaken. This means that, in tobacco, it is now possible to work from a homozygous line to a hetrozygous population and back again in about one year. This provides a sort of 'instant' plant breeding in which a complete breeding cycle is possible in about twelve months. A succession of five breeding cycles which would previously have taken 35-45 years can now be completed in about five years. And if the pollen cell culture method is applied to the technique of breeding for horizontal resistance, the breeding will not only be quick; it will produce permanent resistance and it can probably do this without an unacceptble loss of tobacco quality because no 'good source' of resistance is necessary.

Havana cigars are made from the black tobacco cultivar 'Criollo' (i.e. Creole, native) which is centuries old. Criollo is highly homozygous and no selection is possible within the cultivar. But it can be crossed with related, susceptible, black tobaccos without undue loss of quality. A single, heterozygous, F_1 plant will provide up to 200 flowers with five anthers to each flower. Each anther provides about five haploid plantlets. Thus, one F_1 plant will provide up to 5,000 haploid plantlets which are segregating and which are all different. These can then be doubled with colchicine to produce doubled monoploids which are homozygous at all loci. The intention is to produce up to 100,000 doubled monoploids each year, from some 10-20 different crosses of Criollo with other black tobaccos.

These doubled monoploids will be grown to self-pollination and seed production. Their homozygous seedlings will be screened under field conditions. The best 1% (i.e. about 1000) will be kept on the basis of their blue mold resistance. The lower leaves of these will be harvested for quality tests; the best 1% (i.e. about 10) will be identified and retained. Because tobacco produces up to 150,000 seeds per plant, these selections can rapidly become available as new cultivars. In the interests of urgency, the original doubled monoploids can be used as the parents in new crosses and the whole breeding cycle can then be completed in about one year. With each successive breeding cycle, there should be major improvements in resistance, quality and yield.

The pollen cell culture technique thus eliminates the main time-consuming component of plant breeding; the converting of a heterozygous variant into a homozygous cultivar. All heterozygosity, segregation and variation occur in the pollen. By the time the progeny (i.e. the doubled monoploids) are screened, they are already homozygous. Indeed, the screening work never handles heterozygous plants at all.

This technique has also improved the prospects of producing hybrid tobacco seed because many homozygous lines can be tested for recombining ability.

SUGARCANE IN THE CARIBBEAN

Sugarcane is an Old World species which was taken to Central America some four centuries ago. Most of its parasites were left behind in the Old World and the New World sugarcane industry thrived accordingly. Indeed, this industry became a major historical and

political factor in tropical America. The Old World industry, being
riddled with pests and diseases, never achieved such prominence.

In the course of several centuries, the New World cane
changed genetically and lost resistance to the parasites which were
absent. This led to an increasing crop vulnerability which is
defined as susceptibility in the absence of the parasite.
Buddenhagen (1977) has called such parasites "re-encounter" pests
and diseases. When the absent parasite is re-introduced to its now
susceptible host, it is very damaging.

This kind of damage began to occur in New World sugarcane at
the turn of this century, largely as a result of the activities of
cane scientists who imported cane cultivars from the Old World and,
inevitably, imported various re-encounter parasites with them.
Mosaic, red rot and root rot very nearly destroyed the cane industry
of the Americas during the first two decades of this century. Then
new, resistant cultivars were produced and the problem was solved.
As I have commented elsewhere (Robinson, 1976), this is one of the
classic examples of plant breeding for disease resistance. The
problem was solved and, having been solved, it was forgotten. It is
probably fair to conclude that the only plant disease problems which
have not been forgotten are those which have not been solved. They
have not been solved either because no 'good source' of resistance
could be found or because the resistance to them was temporary.

Sugarcane is somewhat exceptional in that no vertical
resistance occurs in this species. (This is because it is derived
from a wild pathosystem which is continuous; see Robinson, 1979).
There was no breeding choice; the only available resistance was
horizontal resistance. Resistance to sugarcane diseases has proved
permanent. It is perhaps worth emphasising that some famous clones
of sugarcane, such as POJ 2878, have been in cultivation for more
than half a century without any loss of resistance.
More recently, two other cane diseases arrived in the
Americas for the first time and have caused further "re-encounter"
problems. Smut was recorded in the Argentine in the 1940's and
slowly spread northwards, reaching the Caribbean thirty years later.
Rust also appeared in the Caribbean in the 1970's, having apparently
been blown across the Atlantic from Africa. Rust is interesting
because only about 1% of all cane seedlings are susceptible to it;
and all the resistance is horizontal; there is no evidence for any
failures of rust resistance. But when cane breeding is conducted in
the absence of rust, a few highly susceptible cultivars are likely
to be retained.

Cuba is the world's largest producer of cane sugar with an
annual production of some eleven million tons of sucrose. One third
of the Cuban cane acreage was planted to a cultivar which was highly
susceptible to rust. This disease caused an annual loss of one

million tons of sucrose until the susceptible cane was replanted with a resistant cultivar.

By way of contrast, the experience in Barbados was quite different because the arrival of both smut and rust were anticipated. Barbados cultivars were tested abroad for resistance and all susceptible crops were taken out of cultivation before these diseases arrived. In the event, Barbados was lucky; most cultivars were resistant and little replanting was necessary. Nevertheless, when smut arrived it proved to be no more important than a botanical curiosity. Rust arrived a year later and was even less important. Barbados canes are still being tested abroad for other foreign diseases, such as Fiji disease. Eventually, the Barbados cane industry will be permanently and completely resistant to all cane parasites, including the foreign ones. Even the crop vulnerabilities will have been eliminated. Being an optimist, I want to suggest that we can aim for this very desirable situation in all our crops.

WHEAT IN BRAZIL

Working for F.A.O., at the EMBRAPA wheat research station at Passo Fundo, Martinus A. Beek is now conducting a wheat breeding program in Brazil which may well prove to be as influential as Biffen's paper published in 1905. There will be a difference, however. Biffen's work led the whole of plant breeding into the blind alley of vertical resistance; Beek's work will lead it back again. In my opinion, this program is the most important breeding experiment currently being conducted anywhere in the world. But, it is not for me to tell you about it because Beek is here himself and will describe his work later in this conference.

REFERENCES

Biffen, R.H., 1905, Mendel's laws of inheritance and wheat breeding, *J. Ag. Sci.* 1:4-48.
Buddenhagen, I.W., 1977, Resistance and vulnerability of tropical crops in relation to their evolution and breeding, *Ann. N.Y. Acad. Sci.* 287:309-326.
Robinson, R.A., 1976, "Plant Pathosystems", Springer-Verlag, Berlin, Heidelberg and New York, 104pp.
Robinson, R.A., 1979, Permanent and Impermanent Resistance to Crop Parasites; A Re-examination of the Pathosystem Concept with Special Reference to Rice Blast, Z. *Pflanzenzuchtg,* 83:1-39.
Robinson, R.A., 1980, New Concepts in Breeding for Disease Resistance, *Ann. Rev. Phytopath.* 18:189-210.
Robinson, R.A., 1981, In "Breeding Plants Resistance to Insects". Eds. F. Maxwell and P. Jennings, John Wiley, New York and London.

Vanderplank, J.E., 1963, "Plant Diseases; Epidemics and Control.,
    Academic Press, New York and London.
Vanderplank, J.E., 1968, "Disease Resistance in Plants", Academic
    Press, New York and London.

DISCUSSION

BROWNING: In your first figure there was indicated "ESS primitive –
ESS advanced". You went over this rather rapidly. What does ESS
mean?

ROBINSON: ESS refers to Maynard Smith and Price's concept of the
Evolutionarily Stable Strategy. I attach great importance to this
concept, but I no longer use the term because I have modified the
concept very considerably for purposes of pathosystem analysis.

BROWNING: ESS is a powerful concept in population biology. Of
course, ESS means that a given ecosystem has not been able to evolve
a strategy to achieve greater stability than that imparted by the
ESS, or that would be the ESS. This population biology concept
applies to a wild ecosystem. I suggest that we can bring population
biology and plant pathology together by defining a "Stable Agricultural Strategy" (SAS) as the strategy that imparts to a given agro-
ecosystem the maximum stability that we can devise with present
germplasm, knowledge, and technology. SAS implies the management of
different types and amounts of resistance in plant populations well
buffered also by cultural practices that maximize the effectiveness
of beneficial micro-organisms (e.g., antagonists) in the ecosystem.
In short, this becomes synonymous with "durable protection", which
is far more inclusive than the concept of "durable resistance" alone;
genetic protection is only one component of population buffering.
I submit that it is incumbent on us plant protectionists to be able
to say with confidence to a hungry world that "this is a SAS on which
you can rely for food and fiber" or to get on urgently with developing a SAS for each crop.

JOHNSON: You often emphasize that to achieve horizontal resistance
you must begin by intercrossing susceptible cultivars and not start
with a good source of resistance. Did the sugarcane breeders who
were so successful begin their breeding with recognized sources of
resistance or with other susceptible cultivars? Secondly, the use
of haploids from F_1 plants may speed the progress to a homozygous
cultivar, but it will make no difference to the ability to re-
establish the exact genetic combination that provides resistance and
the necessary similarity to the acceptable commercial tobacco
cultivar. This could be established as easily, or perhaps more
easily, from selection in a standard program of diploids through segregating generations. Thirdly, I know nothing in your writing, nor
do I understand any logical reason why you should consider it un-

likely that there should exist incomplete but race-specific resistance in a wild population. Even if there is not, however, that does
not alter the fact that such resistance exists in agricultural crops
and must be dealt with. It is the separation of incomplete resistance that is not durable from other intermediate resistance that is
durable that provides the greatest challenge to the breeder in
achieving durable resistance in new cultivars.

ROBINSON: In sugarcane both situations occurred. There are plenty
of examples of a "good source" of resistance being both useful and
stable. I am not against the use of a good source of resistance so
much as being against the belief that it is essential. Much of the
cane breeding success was due to transgressive segregation by
crossing existing cultivars. Concerning the second part of your
question, yes, I agree but we are in a crisis situation in which
time is the crucial factor. Concerning incomplete resistance, I
consider that the survival value of incomplete vertical resistance
in a wild pathosystem is too low to justify the evolution of the
gene-for-gene relationship.

PARLEVLIET: In 1976, Feeney proposed a model that could explain why
some crops, such as small cereals, often have elusive resistance
while others (sugarcane as you mentioned) seem to have mainly durable
resistances. He said that in natural ecosystems there are dominant
species that cannot escape parasitism. They must, in order to remain what they are, invest in high energy, durable, resistances.
Pioneer plants though, and most of our crops belong to this group,
colonize disturbed habitats for short periods, disappear and pop up
at other places. They cannot so easily be reached and found by
parasites and they can exploit low energy, less durable, resistances.
Sugarcane possible belongs to the first group of species while crops
with elusive resistances might belong to the latter. Would this also
mean that, in crops derived from pioneer species, it is rather
difficult to get truly durable resistance to some pathogens?

ROBINSON: Yes; this is a most useful idea, which we should certainly
bear in mind.

CROSBIE: The doubled monoploid technique was proposed by Chase (1964)
to shorten the time required to produce inbred lines of corn. However, the ability to produce homozygous lines in one step did not
significantly reduce the time needed to produce new corn hybrids.
Most important quantitative traits in maize (including disease
resistance) have low heritabilities and show important genotype x
environment interactions. Lines developed by any method, therefore,
must be evaluated for many traits for several years. Much of this
selection is normally done during inbreeding. Lines developed by
the doubled monoploid technique must also be evaluated in the same
number of environments as lines from other breeding methods, in order
to have the same assurance of the usefulness of the cultivar over

the usual range of environments encountered by the crop. This is especially important to pathogen and pest resistance, yield stability and quality. How does the situation you have described for tobacco improvement differ from that in corn that leads you to believe that your proposal will be faster than other breeding methods?

ROBINSON: This is a crisis situation of great urgency. We hope to overcome the low heritability with large numbers of doubled mono-ploids. But the most important factor is to get results quickly.

WALLER: In a situation where durable resistance is achieved by many vertical pathotypes and vertical resistance genes, do you think that there is a complete lack (or a very low level) of horizontal resist-ance - a situation which could be analogous to a sort of natural "vertifolia" effect?

ROBINSON: Yes, it seems that low levels of HR do occur and are shown by the production of some seed. The more common situation seems to be total susceptibility, in which the host is totally con-sumed and produces no seed or pollen.

THE SOCIO-ECONOMIC IMPLICATIONS OF BREEDING FOR DURABLE RESISTANCE IN DEVELOPING COUNTRIES

C.A.J. PUTTER

Pathosystems Research Unit
University of Natal
PITERMARITZBURG. R.S.A.

INTRODUCTION

The emphasis in this topic of socio-economic conditions suggests that these circumstances may impose qualitatively different constraints on plant breeding programs in developing countries than they do where agriculture is better developed. This presentation will evaluate certain socio-economic constraints on disease resistance breeding programs and discuss some approaches and disease resistance breeding methods that may be inappropriate for developing countries.

A TYPOLOGY OF DISEASE CONTROL STRATEGIES AS INTEGRAL ASPECTS OF RURAL DEVELOPMENT PROGRAMS

Most of the agricultural technology commonly practised in the developed countries is generally thought to be inappropriate for the developing countries. For example, the objective of yield optimization by intensive monoculture permitted by sophisticated mechanization and managerial skills, can readily be shown to be beyond the resources of the average farmer in the developing countries. However, such generalizations obscure much and they could lead to the misunderstanding that "Western" techniques *in toto* are inappropriate for these countries.

Such generalizations do not differentiate between the several sectors of the agricultural community. Furthermore, they obscure the unique role that disease resistance programs may play in the general rural development process; not only insofar as this involves crop protection but perhaps more importantly, as it may facilitate a dovetailing of traditional and foreign concepts.

Disease Resistance Breeding relative to the Structure of Rural
Communities in Developing Countries

The spectrum of agricultural activities in developing countries
may arbitrarily be divided into three major groups. These may be
viewed as forming a continuum because there is considerable overlapping of methods and objectives. This sub-division is also convenient for the purposes of assigning priorities and determining
objectives for rural development programs. The categories range from
those wherein subsistence farming and a barter economy prevail,
through an intermediate sector characterised by a mixed economy, as
may be found for example in smallholder settlement schemes, to those
where intensive cash cropping is practised.

The central assumption of most strategies for the development
of the Third World is that peasant farmers should be encouraged to
traverse this continuum. As they do so, they become increasingly
dependent on sophisticated machine- and processing- technologies
and cash incomes.

Crop protection programs could easily suffer the same fate
(Putter, 1979). To avoid this, certain statistical details charac-
terising these three sectors are pertinent to the theory advanced
earlier that a particular approach to disease resistance breeding
may be valuable in expiditing the rural development process.

Within these three sectors, 60-90 percent of the population
depend solely on subsistence farming; 5-15 percent rely largely on
intensive cash cropping while 20-30 percent are in the intermediate
category. In the statistical categorisation that follows, specific
data will be given and described for the extremes; intermediate
values or descriptions for the intermediate category will be assumed.

The cropping pattern of subsistence farmers is aimed at maximum
diversity with the ecological objective of achieving a sustained
yield. Cash cropping on the other hand emphasises mono-, di- or at
most tri-cropping, under production methods that rely on deterministic manipulation of the agro-ecosystem. The degree of deterministic control associated with intensive farming is responsible for
a ratio between energy put into the system to energy leaving the
system, of 1:3. In contrast, the ratio for subsistence farming is
1:22 (Rappaport, 1967; 1971). These ratios are largely a consequence
of the subsistence peasant often relying on a pre-wheel technology
whereas his cash cropping counterpart uses sophisticated machinery.

Subsistence farming methods undoubtedly constitute an ecologi-
cally stable strategy and there are many valuable lessons to be
learned from them. Unfortunately an appreciation of the positive
aspects of subsistence farming methods is lamentably uncommon

(Putter, 1978; 1980). It is not intended here to romanticise the "simple" life with its lack of penicillin; neither are these remarks made in ignorance of the deleterious pressure that unbridled population increases place on these communities. Instead it is a plea for a greater awareness of the positive aspects of these farming methods and a prelude to the perspective of crop protection methods under these circumstances that will follow. Also these arguments are not raised to obscure the plight of peasant farmers where their agro-ecosystems have succumbed to those pressures arising from socio-economic circumstances. Indeed, from most other "Western" points of view the silent majority of subsistence farmers are worse off than their contemporary, intensive farming associates.

Isolated peasant societies have very limited access to schools and medicine while they are poorly (if indeed at all), served by communication and transport infrastructures. In contrast, cash cropping is practised in areas umbilically dependent on nearby markets and well-developed transport systems which concommitantly afford access to education and health facilities. Even so, urban drift is associated with undernourishment. Famines and epidemics are more commonly experienced by those communities that have tried and failed at implementing development strategies in which "green revolutions" are hailed as panaceas.

The failure of many development projects may be traced to a premature introduction, often as a result of poor extension techniques, of too sophisticated a method or procedure. Commonly, however, insensitivity to the peasant's point of view or lack of credit for a possible scientifically acceptable basis for his highly ritualised and often animistic practises, denies that the peasant has a perception or understanding of the necessity for any change or transition. Such an understanding must preceed any major change in agricultural practises. Therefore, every effort should be made not to rely excessively on solutions which depend on inputs either foreign to the peasant value-system or exotic in terms of their degree of technological sophistication.

This may be achieved readily in the disease resistance breeding aspects of crop protection programs in developing countries and can be demonstrated with the aid of a typology of disease resistance breeding objectives and methods. Such a classification enables the determination of the optimum combination of intervention points that could define the aims of resistance breeding programs.

The first step is to divide the potential list of control measures into two primary groups: those that rely on the manipulation of factors internal to the pathosystem in question and those that require inputs from outside the pathosystem. Each of these groups may be sub-divided further to produce the hierarchical classification in Fig. 1.

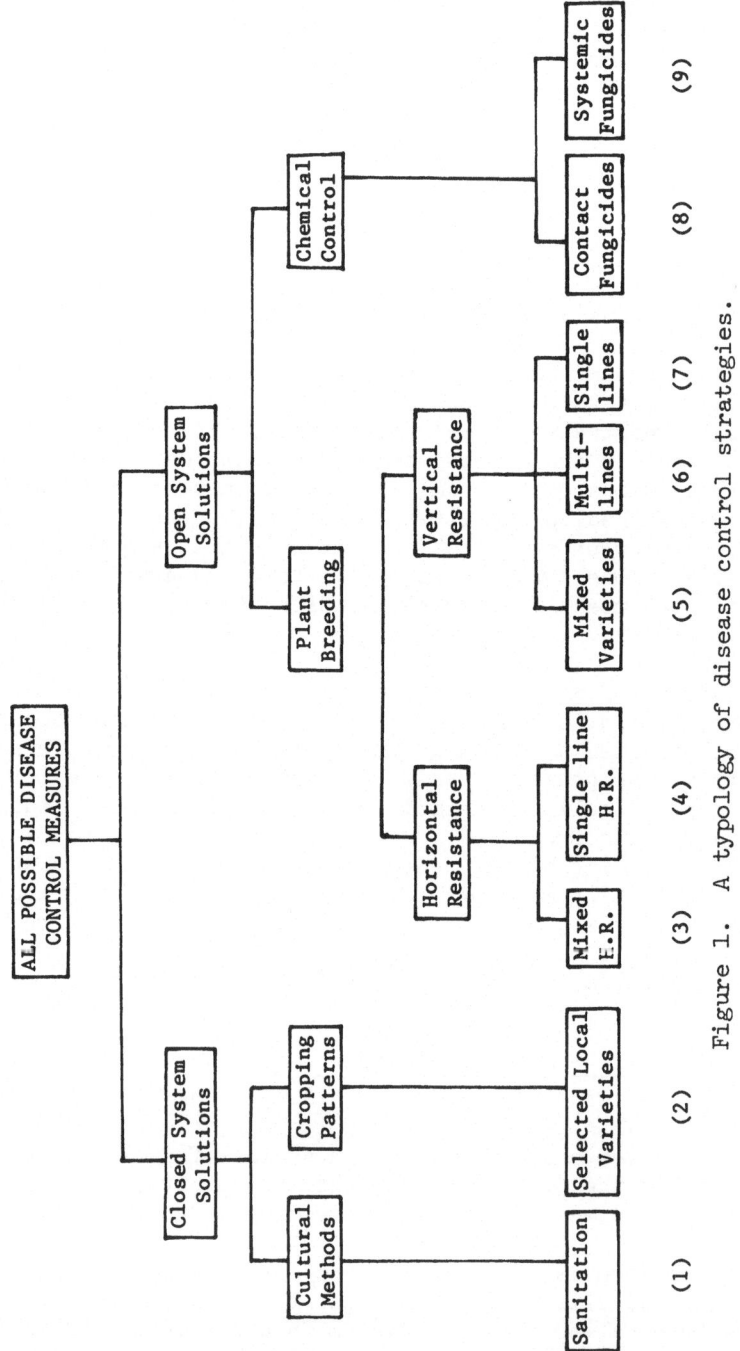

Figure 1. A typology of disease control strategies.

Although not complete, the system in Fig. 1 distinguishes between indigenous disease control possibilities and newer, less familiar approaches and concepts. It also illustrates, from left to right, an increasing order of technological complexity and unfamiliarity of the disease resistance concepts that will have to be faced by peasant farmers in developing countries. Its most important feature, however, is the way in which it reveals the high degree of similarity between strategy numbers (2) and (3) in spite of the fact that the former relies on indigenous possibilities while the latter requires inputs from outside the system. The juxtaposition of these two illustrates how they form a conceptual and technological continuum in spite of the fact that the one is "exotic" while the other is indigenous, thus supporting an earlier argument of this presentation that breeding for disease resistance has a potentially useful role to play in bridging the gap between the known and the unknown - a gap that so often spells the demise of apparently promising solutions to pressing problems.

Comparative Evaluation of Disease Resistance Breeding Strategies

A discussion of sanitation as a disease control measure is beyond the scope of this Advanced Study Institute. Its advantages to tropical subsistence farming have also been discussed by Putter (1980).

Control measure (2) based on the selection of more resistant local varieties, exploits the diversity for which subsistence farming is renowned. Though subsistence gardeners rely largely on one or two staple crop varieties, they are ardent collectors of food plants, both at the species and variety levels. This provides an excellent opportunity to extend the concept of variability, observable within the local landraces selected by the farmer himself for other desirable attributes, to include the concept that plants vary with respect to their response to disease attack. Often peasant farmers do not understand the cause and effect relationship between disease and symptoms (Putter, 1976). The new awareness of variability in disease resistance, could provide an opportunity to establish an interest in disease etiology.

These two factors, viz. an acceptance of the existence of variability and the knowledge of cause and effect, constitute the *sine qua non* of successfully exploiting resistance mechanisms in crop protection programs. They establish a foundation, based on local information that makes it possible to view the next step of introducing foreign concepts, not as something new but as an extension of pre-existing, albeit hidden, knowledge. In developing countries this can be an extremely powerful psychological advantage. There can be no charge of scientific "colonialism", of expatriate domination or imposition of foreign values and objectives.

Having established this foundation, the curiosity and expectations of the peasants may then be used to identify and define how
these principles may be extended to form the basis of a deterministic
strategy in which variable disease resistance is introduced and manipulated. At this stage one aspect of the concept of horizontal resistance, i.e. the *Quantitative* variation in susceptibility to disease, can be introduced as a logical step in the progressive improvement of the cropping pattern. In this way the transition from the
"old" to the "new" will have been accomplished; the scientific basis
of a traditional pattern revealed, acknowledged and praised; that
which was previously "primitive" will now be respectable. There is
no question of imposing the essential value change which Enke (1964)
regarded as an absolute pre-requisite to any development.

Further progression along the sequence of control measures using
disease resistance, as depicted in Fig. 1, is obviously a gradual and
logical process that requires little explanation. Observe, however,
that this progression parallels the progression from mixed diversity
to intensive monoculture, which two factors were previously postulated
as being the opposite extremes in the continuum of agricultural
patterns found in developing countries. Indeed, it does not seem
far-fetched to speculate that this was the pattern of the evolution
of agriculture in the developed countries. After all, not so long
ago we were relying on "field resistance"; on the chance introduction
of a superior variety and we would have regarded the introduction of
"miracle" varieties on which modern green revolutions are based, with
as much superstitious awe and incredulity as would a peasant farmer
today in any developing country.

To complete the conceptually acceptable continuum from the known
to the unknown, the position of fungicides is also depicted in Fig. 1.
However, these will not be discussed here since they are also beyond
the scope of this Advanced Study Institute.

Strategies (2) to (7) in Fig. 1 are directly relevant to the
theme of this Institute and they are offered here on the assumption
that they will serve as a useful guide when disease resistance breeding programs for developing countries are planned. They are also
offered as corroboration of the proposition made earlier that disease
resistance breeding programs can have a general role in facilitating
the rural improvement process rather than being of value only within
the narrow confines of a particular disease problem.

ASPECTS OF DISEASE RESISTANCE BREEDING PROGRAMS INAPPROPRIATE FOR
DEVELOPING COUNTRIES

It is frequently conceded that to start out along the route of
agricultural development, as is the case in developing countries,
has some advantages. Primary among these, is the fact that these
countries can learn from mistakes made elsewhere.

In the developed countries our disillusionment with the boom-and-bust cycle, initiated by an excessive reliance on vertical resistance, would be a case in point. Few plant pathologists employed in developed countries would again rely so heavily on vertical resistance. Thus developing countries were spared the difficulties associated with the boom-and-bust cycle as the vertical resistance approach to plant breeding passed them by.

However, since this Institute is devoted to a discussion of the concept of durable resistance as opposed to temporary resistance, the strategies depicted in Fig. 1 will serve as the basis for the evaluation of some of the potential dangers for developing countries, that may arise from the conceptual vagueness and elusive quality of the definition that describes resistance as being "durable".

Potential Problems that may arise from Defining Resistance as "Durable"

Whatever one's position on definitions of disease resistance concepts, few would disagree that hosts may vary both *qualitatively* and *quantitatively* in their response to disease attack. In passing, we note that both kinds of response to the same pathogen species or race, may occur simultaneously in the same host.

A time scale would be one of the *quantitative* categories wherein the concept of durability is valid, i.e. it would be some stage further along the time scale than temporary resistance. However, since it has not been operationally defined in terms of time scale units, the word durable can be misleading and dangerous for use in developing countries.

During this Institute, Johnson has spoken of resistance which has lasted for five to seven years as being durable. Wolfe & Barrett (1980) used the term implicitly for resistance that lasted three years and these examples support the conclusion that the term durable resistance is entirely a value judgement, imprecisely defined even on a time scale - semantically the only scale to which it can belong.

By defining resistance as durable because it has lasted a few years, several cases of vertical resistances would have been classified initially as being durable. In potatoes the gene R1 for resistance to *P. infestans* is a case in point. In North America the resistance in the potato variety Kennebec conditioned by the gene R1 lasted longer than seven years in spite of the variety's wide cultivation (Van Der Plank, 1963). If, at that time, the differential interaction between the known races of *P. infestans* and potato varieties had been employed, the resistance of Kennebec would have been revealed as being vertical and therefore likely to fall prey to the vagaries of the boom-and-bust cycle. Of course anticipatory searches for differential interactions cannot always be undertaken, as for example

when varieties of the host and necessary races of the pathogen are
not available. However, no such differentiating *procedure* exists
whereby it can be established whether or not a particular resistance
may be described as being durable. Examples in other plant pathosystems of vertical resistance that has lasted long enough to be described as durable, are given by Vanderplank (1982).

It would therefore not be advantageous to base a disease resistance program on the concept of durability rather than on the concepts
of horizontal and vertical resistance. At the very least, the absence of a differential interaction is an indicator that there is a
high degree of probability that such resistance is durable. It has
to be acknowledged that the absence of a differential interaction at
any time is no guarantee that one might not occur in the future.
When this does happen, the resistance hitherto considered to be
horizontal changes its identity to become vertical. However, it is
difficult to see how "durable" as a definition would avoid this
problem. Resistance is durable until it fails, then it becomes
temporary to a greater or lesser extent, depending on how long it
lasted. Ellingboe's (1981) statement that all resistance not yet
shown to be vertical is horizontal, can be paraphrased and applied
to the concept of durable resistance.

There are sound reasons why the word durable is a suitable
adjective that may be used colloquially in the description of the
objectives of a breeding program. However, attempts to elevate
the word into a new population biology term should be opposed
strongly, not only because of their potential to place breeding
programs in developing countries on the wrong track but also because
it is a term that is divorced from any casual mechanism and because
another term has obvious and accepted precedence.

Consider the category of resistance mechanisms that differen-
tiates *qualitatively* between races of the attacking pathogen.
Notice how naturally and comfortably the words "differentiate" and
"qualitative" go together and how naturally they lead to the concept of the differential interaction which in turn leads to the
statistical concept of interaction as measured by the analysis of
variance. In this category of resistance, the word stable is much
more appropriate than is the word durable. A change in quality
implies a change in *state*; which is indeed what happens when vertical resistance breaks down.

Resistance is stable because stabilizing selection makes it so.
This terminology reveals a cause and effect relationship that is not
possible with the use of the term durable resistance. Hence this is
another advantage in adhering to the term stable resistance. It is
patently absurd to think of durable and "durabilising" as having any
meaning whatsoever.

The term stable has precedence over durable not only in plant pathology, but elsewhere in population biology. It is found in such concepts as "evolutionary stable strategy" where it is associated with the process of adaptation. Since adaptation underlies the phenomenon of specific interaction between host and pathogen, it would be logical to preserve the precedence enjoyed by this term elsewhere in biology, in plant pathology as well.

In everyday use for farmers, stable is as suitable as durable and as acceptable when describing the objective of a breeding program. Thus there appears to be every reason to desist from using the term durable and to continue using the more correct term of stable resistance.

However, neither term is a suitable substitute for the use of the terms horizontal and vertical as definitions of resistance. Stable and durable describes an attribute of resistance that both horizontal and vertical resistance may have in common. Even if one uses "general" and "specific" or any other of the many synonyms for horizontal and vertical resistance respectively, the problem is not avoided. Vanderplank's terms are based on a differentiating process that can be stated quantitatively. Their acceptability is further enhanced because they do not contradict basic biochemical processes wherein substrates are catalysed either quantitatively or qualitatively by appropriate enzymes. Although one has to acknowledge that there are some difficulties associated with the differential interaction, the quantitative basis of Vanderplank's definitions are a decided improvement on the descriptive value judgements of non-differentiating attributes on which other definitions are based. The relevance of these comments to this topic emerges when one considers the possible confusion that may arise when descriptive terms are translated, as may be necessary when they are introduced into developing countries. Even if the terms themselves are not translated, there is no guarantee that the subjective bases of the concepts will be understandable or acceptable in cultures different from ours.

Vanderplank (1975) presented and analysed examples of durable vertical resistance. Robinson (1976) also discusses the phenomenon of "frozen" vertical resistance. These data suggest that the growing reputation of vertical resistance being transient, should be reevaluated. In 1978 Vanderplank presented a feasible, albeit speculative biochemical theory that could explain why the resistance in the wheat variety Selkirk, conditioned by the resistance genes Sr6 and Sr9d, has been durable in Canada. Indeed, the extensive converging and corroborating evidence and statistical data presented by Vanderplank (1982), along with the speculative model of the ABC- and XYZ-systems of resistance and virulence genes respectively (Vanderplank, this volume), suggest that some forms of vertical resistance may be permanent rather than merely durable.

If Vanderplank is right, vertical resistance genes have qualities that have hitherto been obscured; qualities which could be manipulated to provide permanent solutions to some of the plant disease problems in developing countries. However, the sophisticated breeding programs that will be required, may be beyond the resources available in developing countries.

Difficulties that may arise from misconceptions about the Quadratic Check and Host:Pathogen Specificity

The phenomenon of highly specific interactions between plant pathogens and their complementary hosts, when these interactions occur on a gene-for-gene basis *sensu* Flor (1971), is especially important to practical pathosystems management. There can be no doubt that Flor used both the quadratic check and a compatible interaction between a pathogen race and a host variety, to indicate when and between which combinations, a specific interaction occurs. Thus, specificity and susceptibility go together; for example a potato variety with the resistance gene Rl can be attacked *only* by a race of *P. infestans* that has the complementary virulence gene rl. All other pathogenic races without the necessary virulence gene are uniformly unable to attack a potato variety with the resistance gene Rl. At this level, *resistance is clearly non-specific* and susceptibility is highly specific. Unless the specific nature of susceptibility at this level is accepted, the very foundation of the concept of physiological races is placed in jeopardy.

In order to avoid temporary or non-durable resistance it will be necessary to avoid genetic mechanisms in the host, i.e. vertical genes, that confer temporary, qualitative, but non-specific resistance to the pathogen population when this population lacks the required specific complementary virulence genes. Specific complementary interaction at this level will only be revealed by a quadratic check involving a minimum of two resistance genes in the host and two virulence genes in the pathogen. In the majority of the approximately 23 pathosystems known to interact on a gene-for-gene basis, separate loci and not merely different alleles at the same locus are involved. In some gene-for-gene interactions different loci have not been identified although these are far in the minority. In such cases, the cautious thing to do is to talk of alleles until such time as segregation studies reveal beyond doubt that different loci are involved.

Ellingboe (1979) very emphatically limits the quadratic check to being an interaction between two alleles at one locus in the host and two alleles at one locus in the pathogen. Within these constraints he further goes on to demonstrate that *resistance is specific; susceptibility not*. When limited to this particular systems level, his conclusion is valid. However, at the level that concerns the practical importance of avoiding specific interactions, the opposite applies, i.e. *susceptibility is specific.*

The quadratic chekc, *sensu* Ellingboe (*loc. cit.*), is only one specific version of a two-way interaction table. To demonstrate a gene-for-gene interaction, i.e. to see if there is a specific, complementary interaction between host and pathogen, requires a two-way interaction table containing two genes (loci) in each of the hosts and pathogens.

Application of the idea of specific resistance at the wrong systems level can cause much confusion. For years it misled us into searching for phytoalexins that could provide this elusive specific resistance-conferring factor.

Specific susceptibility, i.e. compatible interactions between a race of the pathogen and a complementary variety of the host, can be expected to reveal common antigenic surfaces. This is corroborated by the published evidence reviewed by Vanderplank (1978). When confronted with such experimentally confirmed evidence, Deverall (1977) prefers to conclude that: "..the notion of a compatible host and parasite sharing a similar type of molecule is at odds with the simplest hypothesis based on the best-known genetic interactions that the most specific interaction is for incompatibility." Deverall himself presented results which he interpreted as showing that: "..antigens were shared by susceptible cultivars and the races virulent on those cultivars." Though it may be disconcerting when experimental results question hypotheses this is no reason to reject *evidence* and accept *theory.*

It is clearly imperative that a consensus be reached concerning the use of the words specificity, susceptibility and resistance and how these terms are inter-related at the different biochemical, molecular-genetic, plant breeding and disease systems levels. From the point of view of the developing countries it is important to avoid confusing concepts and misleading terms.

In empathy with the plight of the millions of farmers in developing countries, plant breeders and pathologists will do well to put their house in order. Then the goal of finding durable resistance and the efforts of the participants at this Advanced Study Institute might well contribute towards improving the agricultural productivity of the developing countries.

REFERENCES

Deverall, B.J., 1977, Defence Mechanisms in Plants, Cambridge
     University Press, Cambridge, pp. 110.
Ellingboe, A.H., 1979, Inheritance of Specificity: The Gene-for-
     gene Hypothesis, *In:* "Recognition and Specificity in Plant

Host-Parasite Interactions." Eds. J.M. Daly and I. Uritani,
      Japanese Scientific Society Press; Tokyo.
Ellingboe, A.H., 1981, Changing concepts in Host-Pathogen
      Genetics, *Ann. Rev. Phytopath.*, 19:125-188.
Enke, S., 1964, Economics for Development, Dennis Dobson, London.
Flor, H.H., 1971, Current status of the gene-for-gene concept,
      Ann. Rev. Phytopath., 9:275-296.
Putter, C.A.J., 1976, The Phenology and Epidemiology of *Phytophthora
      colocasia* Racib. on Taro in the East New Britain Province of
      Papua New Guinea, M.Sc. Thesis, University of Papua New
      Guinea.
Putter, C.A.J., 1978, Pathosystems management under subsistence
      farming conditions, *3rd International Congress of Plant
      Pathology*, Munich, 16-23 August, 1978.
Putter, C.A.J., 1979, Crop loss quantification in subsistence
      farming systems, *IXth International Congress of Plant Pathology*,
      Washington DC, 5-11 August, 1979.
Putter, C.A.J., 1980, The management of epidemic levels of endemic
      disease under tropical subsistence farming conditions, *In:*
      "Comparative Epidemiology - A tool for better disease management."
      Eds. J. Kranz and J. Palti, PUDOC, Wageningen.
Rappaport, R.A., 1967, Ritual regulation of environmental relations
      among a New Guinea People, *Ethnology*, 6:17-30.
Rappaport, R.A., 1971, The flow of energy in an agricultural
      society, *Scientific American*, 224 (3).
Robinson, R.A., 1976, Plant Pathosystems, Springer-Verlag, Berlin,
      Heidelberg, New York.
Vanderplank, J.E., 1963, Plant Diseases: Epidemics and Control,
      Academic Press, New York, London.
Vanderplank, J.E., 1975, The Principles of Plant Infection, Academic
      Press, New York, London.
Vanderplank, J.E., 1978, The Genetic and Molecular Basis of Plant
      Pathogenesis, Springer-Verlag, Berlin, Heidelberg, New York.
Vanderplank, J.E., 1982, Host-Parasite Interactions in Plant Disease,
      Academic Press, New York.
Wolfe, M.S., and Barrett, J.A., 1980, Can we lead the pathogen astray?
      Plant Disease, 64:148-155.

DISCUSSION

JOHNSON: Durable resistance is not equivalent to horizontal resistance because some vertical resistance has been highly durable.
Green & Campbell (1979) quote as examples the wheat cultivars Selkirk,
Pembina, Manitou, Neepawa, Napayo and Glenlee which have been durably
resistant to stem rust, although the resistance is essentially vertical. We should be asking what stabilizing selection is needed to
preserve the durability of vertical resistance.

There is limited merit in trying to identify vertical and

horizontal resistance through their durability. What are needed
are biochemical or physiological systems of recognizing the sort of
resistance. The definitions of the two sorts of resistance are
guides. Vertical resistance implies a disease system with both a
specific elicitor in the pathogen and a specific receptor in the host.
Horizontal resistance implies a disease system in which either the
elicitor or the receptor or both are unspecific.

Reference

Green, G.J., and Campbell, A.B., 1979, Wheat cultivars resistant to
    Puccinia graminis tritici in western Canada: their development,
    performance and economic value, *Can. J. Pl. Path.*, 1:3-11.

PUTTER: The definitions of horizontal and vertical resistance given
in this paper were very clear. However, as stated they may be assumed
to occur together in a cultivar. The difficulty that is not answered
is how, in practical terms, the supposedly horizontal component can
be distinguished and separated. Race-specific components of resistance
can be identified when pathogen isolates with specific pathogenicity
for them can be found, but how long must a cultivar be grown before
we can be sure that it possesses no further race-specific components?
I believe it is impossible to state with conviction that a given level
of *durable* resistance is, and will remain race-non-specific. It may
perhaps be possible to identify a genetic component that has apparently
controlled durable resistance but not to claim that it, or components
of it, will remain permanently effective and therefore classifiable
as race-non-specific.

WOLFE: There are at least two meanings of the word specificity, and
they cannot be evaded. Within a 2-variable system there is specificity
implying interaction, i.e. vertical resistance. Within a 3-variable
system there can be specificity or pseudospecificity arising from
the inadequacy of the resistance. For example, in tomato the gene
I gives quantitative resistance to *Fusarium oxysporum* f.sp.
lycopersici, a 2-variable system with no specificity. Race 2 of
Fusarium that overcomes the resistance of gene I is, in the absence
of this gene, less fit to survive than race 1. The availability of
gene I has enabled tomato growers to shorten their crop rotation
thereby allowing race 2 to accumulate in soil. In this 3-variable
system, shortened crop rotation being the third variable, there would
be a pseudospecificity of race 2 for gene I.

PUTTER: In the simple case of a single resistant variety overcome
by the pathogen, the problem for the grower, breeder, etc. is highly
specific: it apparently concerns the reaction of a pathogen population specifically to a key variety. Simply at that level, however, it cannot be stated to be a specific interaction involving a
matching of the host resistance genes by the appropriate pathogen
virulence genes.

DURABLE RESISTANCE IN PERENNIAL CROPS

N.A. Van der Graaff

Food and Agriculture Organization of the United Nations

SUMMARY

In this paper, pathosystems are considered with the aim of
determining why particular diseases (and pests) are important and
how to obtain a system in which they become unimportant. The
crops under consideration are mostly "traditional" and "unimproved". Crops considered are fig, grape, coffee and coconut.
Their various diseases are discussed and, where possible, reasons
for recent outbreaks of diseases in these crops are given. Most
outbreaks appear to be "man made": in the traditional situations,
diseases and pests are relatively unimportant. The diseases and
the pests that are important in these crops were either new to the
crop, became more important through new cultural practices or
resulted from the use of material that was not well tested with
regard to the local disease and pest situation. The latter condition
may nearly be intractable when diseases have very long latent
periods, as in coconuts. In other perennials, disease problems
can often be related to the above causes. In temperate deciduous
fruits, the history of the crop is usually too complicated for identification of such causes to be possible. Concerning traditional
diseases in perennial crops, vertical resistance is mostly absent
or not effective and various levels of polygenetically inherited
disease resistance can be found within the range of cultivars.
Monogenetic vertical resistance is chiefly found in wild related
species; its use in breeding programmes will result in a differentiation in vertical pathotypes. With regard to diseases that are
new to a crop, vertical resistance will hardly ever occur and all
resistance found within the crop species can probably be used.
The study of the balance between host and pathogen in "traditional"
and "unimproved" crops indicates that a situation of stable and
durable resistance can be reached.

263

INTRODUCTION

This paper is concerned with durability of resistance and resistance breeding in perennial crops. Although in this Study Institute some definitions on resistance have been already given and discussed, I wish to define some terms as they will be used within the context of this paper.

Horizontal resistance is quantitative; its inheritance is poly- or oligogenic, and there is a negligible level of specificity between elements of the host and pathogen population.

Vertical resistance is mostly qualitative, its inheritance is monogenic or oligogenic, and there is a specificity between elements of the host and pathogen population caused by a gene-for-gene relationship.

Durability concerns the ease with which the pathogen can adapt to the host's resistance. Each resistance and type of resistance has been part of a "durable" pathosystem (except in the case of newly encountered diseases). Thus its original function in that system will mainly determine its durability. In such a wild pathosystem, vertical resistance will only be functional through multiline effects.

To obtain "durable" resistance it is necessary to have an insight in the history of the crop and its diseases: the "pathosystem" has to be analysed. Most important is the realization that the great majority of plant diseases are not important and that resistance to these diseases is durable.

In this paper we shall consider a number of pathosystems, determine, where possible, why diseases are important and indicate ways to obtain durable resistance. For this pathosystem analysis, I used the excellent publication of Buddenhagen (1977).

PATHOSYSTEM ANALYSIS AND MANAGEMENT

Some crops have been only recently taken into cultivation; this will have often resulted in an unbalanced pathosystem. An obvious case is the cultivation of rubber in South America and the serious attacks of South American Leaf Blight. Little can be said concerning the possibilities of obtaining adequate and durable resistance against this disease.

To obtain an insight into durable resistance in other crops, it is necessary to study those that are still "unimproved": that is to say, those crops having cultivars that have been selected only by many generations of farmers. Through local selection, a

wealth of local cultivars has been developed, which are well
adapted to local growing conditions and, under the traditional
conditions of crop husbandry, have disease levels that cause
insignificant losses. Owing to the nature of this type of selection,
no rapid adjustment of the pathogen to the resistance level of the
host is to be expected. Depending on the nature of the crop,
resistance will be mostly horizontal, although there may be a
possibility of the presence of vertical resistance. During the
traditional selection process, heterogeneity decreased and, therefore, the role of vertical resistance gradually diminished. As a
consequence, in fairly homogenous crops, effective vertical resistance is rare and the vertical genes will be mostly found among
wild progenitors.

Examples of perennial crops in which diseases play or have
played a minor role in the course of their domestication will be
given. In reviewing these crops, we shall also identify the
reasons why some diseases have instead become important.

Fig

The fig is usually a clonally propagated crop. As such, it has
a considerable range of pathogens. Weber (1973) lists as many
as 22 fig diseases, including two rusts, leaf spots caused by
Ascochyta, Alternaria, Cercospora, Cephalosporium and others.
Nevertheless Forkner (1919, cited by Condit, 1947) stated on the
fig:

I live for ever
I have no disease
And for six thousand years
I have not failed to produce a crop each year

Nearly all fig varieties are traditional and some are very old.
Storey (1976) indicated that: "Sari lop for example has been grown
in the Meander Valley of Turkey for possibly as long as 2,000 years.
Dottato of Italy was praised by Pliny (A.D. 23-79); Verdone has
been grown in the Adriatic region for hundreds of years."

Although these varieties are very old, resistance to their
pathogens has remained adequate and unchanged. Thus, it can be
concluded that within this crop, selection by farmers has been
efficient enough to maintain diseases at insignificant levels.

Grapes

Vitis vinifera is a traditional crop in Europe. The species
probably originates from central Asia (Olmo, 1978). Cultivation

gradually spread westward from an early domestication in the Near East. Multiplication used to be through cuttings; presently, the normal type of grape multiplication is through grafting on rootstocks. At present, in Europe, the majority of grown cultivars have been obtained through selection by farmers. However, looking at today's disease situation in Europe, one would conclude that the pathosystem is very unbalanced. Regular fungicide treatments are needed to control downy mildew *(Plasmopara viticola)*, powdery mildew *(Uncinula necator)* and black rot *(Guignardia bidwellii)*. However, it should be realized that these diseases (and the root aphid, *Phylloxera virifolia)*, were not part of the original pathosystem where the grape cultivars were selected. Root aphids were introduced from America, after which the European grape culture could only be saved by grafting *Vitis vinifera* on resistant American rootstocks. However, with the rootstocks, powdery mildew and black rot were also introduced into Europe, after which only the chance discovery of Bordeaux mixture saved the grape culture from disappearing. The only old world disease that sometimes damages grapes is anthracnose *(Elsinoe ampelina)*.

Resistance to the root aphid is polygenetically inherited (Boubals, 1966; Pouget and Kim, 1977) and has proved to be durable. Resistance to powdery and downy mildew varies in European grape varieties but is, in general, too low for satisfactory control. Both pathogens co-evolved with wild grape species in North America; consequently, sufficient resistance is found in North American grape species. Boubals (1959, 1961) studied the inheritance of non-necrotic, lesion-limiting resistance in grapes. He found that polygenetic systems existed for resistance to downy mildew and powdery mildew. As long as these types of resistance are used in grape breeding, a "breakdown" of resistance is unlikely. Only recently, Shtin and Filippenko (cited by Olmo, 1978) found monogetic resistance to *Plasmopara viticola*; surprisingly, this resistance was derived from *Vitis amurensis*, a Russian species.

In summary, traditional grape diseases are of little importance. Problems arose when new encounters occurred with pathogens that evolved in a different, but related, pathosystem and to which resistance levels in the original crop were insufficient.

Coffee

The main species grown for coffee production are *Coffea arabica* and *C. canephoro*. Both species are of African origin; *C. arabica* is a species from the tropical highlands. It originated in Ethiopia, where it has been in culture in the southwestern part of the country for an undefined period. *C. canephora* was culti-

vated in Uganda before the time of the European explorations.
C. arabica is self fertile and is normally propagated by seed.
C. canephora is self sterile, and is often vegetatively propagated.

Coffea arabica is considered to be a crop with many diseases.
In Ethiopia, the centre of domestication of the crop, diseases have
been of minor importance until recently. The crop has been traditionally grown under practically the same conditions under which
it was originally domesticated. Coffee populations are very heterogeneous and the crop is grown under heavy shade and with high tree
densities to suppress weed growth. Fertilizers have never been
applied; yields per hectare are low but so are all inputs. Under
such conditions, the disease situation in Ethiopia is unique and
provides a rare opportunity to reveal a well-balanced pathosystem
situation. These will be examined for some individual pathogens.

Leaf rust, *Hemileia vastatrix*, is one of the major coffee
diseases outside Ethiopia. In that country, the disease hardly
warrants control. It is influenced by altitude: at higher altitudes
coffee rust is less prevalent; nevertheless at some locations, *C.
arabica* is grown at low altitudes without major rust damage.
Differences in rust susceptibility within the Ethiopian coffee population exist among geographical areas. The coffee from Eastern
Ethiopia (Hararge) is more susceptible to rust than coffee from
the western part of the country; within the latter coffee population,
geographical differences in susceptibility can also be found (Van
der Graaff, 1981). These differences probably reflect disease
selection pressure in the various locations, chiefly due to clima-
tological factors. Comparable differences exist with regard to
Phoma tarda, *Cercospora coffeicola* and a leaf miner (Van der
Graaff, 1981). Thus, the Ethiopian *C. arabica* population is
composed of a patchwork of landraces that are adapted to fairly
specific local disease and pest conditions.

It is known that vertical resistance to leaf rust occurs in *C.
arabica* (Rodrigues et al., 1975). Also, a quantitative variation
in resistance exists (Eskes, this volume). It may be assumed
that horizontal resistance accounts for the major differences
between locations in Ethiopia. Within each location, effects of
vertical resistance may increase the effect of horizontal resistance, although the effect of vertical resistance has probably
diminished in the process of domestication. Outside Ethiopia,
coffee types are grown in which horizontal resistance to leaf rust
is not at its maximum. These coffee types did not originate
directly from Ethiopian coffee types. Coffee was instead first
taken to Yemen from Ethiopia in the first half of the second millen-
nium. In Yemen, conditions were adverse for leaf rust development and, thus, there was no selection pressure for disease resistance. From Yemen coffee was taken to other countries, and the
pathogen caught up with the crop only after large cultivation areas

were established. In most cases, leaf rust resistance is insufficient and protection with fungicides is needed. In the past, the
selection for resistance was only concerned with vertical resistance, but this has not led to any lasting solution. A programme
concerned with horizontal resistance is presented by Eskes elsewhere in this volume.

As indicated previously, disease resistance toward *Ascochyta*
(Phoma) tarda varies among populations from geographically different origins in Ethiopia. Under normal conditions of cultivation,
the disease presents no problem. However, changes in the agricultural system increase its economic significance: rejuvenation
of coffee trees through "stumping" is a newly proposed practice;
under this system, old bearing stems are removed and replaced
by new shoots. However, growth of these new shoots is often
retarded by the pathogen, thus increasing the time span between
rejuvenation and first crop.

Brown eye spot, *Cercospora coffeicola*, causes no major
damage. In Ethiopia, landraces from dry areas show susceptibility in wet locations. The pathogen has a worldwide occurrence
and resistance appears to be sufficient except in locations where
unsuitable varieties have been inadvertently chosen. For example,
Cercospora coffeicola causes serious damage to berries of susceptible cultivars in Zambia and the same damage is reported from
Malawi (Siddiqi, 1970).

Vascular wilt, *Gibberella xylarioides*, has a very long incubation period (at least 9 to 10 months). On Arabica coffee, this
disease is only reported from Ethiopia (Kranz and Mogk, 1973).
The system host-pathogen is a prototype of an endemic disease
(Van der Plank, 1975). Quantitative differences in susceptibility
to the disease have been found in Ethiopia (Van der Graaff and
Pieters, 1978; Pieters and Van der Graaff, 1980). In the dense
stands of coffee in western Ethiopia, damage by the disease is
minimal. However, when "modern" cultural practices are introduced, conditions change and influence the relation between the
pathogen and its host. Outbreaks of vascular wilt have often been
associated with the "modernization" of coffee growing practices.
These practices include removing of shade, and regulating and
limiting number of trees to increase production. This also
increases weed growth. Consequently, soil cultivation, including
digging and slashing of weeds, is required. Trunks and roots of
trees are often wounded in the course of these procedures and this
increases an effective transmission of the pathogen.

Coffee Berry Disease (CBD), *Colletotrichum coffeanum*, is
probably an insignificant pathogen of a related coffee species. On
C. arabica, the pathogen causes an anthracnose of green berries.
The disease is very damaging; losses of the complete crop are not

unusual. The first observations of the disease date back to 1922, when it was observed in western Kenya (Macdonald, 1924). The disease gradually spread over the main Arabica coffee producing areas in Africa and, at present, only some southern African coffee producing locations are outside the infested area (southern Tanzania, Zambia, Malawi, Zimbabwe, South Africa). In 1971, the disease was found in Ethiopia (Mulinge, 1973) and spread very rapidly. By about 1975, most areas were infested; in 1978, the disease was also found in the geographically isolated area of Harerge. Damage is considerable; it is estimated that 20 to 25% of the crop is lost annually (Van der Graaff, 1981). Variation in resistance was observed early; the high level of resistance of the Rume Sudan coffee type was reported by Firman in 1964. In another paper in this volume the CBD resistance programme in Ethiopia and the nature of the resistance is discussed by Van der Graaff.

Other diseases: In Ethiopia, many insignificant *C. arabica* pathogens probably exist. Outside Ethiopia, there are still *Coffea arabica* diseases that have not been redistributed to the main growing areas; they include grey rust *(Hemileia coffeicola)* and *Nematospora coryli* , a yeast transmitted by the *Antestiopsis* bug. Both these diseases are probably part of a related host-pathogen system. American leaf spot *(Mycene tricolor)* is probably a new-encounter disease and is still restricted to South America.

Coffea canephora can be attacked by the same pathogens as *C. arabica;* however, its resistance to most diseases is higher. The only major problem consists of a vascular wilt caused by *Gibberella xylarioides* . Varietal differences in resistance were found and resistant material was distributed. (Meiffren, undated; Saccas 1951, 1956). The disease appears to have gradually disappeared.

Conclusions: Arabica coffee grown under traditional conditions is a crop that is in good balance with its pathogens. Where major diseases have occurred this has resulted from the following situations:

(a) Crop selection and breeding has occurred outside the range of a potentially major pathogen *(Hemileia vastatrix)* (re-encounter).
(b) The crop comes in contact with pathogens of related species (Coffee Berry Disease) (new-encounter type a).
(c) Omnivorous pathogens of unrelated pathosystems are able to attack the crop *(Mycene tricolor)* (new-encounter type b).
(d) Changes are introduced in cultural practices *(G. .xylarioides, Phoma tarda)* .
(e) Cultivars are chosen that are unsuitable under certain ecological conditions *(C. arabica* varieties susceptible to

Cercospora coffeicola).

Coconuts

"Most coconut breeding programmes have been concerned
with improving yield. The increasing seriousness of such diseases
as cadang-cadang in the Philippines, Kaincopé in West Africa, and
lethal yellowing in the West Indies has highlighted the need to
consider breeding for disease resistance. In such cases, indeed,
the production of resistant genotypes may be the only practical
answer to disease problems." (Child, 1974).

The low concern of breeding programmes with disease resistance is understandable: diseases and pests have not been a limiting
factor in coconut culture in most countries. However, in the after-
math of the expansion of the coconut culture in the last century,
disease problems have gradually intensified. The most important of
these diseases are: lethal yellowing, a killing disease caused by a
mycoplasma; cadang-cadang, a slow decline disease caused by
a viroid; and Kerala wilt, a decline of an unknown etiology. In
Central America, chief problems also arise from red ring caused
by the nematode *Rhadinaphelenchus cocophilus* and phloem necrosis
caused by flagellate protozoans.

The disease situation concerning coconuts can be characterized by two major situations: (a) The insular character of the crop
offering many possibilities of genetic drift and separate host-
pathogen evolution; (b) The fact that most important diseases
are of the slow-decline type.

The possible origin of lethal yellowing has recently been
proposed by Chiarappa (1979). He presented evidence that lethal
yellowing, a disease found in the Caribbean and possibly West
Africa, may be identical to other wilts that are observed in coconuts in other areas. In resistance trials in Jamaica, tall coconuts
from the mainland of South East Asia and Sarawak showed an intermediate degree of resistance. Mixed resistance and some susceptibility occurred in coconuts from New Guinea and the Solomon
Islands. Material from the Pacific was fairly susceptible, while
material from the New Hebrides and from India, Sri Lanka and
Africa was highly susceptible (Harries, 1978). This indicates a
possible co-evolution of host and pathogen on the Malaysian penin-
sula. Cases of wilt-type diseases in southeastern Asia are reported
sporadically (Chiarappa, 1979).

More cases of isolated co-evolution of host and pathogen are
to be expected in coconuts. In the New Hebrides, all imported
coconut material dies off rapidly but local coconuts are resistant
(Grylls cited by Chiarappa, 1979). Comparable cases have been

observed in the Solomon Islands and Indonesia.

Cadang-cadang, a slow decline caused by a viroid, occurs on Guam (Boccardo et al., in press) and part of the Philippines. Trees gradually decline although it may take 20 years for a tree to die. The history of cadang-cadang is insufficiently documented to provide evidence on the origin of the disease. It is, however, likely that the pathogen co-evolved with coconuts somewhere else in the Pacific and later came into contact with more susceptible material. Differences in resistance have not yet been proven.

Kerala wilt is a slow decline of unknown etiology. Very little is known on the history of the disorder. A critical examination of data published by Rawther and Pillai (1972) shows that differences in resistance may occur.

Red ring and heart rot are new encounter diseases in the new world. For both diseases, variation in susceptibility is reported (Alexander, 1979; Martinez Lopez et al., 1979).

The important diseases in coconuts are mostly of an endemic nature (Van der Plank, 1975). They probably co-evolved in a part of the host-pathogen system and then were brought in contact with more susceptible material. It is extremely difficult to avoid susceptibility to such diseases. Firstly, the pathogen may only be present at an insignificant level in the original population so the epidemic needs time and sufficiently susceptible material to start; secondly, owing to slow disease progress, it may take years before it is possible to identify a diseased tree. In coconuts, the situation is very complicated due to the elusive nature of the pathogens involved. Resistance to the various diseases can be found in the locations where the disease and pathogen co-evolved. Such resistance is expected to be durable.

The new-encounter diseases in the new world are of the same nature as the other major coconut diseases. Quantitative variation in resistance has been found and there is little reason to believe that resistance would not be durable.

The nature of diseases makes each coconut improvement programme extremely difficult. Introduction of new germplasm can result in the introduction of pathogens new to local varieties or may result in the introduction of new genetic susceptibilities, the effects of which will only be seen many years later. Therefore, improvement programmes can only avoid these problems if some sort of an international network could be established in which:

(a) Material is identified that is susceptible or resistant to the main diseases that are known at present. This can be done by on-site testing in major disease areas and by the develop-

ment of rapid testing methods.
(b)   Varieties are tested in various traditional coconut areas as
       a "trap crop" to identify potentially dangerous diseases.
(c)   Material is exclusively used for local improvement, that
       has been grown in the area for a considerable time and
       has remained disease free.

Other perennials

The list of perennials with major disease problems may
easily be extended. Cacao swollen shoot is a major new-encounter
disease in West Africa while witched broom (*Crinipellis
perniciosus*) is a major problem in South America. Types of
cocoa resistant to swollen shoot have been selected (Legg and
Lockwood, 1981). Date palm has a major, probably new or re-
encounter, disease: Fusarium wilt (Bayoud). This disease occurs
in Morocco, is gradually spreading into Algeria, and is said to
have originated in southwestern Morocco (Toutain, 1965). Resistant varieties exist in that country and a very closely related species,
which occurs in southwestern Morocco, is also resistant (Oudejans,
1969). This may indicate a co-evolution of host and pathogen in
that area.

Other disease problems concern clove (sudden death in
Zanzibar), and tea (blister blight caused by *Exobasidium vexans*).

All major disease problems in the above crops, when more
closely scrutinized, may have one or more of the previously
cited causes.

So far, mainly tropical, industrial, perennials have been
cited. Temperate crops such as apples, pears, plums and
smaller fruits have been omitted. The main reason for this
omission is that the history of these crops and its relation to
diseases is more complicated. In most of the major deciduous
fruit cultivars, vertical resistance is absent or not functioning
and varying levels of polygenetically inherited disease resistance
can be found within the range of cultivars. Vertical resistance is
chiefly found in wild progenitors and related species. For example,
monogenic resistances have been found against apple scab and apple
mildew in wild progenitors. In the wild populations these genes
may have functioned through multiline effects. Vertical resistances
to "scab" of apples have been used in breeding programmes and
some have broken down (Williams and Kuc, 1969). Although
Russell (1978) stated that the gene Vf was durable, the use of
monogenic resistance against scab will undoubtedly result in
further widening of the spectrum of vertical pathotypes. The
observation of Keep (1975) in part citing Knight and Alston
there is little evidence for the rapid breakdown of major gene

resistance to diseases in fruit crops is true. However, the reason
is probably not the durability of vertical resistance but the avoidance of it in fruit crop breeding programmes. For example, the
use of vertical resistance in black currants to *Sphaerotheca mors-
uvae* has quickly resulted in the appearance of vertical pathotypes
(Trajkovski and Pääsuke, 1976). However, some, probably
monogenic, resistances appear to be durable.

CONCLUSIONS

In order to understand durable resistance one must study
unimportant diseases. In "traditional" perennial crops most
diseases are unimportant. Their domestication has occurred
with a wealth of local cultivars which have, for a given location,
ample resistance to local diseases and pests. In the process of
domestication, heterogeneity of perennial crops has decreased
and the importance of vertical resistance (multiline effects) has
disappeared. Conditions causing or aggravating disease situations
are:

(a) Re-encounters of host and pathogen after prolonged geo-
     graphical separation.
(b) Introduction of pathogens from related pathosystems that
     have, however, co-evolved separately (new-encounter type a).
(c) Encounters with pathogens of unrelated species that are able
     to attack the crop (new-encounter type b).
(d) Changes in agricultural practices that disturb, in some way,
     the balance of the pathosystem.
(e) The use of susceptible material in selection, breeding or
     planting programmes when the susceptibility is not timely
     recognized, or is considered to be unimportant for reasons
     of yield or quality. Special problems are caused by systemic
     diseases with a long disease cycle; susceptibilities will only
     show up after many years and can, therefore, hardly be
     avoided.

Resistance to new-encounter pathogens (if sufficient resistance
can be found) is probably always durable. Durable resistance to
"traditional" diseases can be found in locations where host and
pathogen have developed together and have remained in close
association for a long time. In perennial crops, effective resistances can often be found in cultivars that have not been subjected
to determined plant breeding. This resistance is mostly horizontal.
Vertical resistance has only been found in some cases. The best
example is *Hemileia vastatrix* in Arabica coffee, a very heterogeneous crop in its country of origin. Until now, vertical resistance
has seldom been used in perennial crop breeding; some examples
are given in Table 1. In apples, a gene against *V. inaequalis* is
said to be "durable" (Russell, 1978). In black currant, some

Table 1.   Some perennial crops in which monogenic
             resistance has been used.

Crop	Disease	Vertical Pathotypes reported
Arabica coffee	Leaf rust	+
Apples	Scab	+
	Mildew	
Black currant	Mildew	+
Rubber	South American leaf blight	+
Grape	Downy mildew	

monogenic resistances also appear to be "durable" (Trajkovski
and Pääsuke, 1976). The genes for resistance against mildews
in apples and grapes have hardly been challenged. Horizontal
resistance in perennials appears to be "durable". Reports of
"breakdown" concern Sigatoka negra on banana in Central America
and Dutch elm disease. In both cases strains appeared with a
higher level of horizontal pathogenicity. Both diseases are,
however, introduced diseases (re-encounter type a) and it might
be speculated that strains with the maximum horizontal pathogenicity had not been previously introduced (Brazier, 1979). Nevertheless, some increase in horizontal pathogenicity cannot be
excluded when high levels of horizontal resistance are used. In
any event, the careful study of the pathosystems of traditional
"unimproved" crops, indicates that a situation of stable and
durable resistance can be reached.

REFERENCES

Alexander, V.T., 1980, Varietal resistance studies for hartrot
     of coconut, *in:* "Proceedings fourth meeting of the inter-
     national council on lethal yellowing", D.L. Thomas, F.W.
     Howard, and H.M. Donselman, ed., University of FLorida,
     Fort Lauderdale.
Boubals, D., 1959, Contribution à l'étude des causes de la
     résistance des vitacées au mildiou de la vigne, *Ann. Amelior.
     Plant.*, 9:5-233.
Boubals, D., 1961, Etude des causes de la résistance des vitacées
     à l'oidium de la vigne - *Uncinula necator* (Schw.) Burr.- et
     de leur mode de transmission hereditaire, *Ann. Amelior.
     Plant,* 11:401-500.
Boubals, D., 1966, Hérédité de la résistance au *Phylloxera
     radicicole* chez la vigne, *Ann. Amelior. Plant.*, 16:327-347.

Brasier, C.M., 1979, Dual origin of recent Dutch elm disease outbreaks in Europe, *Nature*, 281:78-80

Buddenhagen, I.W., 1977, Resistance and vulnerability of tropical crops in relation to their evolution in breeding, *in:* "The Genetic Basis of Epidemics in Agriculture", *Ann. N.Y. Acad. Sci.*, 287:309-326.

Chiarappa, L., 1979, The probable origin of lethal yellowing and its co-identity with other lethal diseases of coconut, Fifth Session of the FAO Technical Working Party on Coconut Production, Protection and Processing, Manila, Philippines, FAO,AGP:CNP/79/4.

Child, R., 1974, "Coconuts", Longman, London.

Condit, I.J., 1947, "The fig", Waltham, Mass.

Firman, I.D., 1964, Screening of coffee for resistance to Coffee Berry Disease, *East Afr. Agric. For. J.*,29:192-194.

Harries, H.C., 1978, Evolution, dissemination and classification of *Cocos nucifera*, Bot. Rev., 44:265-319.

Keep, E., 1975, Currants and Gooseberries, *in:* "Advances in fruit breeding", Janick, J., and Moore, J.N. ed., Purdue Univ. Press, West Lafayette.

Kranz, J., and Mogk, M., 1973, *Gibberella xylarioides* Helm & Saccas on arabica coffee in Ethiopia, *Phytopathol. Z.*,78:365-366.

Martinez Lopez, G., Jimenez, O., and Mena-Tascon, E., 1980, Flagelated protozoans in coconut palms in the south-west of Colombia, *in:* "Proceedings, fourth meeting of the international council on lethal yellowing", D.L. Thomas, F.W. Howard, and H.M. Donselman, ed., University of Florida, Fort Lauderdale.

McDonald, J., 1924, Annual report of the mycologist for the year 1922, *in: Kenya Dep. Agric. Ann. Rep.*, 1922:111-115.

Meiffren, M., undated, Contribution aux recherches sur la trachéomycose du caféier en Côte d'Ivoire, Inst. Fr. Café Cacao, mimeographed, 19 p.

Mulinge, S.K., 1973, Coffee Berry Disease in Ethiopia, *FAO Plant Prot. Bull.*, 21:85-86.

Olmo, H.P., 1978, Genetic problems and general methodology of breeding, *in:* "Génétique et amélioration de la vigne, Ile Symposium International sur l'Amélioration de la vigne, Bordeaux, 14-18 juin 1977", INRA, Paris.

Oudejans, J.H.M., 1969, Date palm, *in:* "Outlines of perennial crop breeding in the tropics", F.W. Ferwerda, and F. Wit, ed., Misc. Pap. Landbouwhogesch. Wageningen, 4.

Pieters, R., and Van der Graaff, N.A., 1980, Resistance to *Gibberella xylarioides* in *Coffea arabica*: evaluation of screening methods and evidence for the horizontal nature of the resistance, *Neth. J. Plant. Pathol.*, 86:37-43.

Pouget, R., and Kim, S.K., 1978, Etude methodologique de la résistance au Phylloxera: application a quelques croisements interspécifiques, *in:* "Génetique et amélioration de la vigne,

IIe Symposium International sur l'Amélioration de la Vigne, Bordeaux, 14-18 juin 1977", INRA, Paris.

Russell, G.E., 1978, Plant breeding for pest and disease resistance, Butterworths, London.

Rawther, T.S.S., and Pillai, R.V., 1972, Note on field observations on the reaction of coconut to root wilt, *Indian J. Agric. Sci.*, 42:747-749.

Rodrigues Jr., C.J., Bettencourt, J.A., and Rijo, L., 1975, Races of the pathogen and resistance to coffee rust, *Ann. Rev. Phytopathol.*, 13:49-70.

Saccas, A.M., 1951, La trachéomycose (carbunculariose) des *Coffea excelsa, neo-arnoldiana* et *robusta* en Oubangui-Chari, *Agron. Trop.*, Paris, 6:453-506.

Saccas, A.M., 1956, Recherches expérimentales sur la trachéomycose des caféiers en Oubangui-Chari, *Agron. Trop.*, Paris, 11:7-38.

Siddiqi, M.A., 1970, Incidence, development and symptoms of *Cercospora* disease of coffee in Malawi, *Trans. Br. Mycol. Soc.*, 54:415-421.

Storey, W.B., 1976, Fig (*Ficus carica*), *in* "Evolution of crop plants", N.W. Simmonds, ed., Longman, London, 205-208.

Toutain, G., 1965, Note sur l'épidemiologie du Bayoud en Afrique du Nord, *Al Awamia*, 15:37-45.

Trajkovski, V., and Pääsuke, R., 1976, Resistance to *Sphaerotheca mors-uvae* (Schw.) Berk. in *Ribes nigrum* L. 5. Studies on breeding black currants for resistance to *Sphaerotheca mors-uvae*(Schw.) Berk., *Swedish J. Agric. Res.*, 6:201-214.

Van der Graaff, N.A., 1981, Selection of Arabica coffee types resistance to Coffee Berry Disease in Ethiopia, Med. Landbouwhogesch., Wageningen, 81-11, 110 p.

Van der Graaff, N.A., and Pieters, R., 1978, Resistance levels in *Coffea arabica* to *Gibberella xylarioides* and distribution pattern of the disease, *Neth. J. Plant Pathol.*, 84:117-120.

Van der Plank, J.E., 1975, Principles of plant infection, Academic Press, New York, 216 p.

Weber, G.F., 1973, Bacterial and fungal diseases of plants in the tropics, Univ. of Flor. Press, Gainesville, 673 p.

Williams, E.B., and Kuč, J., 1969, Resistance in *Malus* to *Venturia inaequalis*, *Ann. Rev. Phytopathol*, 7:223-246.

BREEDING ELM, PINE AND CYPRESS FOR RESISTANCE TO DISEASES

L. Mittempergher and P. Raddi

Centro di studio per la patologia delle specie
legnose montane del C.N.R.
Firenze, Italia

INTRODUCTION

At our Laboratory of Forest Pathology in Florence, we have
been working for nearly ten years on three breeding programs for
disease resistance in trees which are of interest for Italy.

In this paper, we give more emphasis to aspects of the
durability of resistance than to the methodology of our work.

The three pathosystems we are dealing with are each different
from the other and interesting for some aspects. In the first one
(elm - *Ceratocystis ulmi*), where the host resistance is regarded as
of the horizontal type, new strains of the fungus have been found
that are definitely more pathogenic than the preexisting ones. The
second pathosystem (pine - *Cronartium flaccidum*) appears to be
mainly vertical, but no cases of breakdown of simple mechanisms of
resistance have been found so far. However, a case of breakdown of
a monogenic resistance was observed in sugar pine attacked by *C.
ribicola*. The third pathosystem (cypress – *Seiridium cardinale*) is
still little known and resembles a pure horizontal pathosystem.

ELM

The pathosystem elm-*Ceratocystis ulmi* is one of the best known
in the field of breeding for disease resistance in forest trees.
Selection of elms resistant to *Ceratocystis ulmi* (Buism.) C. Moreau
was begun by Christine Buisman in Holland in 1929. Heybroek, who
continues Buisman's work, established that resistance in elm is
polygenically inherited because it was built up step by step

277

through many crosses during several generations. Hypersensitive
reactions are not known and the ranking of elm clones for susceptibility to the disease varies little after inoculation with isolates
of different pathogenicity. This leads us to conclude that we are
dealing with a system resembling horizontal resistance prompting
the idea that there would be no danger of resistance breakdown.
The pathosystem seemed to have found a natural equilibrium in Europe,
until approximately 1970 when a new epidemic started which now
threatens the survival of all the European species of elm. Gibbs
et al. (1975) concluded that a more aggressive strain of the fungus
was causing the new epidemic: the more aggressive strain is characterized by different cultural features as well as by a higher
pathogenicity. The Dutch selection of elm, "Commelin", resistant
to the old strain was killed by the new one. To complicate things
even more, Brasier (1980) recently recognised two races within the
more-aggressive strain, which were named North American (NAN) and
Eurasian (EAN) according to the geographical areas in which they are
prevalent. These races can also be recognized by differences in
cultural features and in the level of pathogenicity. The English
elm (*Ulmus procera* Salisb.), which is moderately resistant to the
disease, is usually killed in the first year of infection by the
NAN race; if the tree does not die in the first year, disease recurs
in the second year and death follows. The EAN race of the aggressive
strain shows a wider range of variation, with some isolates significantly less pathogenic than the NAN race. On the English elm,
the isolates with the highest pathogenicity behave similarly to the
NAN isolates, but with less pathogenic isolates some recovery may
take place and disease recurrence in the following year seems to be
rare.

This appearance of strains of the pathogen much more aggressive
than the preexisting ones in a horizontal pathosystem seems in
contrast with what has been found and theorized on horizontal resistance, as in this system the pathogenicity should be at its highest
level. To understand the situation better, it appears useful to
recall the results of Brasier's work (1980) on the genetics of
pathogenicity in *C. ulmi*. Numerous isolates varying in aggressiveness and collected from three continents were crossed in the
laboratory. The distribution of pathogenicity in the progeny
indicated that, together with some additive components, there was
a large negative interaction between parents, as the mean of the
progeny fell significantly below the mid-point of the two parents.
Data from reciprocal crosses excluded the possibility that cyto-
plasmic effects could account for the interaction. The segregation
of the cultural characters such as growth rate and colony morphology
also showed extensive genetic differences between the parents.
Thus, Brasier proposed that the genetic system governing pathogenicity is complex, polygenic and quantitative and that the high
level of pathogenicity in the more-aggressive strain is conferred
by special gene combinations which on hybridization with the less-

aggressive strain are dispersed. In nature, there is no indication
of hybridization between the two strains, although Brasier examined
thousands of wild isolates. Thus the two strains are genetically
isolated and evolved separately, and could be regarded as different
subspecific taxa. As they are not elements of the same population,
we are no longer dealing with a horizontal pathosystem in which the
pathogen varies widely in pathogenicity, but with at least two
different taxa of pathogen. How could these taxa arise? We suspect
that the pathosystem is not purely horizontal as Brasier found qualitative differences in pathogenicity by crossing the two main strains
of pathogen. He suggested a molecular explanation of the difference
in terms of operon-type genetic system, i.e. in terms of major and
minor genes. "Substitution of an operator allele in the more-
aggressive strain with equivalent locus in the less-aggressive strain
might lead to half of the progeny being non-functional at the operon
concerned and substitution of alleles at the structural loci might
also impair the efficiency of the metabolic pathway involved"
(Brasier, 1980).

Secondly, Holmes (1965), Heybroek (1975), Gibbs *et al.* (1975)
and ourselves found interaction between host clones and pathogen
isolates indicating elements of specificity in the resistance of
the host. It must be stressed that several authors found that the
ranking of elm clones did not vary significantly when they were
inoculated with isolates of the two strains of pathogen. Particularly, no elm clone was ever found to exhibit more disease symptoms
after inoculation with the less-aggressive strain than after inoculation with the more-aggressive one. Nevertheless, for a couple
of years, we have noted some changes in the ranking of the hosts
(Table 1). The interaction between host clones and pathogen
isolates was significant at the 5 percent level four weeks after
inoculation, but only at the 10 percent level ten weeks after inoculation. Moreover, Townsend (1979) in his breeding work pointed
out the existence of specific combining ability in the transmission
of resistance in elm.

The aim of the breeding work in Florence is to introduce
resistance in *U. carpinifolia* by hybridization with the oriental
species *U. pumila, U. parvifolia,* and *U. wilsoniana*. Selection is
on the basis of disease symptoms that develop following inoculation
with a suspension of propagules of the more aggressive strain into
the xylem vessels of the tree. Variation in the amoung of inoculum
does not seeem to influence disease development. A defoliation and
dieback that does not exceed 15-20 percent of the crown and a
possible recovery of the diseased branches are considered to be
expressions of a good level of resistance. These trees should not
show disease recurrence in the following year. The speed of work
is restricted by the juvenile resistance of the elm which compel
us to delay inoculation and selection until the progeny from crosses
are three or four years old.

Table 1. Mean disease symptoms (transformed to arcsin % incidence) recorded on 5 Siberian elm clones four and ten weeks (in italics) after inoculation with 5 isolates of *C. ulmi* are reported along with their coefficients of variability. Between 7 and 10 self rooted 6-year-old trees were inoculated for each clone on May 14, 1981. Interactions between elm clones and isolates is significant at 5 percent level at the first date, but only at 10 percent level at the second date.

Clones of Siberian elm (*Ulmus pumila*)

Isolate	S 1 Mean	CV	S 13 Mean	CV	S 12 Mean	CV	S 2 Mean	CV	S 14 Mean	CV	Total row Mean	CV
239	32	14.7	42	13.8	28	16.6	21	15.7	40	9.3	32.5	6.8
	32	*13.5*	*46*	*14.7*	*28*	*15.4*	*26*	*14.3*	*40*	*9.3*	*34.5*	*6.5*
182	22	11.8	35	16.1	24	17.0	28	18.6	42	13.2	31	7.7
	22	*11.0*	*40*	*15.6*	*33*	*15.0*	*32*	*13.8*	*42*	*14.3*	*34*	*7.0*
65	22	11.3	20	34.5	10	34.3	16	27.0	15	36.5	16.5	16.4
	24	*11.2*	*21*	*32.5*	*12*	*23.6*	*22*	*18.8*	*16*	*28.4*	*19*	*16.6*
215	1	211.8	0	–	0	–	0.7	70.5	0.1	28.5	0.3	41.6
	2	*94.9*	*0*	*–*	*0*	*–*	*3*	*47.1*	*0*	*–*	*1*	*43.8*
166	24	10.2	28	19.1	26	14.0	38	9.68	30	11.3	30	5.8
	24	*9.8*	*38*	*15.6*	*28*	*17.9*	*40*	*9.8*	*32*	*9.3*	*32.5*	*6.1*
Total	20	10.1	25	13.3	18	13.3	21	11.6	25.5	11.3	22	5.4
Column	*21*	*9.6*	*29*	*12.9*	*20*	*12.5*	*25*	*9.6*	*26*	*10.8*	*24*	*5.0*

ANOVA

Sources of varieties	df	MS	F	MS	F
Among blocks	24	1712.68	12.43[++]	*1860.05*	*12.10[++]*
Among isolates	4	8494.99	61.65[++]	*9454.68*	*61.51[++]*
Among clones	4	495.79	3.60 N.s.	*515.43*	*3.35 N.s.*
Isolates x clones	16	321.33	2.33[+]	*297.54*	*1.93 N.s.*
Error	195	137.78		*153.71*	

Testing of selected clones and provenances of oriental species of elm is performed in several European countries in the context of EEC project for Dutch Elm Disease Control.

PINE

Blister rust of two-needled pines is caused by *Cronartium flaccidum* (Alb. et Schw.) Wint. and is widespread in several countries of Europe. In the past 30 years epidemic outbreaks of this disease have caused heavy damage in central and southern Italy in plantations and nurseries of Italian stone pine, maritime pine and Austrian pine. Infection occurs through cotyledons, primary and secondary needles and, possibly, tender stem tissue. The first signs of infection are yellow and red spots on the needles. The fungus can invade the main stem from infected branches. In experiments, diseased seedlings usually die within 2 or 3 years after inoculation.

Most recent studies indicate that each pathogen within the genus *Cronartium* that has been investigated with regard to variability shows sharp differences in levels of virulence. This variability has been demonstrated on both the pine host and the various alternate hosts (Powers, 1980). In the species *C. quercuum, formae speciales* have been found and the presence of "physiological races" has been suggested although the use of the term may be premature, since the conditions relating to rust collections and pine seedling families are very different from those used with the wheat rust pathogen.

In some of the hosts, simple mechanisms of resistance, controlled by oligogenes, have been found in the needle and in bark tissue. Kinloch (1980) found a hypersensitive resistance reaction in sugar pine 14 years ago, but recently this resistance gene has been matched by a race with a complementary virulence (Kinloch & Comstock, 1981). Although the study of the heritability of resistance and pathogenicity in these pathosystems is more difficult than that in cereals and their rusts because we deal with seedling populations instead of pure lines and with basidiospores coming from meiosis instead of urediospores, it can be said that a great part of the variability of resistance and pathogenicity is vertical. This is also confirmed by the interaction between populations of host and pathogen. On the other hand, mechanisms of resistance exist that are quantitatively inherited as, for example, "slow rusting" and resistance associated with aging. It is evident that a chance selection for a non-durable type of resistance to pine would have worse consequences than in annual crops. To avoid this danger, it has been suggested that the number of major genes for resistance should be identified and the greatest diversity of them should be accumulate in selections. In current tree improvement programs, the main emphasis has been on progeny testing for relative levels of resistance.

The research work in progress at our Laboratory deals with:
(a) some aspects of fungus biology; (b) methodology connected with
the breeding program particularly with volumes of inoculum; (c)
breeding rust resistant pines. From the work already performed on
two-year-old seedlings of the three main species of Mediterranean
pines, it has emerged that the Italian stone pine did not show any
appreciable level of resistance to rust, the maritime pine rated
intermediate while the Austrian pine was the most resistant (Raddi
et al., 1979). It is interesting to note that the habitat in which
Austrian pine lives is the most favorable for the development of
the disease. Concerning pathogenicity, we found that rust collections from the respective habitats of the three hosts (seaside,
hill and mountain) were all able to attack the three species but
were more adapted to the species prevalent in the zone of rust
collection (Mittempergher & Raddi, 1977).

The most common mechanism of resistance we found is what Hoff
& McDonald (1980) called "spotted-only-seedlings", i.e. seedlings
which became infected but did not show any disease symptoms on the
stem. In western white pine infected by *C. ribicola*, this mechanism
seems to be controlled by a recessive gene.

Recently, Raddi & Ragazzi (1980) studied the behavior of the
basidiospores of *C. flaccidum* on the needles of several species of
pine by a fluorescent labelling technique. The first observations
showed significant differences among the tested pine species for:
(a) basidiospore germination after 40 hours (from 44 to 93 percent);
(b) beginning of germination (from five to 20 hours); (c) germ tube
length after 40 hours (from 10 to 400 µm); (d) germ tube branching
(from 13 to 71 percent); (e) germ tube penetration between the
guard cells (from seven to 64 percent). From these observations,
it appears that the prevention of the needle infection may be
another resistance mechanism. Hoff & McDonald (1980) consider this
to be a threshold trait because it varies with the intensity of the
inoculation. In our studies we observed a few maritime pine
seedlings that developed the first fructifications of the fungus,
the spermagonia, but not the second ones, the aecia. We only
found one seedling of maritime pine that was able to recover completely from a well developed canker with aecia. Several infected
seedlings of maritime pine are still alive many years after inoculation. This "slow rusting" mechanism, as well as the restricted
development of spemagonia and complete recovery, are presumably
complex and under polygenic control.

At present, the most important objectives of our research work
are: (a) to understand the genetic control of the blister rust
resistance mechanisms found in maritime pine seedlings; (b) to combine all the heritable types of resistance by intraspecific crosses.

To reach these two goals we cross pines with these mechanisms following a North Carolina Model 2 mating scheme.

CYPRESS

The Italian cypress (*Cupressus sempervirens* L.) is a middle-eastern species naturalized in Italy where it has found several uses. In central Italy, it is one of the few species adapted to the reforestation of calcareous, clayey, dry, and shallow soils and is widely used as a windbreak. For centuries, the Italian cypress has been a characteristic component of the landscape in Tuscany and is one of the main elements in the constitution of parks, gardens, and avenues. The nursery production of cypress generates a conspicuous internal and external trade.

Presently, the cypress canker, caused by *Seiridium (Coryneum) cardinale* (Wag.) Sutton & Gibson, is the most destructive disease of cypress in the Mediterranean area. A survey carried out in the province of Florence in 1979 has shown that 25 percent of 4 million cypresses are dead or heavily damaged by canker and that another 25 percent shows the first stages of the disease. Eradication and chemical preventive measures may be suggested only in certain situations, i.e. in nurseries and for certain ornamental plantations, but they cannot be taken into consideration in forestry. Curative measures are not really effective (Panconesi, 1980).

In recent years, research has therefore been directed toward genetic improvement of the Italian cypress for resistance to canker. Local authorities in Tuscany are sponsoring our breeding program for five years and the EEC is supporting an international research program for four years.

First results of artificial inoculations carried out six years ago show that (Raddi & Panconesi, 1977): (a) 88 percent of the inoculated trees are severely damaged by the disease; (b) 10 percent have cankers of reduced size or are in process of recovery; (c) two percent recover completely from the canker; (d) there is ample susceptibility to the disease among the various cypress species; (e) immunity does not exist and the mechanism of recovery in two percent of the Italian cypresses is similar to that found in *C. glabra* and *C. bakeri*, which are the most resistant species; (f) so far, artificial inoculations carried out using numerous wild isolates did not show any consistent variability of pathogenicity (Table 2) (Raddi & Panconesi, 1981).

Although we are not sure that a substantial part of the pathogen population was introduced from America, for the moment, we may assume to be dealing with a horizontal pathosystem and durable resistance. All the trees which recovered from the canker after artificial inoculation, if reinoculated, were still able to

Table 2. Canker lengths in cm eight months after inoculation
          recorded on eight trees of Italian cypress inoculated
          with the following isolates of *Seiridium cardinale*:
          isolate No. 1 from *C. arizonica* (Firenze); isolate No.
          2, 3, 4, 5, 6 from *C. sempervirens* (Bari, Palermo, Firenze
          locality Monte Morello, Panzano, and Serpiolle
          respectively); isolate No. 7 from *Juniperus communis*
          (Firenze locality Capannuccia); isolate No. 8 from
          Cupressocyparis leylandii (Pistoia)

Tree	Isolate								%	Err.
	No. 1	No. 2	No. 3	No. 4	No. 5	No. 6	No. 7	No. 8		
1	9.5	6.5	4.0	9.8	8.5	9.5	6.3	9.2	7.9	0.7
2	7.2	8.1	6.3	8.0	6.4	7.0	8.5	6.0	7.2	0.3
3	12.5	2.5	8.0	5.0	4.5	6.5	8.0	7.3	6.8	1.1
4	5.3	6.9	5.2	2.2	7.6	8.2	8.6	6.7	6.3	0.7
5	10.4	4.5	6.5	8.1	8.4	6.5	7.5	8.0	7.5	0.6
6	10.0	10.5	4.7	6.5	8.8	8.6	6.9	5.0	7.6	0.8
7	16.0	7.1	6.2	8.2	9.0	5.3	10.1	9.2	8.9	1.2
8	8.2	5.0	5.0	6.0	7.2	5.5	7.0	7.2	6.5	0.4
X	9.9	6.4	5.8	6.7	7.5	7.1	7.9	7.3	7.4	0.3
Err.	1.2	0.9	0.4	0.8	0.5	0.5	0.4	0.5		

ANOVA

Source of variation	df	SS	MS	F
Among isolates	7	39.09	5.58	1.48 NS
Among plants	7	83.13	11.88	3.15 NS
Isol. x plants (sampling error)	49	184.63	3.77	
Total	63	306.85		

recover. In this situation, our program of genetic improvement is
based on: (a) identification of resistant cypresses by mass
artificial inoculations; (b) concentration of resistance genes through
intraspecific crosses and mass selection of the progeny; (c)
introduction of resistance genes through interspecific crosses.

Another aim of our research is to obtain information on the
genetic nature of the resistance mechanisms. To this purpose, in
1977, a program of controlled crosses was started according to the
North Carolina Model 2. Thirty-two Italian cypresses grouped in
four classes of susceptibility were crossed in all possible combinations with four testers. Each of the testers had a different degree
of canker susceptibility. The results of artificial inoculations
of the progenies in two localities will present information on the
value of h^2 of the resistance, of GCA and SCA and of the genetic
gain per generation. With the results of this program we will have
a better basis on which to redesign our breeding program.

REFERENCES

Brasier, C.M., 1980, Genetics of pathogenicity in *Ceratocystis ulmi*
      and its significance for elm breeding, Workshop on the Genetics
      of Host-parasite interactions in forestry, Wageningen, Sept.
      14-21.
Gibbs, J.N., Brasier, C.M., Nabb, H.S., and Heybroek, H.M., 1975,
      Further studies on pathogenicity in *Ceratocystis ulmi*. *Eur.
      J. For. Path.*, 5:161-174.
Heybroek, H.M., 1975, Dutch elm disease. Proceedings of IUFRO
      Conference, Minneapolis-St. Paul, USA, Sept. 1973 (Discussion
      on page 72).
Hoff, R.J., and McDonald, G.I., 1980, Improving rust-resistant
      strains of inland western white pine, *USDA Forest Service Res.
      Pap.* Int-245.
Holmes, F.M., 1965, Virulence in *Ceratocystis ulmi*, *Neth. J.
      Plant Pathol.*, 71:97-112.
Kinloch, B.B., 1980, Mechanisms and inheritance of rust resistance
      in conifers, Workshop on the Genetics of Host-parasite
      interactions in forestry, Wageningen, Sept. 14-21.
Kinloch, B.B., and Comstock, M., 1980, Virulent race of *Cronartium
      ribicola* to major gene resistance in sugar pine confirmed,
      Plant Dis., 65 (in press).
Mittempergher, L., and Raddi, P., 1977, Variation of diverse
      sources of *Cronartium flaccidum*, *Eur. J. For. Path.* 7:93-98.
Panconesi, A., 1980, Il cancro del cipresso in Toscana: aspetti
      biologici, *in:* "Il cipresso: malattie e difesa", V. Grasso
      e P. Raddi, ed., Firenze: 127-133.
Powers, H.R., 1980, Pathogenic variability within the genus
      Cronartium, Workshop on the Genetics of Host-parasite inter-
      actions in forestry, Wageningen Sept. 14-21.

Raddi, P., Mittempergher, L., and Moriondo, F., 1979, Testing of
    Pinus pinea and *P. pinaster* progenies for resistance to
    Cronartium flaccidum, *Phytopathology* 69:679-681.
Raddi, P., and Panconesi, A., 1982, Cypress canker disease in Italy:
    biology, control possibilities and genetic improvement for
    resistance, *Eur. J. For. Path.* (in press).
Raddi, P., and Ragazzi, A., 1980, The current status of pine blister
    rust (*Cronartium flaccidum*) disease resistance in two-needle
    pines in Italy, Workshop on the Genetics of Host-parasite
    interactions in forestry, Wageningen Sept. 14-21.
Townsend, A.M., 1979, Influence of specific combining ability and
    sex gametes on transmission of *Ceratocystis ulmi* resistance in
    Ulmus, *Phytopathology* 69:643-645.

RESEARCH FOR DURABLE RESISTANCE TO VASCULAR WILT DISEASE
(*Fusarium oxysporum* f.sp. *elaeidis*) OF OIL PALM (*Elaeis guineensis*)

J. L. Renard and J. Meunier[*]

IRHO Phytopathology Dept.
Dabou, Ivory Coast
*IRHO Plant Breeding Dept.
Montpellier, France

INTRODUCTION

Vascular wilt is the most serious oil palm disease in Africa. The fungus responsible, which develops in the xylem, is *Fusarium oxysporum* f.sp. *elaeidis*. In the ten years following planting, 50 percent of the trees can be affected in certain zones, damage being much greater in replanting than in extension to new areas.

The spread of plantations over several thousand hectares and the perennial nature of the plant rule out any direct action against the parasite, either in the soil or in the palm. By 1955, very marked differences in performances between families had been observed, and it was decided to carry out selection for crosses with durable resistance to Fusarium wilt under plantation conditions (Renard *et al.*, 1972). This character was therefore included in the general program of recurrent reciprocal selection set up to improve yield (Meunier & Gascon, 1972).

PRINCIPLE

The disease can be induced in the pre-nursery stage by artificial inoculation (2 x 10 propagules/plant applied to the soil around the collar). The test is read five months after inoculation. In a group of families tested at the same time, called a "series", each family is characterised by an index calculated as follows:

$$I = \frac{\% \text{ disease in family concerned}}{\text{Mean } \% \text{ disease in all families tested}} \times 100$$

The lower the index, the more resistant the family. In addition, parent trees will be characterised by the mean of the indices
of the crosses into which they enter, those with low indices being
retained for later crosses.

RESULTS

In a series comparing the value of nine types of cross between
progenies by selfing or by crossing between four parents, the index
of the crosses was in total agreement with the index calculated on
those of the parents on the basis of additivity (Table 1).

Table 1. Indices observed (and calculated) of the crosses between
progenies of four E. *guineensis* parents.

Parents	Indices	L2T Self	L7T Self	L2T x L7T
		70	130	100
D1OD	60	62 (65)	97 (95)	82 (80)
L4O4D	142	109 (106)	137 (136)	116 (121)
D1OD x L4O4D	101	87 (86)	116 (116)	92 (101)

This suggests, therefore, that additive factors are prepon-
derant in the inheritance of tolerance. A diallel test was carried
out with eight parents, and the crosses obtained were inoculated,
using two strains of the pathogen. No differential interaction between strain and families was observed. Moreover, the same applied
to other trials. Analysis of variance shows that there are highly
significant differences between parents as regards general combining ability, total variability being due mainly to additive
genes (67 percent of total variance); a small but significant
maternal effect was detected (Meunier et al., 1979). These results
show clearly that resistance is of the horizontal type, and that
selection based on the test using a sufficiently large dose of
inoculum will enable resistance characters to be distinguished.
Using the results of the tests, seed production is assured by reproduction of the best crosses, trees descended from selfs of both
parents being crossed between themselves. The test thus makes it
possible to classify each type of hybrid, characterise it by an
average index and by the distribution of the indices of the crosses
on either side of the theoretical mean value of 100 (Table 2).

The most tolerant crosses, with an index below 100, are field
planted in an infested zone on the IRHO Experimental Station, Dabou
(Ivory·Coast). Over a five year period, the mean annual percent of

Table 2.   Classification of hybrids reproduced on the basis of the
            test, and performance in the field.

Hybrids reproduced	No. of crosses tested	% crosses Index			No. of trees planted	Mean % diseased palms/yr	Mean Index of lines
		100	100	Mean			
D115D x L2T	92	74	26	79	7 869	1.0	64
D5D x L5T	33	55	45	95	5 048	1.1	81
D118D x L2T	55	67	33	87	9 922	1.9	75
D10D x L2T	428	60	40	91	31 603	2.0	78
D118D x L451T	44	57	43	100	8 987	2.1	81
L407D x L451T	65	15	85	132	1 469	3.8	90

wilted palms per hybrid is related to the mean indices observed
either on test of in the field (Table 2).

The distribution of crosses according to index categories,
explained both as those planted and as the percentage of families
(F) in each category with less than one percent disease per year is
shown in Table 3.

Table 3. Distribution of crosses according to index category.

Index of category	0 - 40	41 - 80	81 - 100	100
F (1%/year)	52.6	34.2	25.6	0
Total crosses planted	19	161	125	5

All these results show that it is possible to achieve notable
and durable improvement of the resistance of oil palm to vascular
wilt on the basis of the disease indices obtained in the pre-nursery
tests of potential parents and their hybrids.

DISCUSSION - CONCLUSION

There are families, however, in which the occurrence of wilt
is more frequent than would have been expected from the value of

the index.   One explanation, and the most likely, seems to lie in the
incubation period of the disease, as the test is in fact only
carried out over five months and consequently does not allow slow
incubation to manifest itself.   The physiological stage of the palm
may also be a factor, as well as the complexity of the natural
inoculum in the soil.   These hypotheses are being studied.

Apart from these exceptions, which remain to be explained, the
observations made for about 15 years on numerous crosses have shown
no sudden modification of degrees of resistance, and the qualifying
adjective "durable" can be applied to oil palm resistance.   Different
factors can affect the level of resistance, such as the potential
infectivity of the soil inoculum, fertilization, methods of cultivation, but they do not change the classification of the crosses in
relation to each other; and this is important, since not only must
this resistance last for a generation (about 20 years) but it must
be transmitted to the following generation.

The pre-nursery tests are carried out on all material collected
in prospections.   All populations have similar performances and show
great variability (Renard *et al.*, 1980).   However, it should be
mentioned that characters imparting total resistance (index = 0)
have been found in *Elaeis melanococca* populations from Central
America and Colombia, where vascular wilt is unknown.   These
characters are maintained in the F1 hybrid with *E. guineensis*; this
resistance is doubtless due to a single gene, or very few.

REFERENCES

Meunier, J., and Gascon, J.P., 1972, Le schéma général d'amélioration
    du palmier à huile a l'IRHO. *Oléagineux*, 27:1-12.
Meunier, J., Renard, J.L., and Quillec, G., 1979. Hérédité de la
    resistance à la fusariose chez le palmier à huile *Elaeis
    guineensis* Jacq., *Oléagineux*, 34:555-561.
Renard, J.L., Gascon, J.P., and Bachy, A., 1972, Recherches sur la
    fusariose du palmier à huile, *Oléagineux*, 27:581-591.
Renard, J.L., Noiret, J.M., and Meunier, J., 1980, Sources et
    gammes de resistance à la fusiose chez les palmiers à huile
    Elaeis guineensis et *Elaeis melanococca*, *Oléagineux*, 35:387-
    393.

INCOMPLETE RESISTANCE TO COFFEE LEAF RUST

Albertus B. Eskes

Food and Agriculture Organization of the United Nations
Instituto Agronomico of Campinas
13.100 Campinas, SP, Brazil

SUMMARY

Incomplete resistance to *Hemileia vastatrix* Berk. et Br. has been studied in Brazil under field, greenhouse, nursery and laboratory conditions. The following coffee populations were studied: breeding lines and cultivars of *C. arabica* , and *C. canephora* cultivars Kouillou, Icatu and Catimor. Results are presented on methodology, on the effect of environment, on leaf age and productivity, on the inheritance of resistance and on host-pathogen interaction studies. Indications for incomplete vertical resistance were obtained for the Icatu and Kouillou populations. Incomplete resistance to coffee leaf rust is often expressed by heterogeneous reaction types. The relation of the results to resistance theories is briefly discussed.

INTRODUCTION

Coffee leaf rust caused by *Hemileia vastatrix* is a major disease of *C. arabica* . Monogenic, specific, resistance of the hypersensitive type has been used to control the disease, but the durability of such resistances has been disappointing. Although strategies have been suggested to improve the efficiency of vertical resistance (Nelson, 1973), the perennial nature of the crop makes such strategies difficult to implement. Partial resistance or some types of partial resistance may possess a better durability. There-fore, in 1976, the Instituto Agronomico of Campinas, Brazil and the Food and Agriculture Organization of the United Nations begun a project aimed at the obtention of high levels of partial resistance (sensu Parlevliet, 1979). In this paper, the results of this project are described.

THE PATHOSYSTEM

The Natural Pathosystem

The center of origin of *C. arabica* is most likely Ethiopia, where the species grows at an altitude from 1200 to 2100 meters (Van der Graaff, 1981). The origin of coffee leaf rust is possibly in Central or North East Africa (Chaves et al., 1970).

Coffee leaf rust is only a minor disease in Ethiopia (Van der Graaff, 1981). However, when *C. arabica* is grown in the tropical lowlands, it may become severely diseased. The apparent balance between host and pathogen in Ethiopia may be due to: (a) a low adaptive capacity of coffee leaf rust to Ethiopian conditions, (b) control by natural enemies, and/or (c) control by efficient host resistance. No adequate information is available on the adaptive capacity of *H. vastatrix* to Ethiopian conditions.

Concerning biological control, a hyperparasite, *Verticillium hemileia*, occurs in Ethiopia. It is however also found in other countries, including Brazil (Locci et al., 1971). For example, in Brazil, *Verticillium* appears to be of little importance in the control of the disease, although this is not necessarily the case in Ethiopia.

Concerning resistance, four genes for race-specific resistance have been identified in *C. arabica* accessions from Ethiopia: SH_1, SH_2, SH_4 and SH_5 (Rodrigues et al., 1975). The most common genes present in Ethiopia seem to be SH_1 and SH_5 (Meyer et al., 1968). However, many plants do not have any recognizable major resistance gene. Therefore, the resistance of *C. arabica* in Ethiopia may be mainly due to minor genes.

The second economically important coffee species is *C. canephora*. It is a broad coffee species with several varieties and subvarieties (Chevalier, 1947) that grow wild in West and Central African countries. The Robusta variety, which originates from Zaire, is one of the most resistant varieties of *C. canephora*. The West African Kouillou variety has a lower resistance level to coffee leaf rust. In Indonesia, Robusta proved to be resistant from the beginning, although not completely immune. Although in this country some adaptation of the rust to *C. canephora* may have occurred (Cramer, 1947), the resistance of Robusta has apparently remained sufficient over many decades.

The products of natural and artificial hybrids between *C. canephora* and *C. arabica*, named Catimor and Icatu, possess several major genes for resistance (Rodrigues et al., 1975). In addition, incomplete types of resistance are frequent in these hybrid populations. These observations suggest that major as

well as minor genes may be important for an efficient genetic
control of coffee leaf rust in *C. canephora*.

The Agricultural Pathosystem

Severe epidemics of coffee leaf rust may occur in coffee
plantations of *C. arabica*, especially in the tropical lowlands (up
to about 1000 to 1200 metres above sea level). The disease was
found in Brazil in 1970 and has now spread to most South and
Central American countries. The coffee cultivars grown in
America possess little genetic variation and all of them are
susceptible to coffee leaf rust race II. The most widely distributed
race of *H. vastatrix* is race II, possessing the $v5$ virulence gene.
This gene matches the $SH5$ resistance gene, which is present in
almost all coffee cultivars of *C. arabica*.

In Brazil, yield losses due to coffee leaf rust are estimated
to be about 30 per cent. The epidemic follows a well-defined
annual pattern for most of the Brazilian coffee growing areas.
Maximum disease incidence occurs at harvest time (June, July),
which is at the beginning of the dry season. At this point, a
considerable natural or pathogen-induced leaf fall occurs and the
epidemic declines. Its lowest point is reached in November and
December, two months after the onset of the rainy season (Monaco,
1977).

The use of the traditionally known resistance genes SH_1,
SH_2, $SH3$, SH_4 and $SH5$ has not efficiently controlled the disease.
In Brazil, seven different races were detected between 1970 and
1980, and only the SH_3 gene, which is probably derived from a
hybridization with *C. liberica* (Rogrigues et al., 1975), is still
effective. However, in India this gene apparently lost its effectiveness after seven years of large scale commercial use
(Visveswara, 1974). Three races with virulence to *C. canephora*
resistance genes have been detected so far by us at the Campinas
research station (Eskes et al., 1981b).

Recent breeding efforts for resistance to *H. vastatrix* are
mainly directed to the hybrid populations Catimor and Icatu. Many
plants of these populations have resistance to all known rust races,
while other plants are only resistant to some races (Rodrigues et
al., 1975). So far, in Brazil, only one race has been found that
matches the resistance of a few Icatu and Catimor plants.

STUDIES ON METHODOLOGY

<u>Assessment of Resistance</u>

Incomplete resistance to coffee leaf rust was estimated by observations on natural disease incidence in the field and through artificial inoculations in the field, nursery, greenhouse and laboratory. For each method, illustrated assessment scales were designed permitting a rapid evaluation of disease incidence or severity (Eskes and Toma-Braghini, in press). A 0 to 9 scale was also proposed to evaluate the infection type. This scale includes a wide variation for heterogeneous infection types, which were observed with a certain frequency in our experiments. The scale values 0 to 3 indicate a variation from immune to resistant, 4 and 5 indicate heterogeneous infection types with a low sporulation intensity, 6 and 7 indicate heterogeneous infection types with increased sporulation intensity, and 8 and 9 indicate a susceptible reaction, with mostly profusely sporulating lesions.

A laboratory test for resistance, using leaf disc inoculations (leaf disc inoculation test), has been standardized and amply tested. In the tests, leaf discs of 1.7 cm. diameter were inoculated with a 0.025 ml droplet of a uniform uredospore suspension in distilled water. After incubation of the leaf discs in the dark for 24 h, at $22 \pm 2^{\circ}C$ and in water saturated air, the droplets were allowed to dry within a few hours. Afterwards the discs were maintained at $22 \pm 2^{\circ}C$, in water saturated air, and under moderate light conditions. Observations were made on resistance parameters like latency period (number of days from inoculation to the date that 50 per cent of the leaf discs that ultimately sporulate, are sporulating), infection frequency, infection intensity, and infection type.

The method gave reliable results for the determination of complete as well as incomplete types of resistance. It was also successfully applied for race identification and for the determination of the effect of leaf age and light intensity on resistance. When individual plants of different coffee populations were tested, significant coefficients of correlation were obtained between the results of the leaf disc inoculations and those of simultaneous field or greenhouse inoculations (Table 1).

The advantages that make the leaf disc test attractive for research purposes and large scale resistance screening are:

(a)   the method is fully quantitative,
(b)   little time and space are needed,
(c)   little pathogen and host material is needed,
(d)   inoculations are mostly successful,
(e)   many host-pathogen combinations can simultaneously be studied in one experiment, and

Table 1. Components of resistance: correlation coefficients between the resistance parameters obtained in three experiments that were conducted in the field, greenhouse and laboratory.

	Field inoculations			Average field disease	Greenhouse inoculations		
	infection type	infection frequency	latency period		infection frequency	latency period	leaf retention
Lab inoculations — infection type	.75**			.58*			
infection frequency		.94**	.01	.59*			
latency period		.76**	−.47	−.71*	.66**	.80**	
Average field disease incidence	.68*	.50*	−.38		.38	.68*	.75**

* P (r = 0) ≤ 0.05
** P (r = 0) ≤ 0.01

no underscoring: results from 15 *C. canephora* cv. Kouillou plants,
underscored once: 70 F_3 plants of Agaro x Catuai,
underscored twice: P_1, P_2, F_1, 10 F_2 plants and their F_3 progenies of Mundo Novo x Ibaarê.

(f)   the resistance of field plants grown far away from a
        laboratory can be determined by remittance of leaves
        through the mail.

PARTIAL RESISTANCE OF *C. ARABICA*

Commercial Cultivars

The Brazilian *C. arabica* cultivars generally show little
genetic variability. The cultivars Mundo Novo and Catuai possess
some variability, because they have been derived from varietal
crosses. These two cultivars are genetically closely related and
have a high productivity (Carvalho et al., 1969).

Greenhouse inoculations of 75 single tree progenies of the
Catuai cultivar showed relatively little variation for latency period.
Significant differences were found for infection frequency (number
of lesions per leaf) with the most susceptible and most resistant
progeny showing a two-fold difference. These results, however,
were not completely repeatable and correlation with the field incidence of rust on these progenies was poor. Furthermore, field
observations indicated that variability for partial resistance in the
Catuai, and also the Mundo Novo cultivar, is too low to obtain
significant selection progress within these cultivars (see also
under the effect of leaf age, etc.).

The Ibaarê Coffee Type

At the Campinas Research Station, the most susceptible variety
of *C. arabica* is a selection of the Harar coffee type, called Ibaarê.
In Africa, the Harar coffee type is also highly susceptible to coffee
leaf rust (Sylvain, 1955; Van der Graaff, 1981). Under Brazilian
conditions, the Ibaarê selection consistently shows a much higher
incidence of rust lesions than commercial cultivars. In the absence
of the rust, the yield potential of the Ibaarê type is equal to that of
the Mundo Novo cultivar.

The components of partial resistance were analyzed to explain
why Ibaarê is more susceptible than Mundo Novo (Tables 2 and 3).
In greenhouse inoculations in the cool and dry winter months,
Ibaarê showed a consistently longer leaf retention period, a shorter
latency period, and a higher infection frequency than Mundo Novo.
In experiments in summer, these components did however not
differ. In leaf disc tests, the latency period was more consistently
shorter for Ibaarê (Table 3).

Table 2. Components of resistance in *C. arabica*: summarized results of several experiments in the field, greenhouse and laboratory.

Coffee type	Field disease incidence	Components of resistance				
		infection frequency	latency period	sporulation intensity	leaf retention period	early necrosis
Agaro	low/medium	+(a)	+	+/++	+/++	+
Mundo Novo	medium	+	+	++	++	++
Ibaarê	low	+/++	+/++	++/+++	+++	++

(a) Increasing numbers of "+" indicate increasing susceptibility. The denotation "+/++" or "++/+++" indicates variation in the results between experiments.

Table 3. Components of resistance in *C. arabica* coffee
as measured in the greenhouse.

Coffee type	Disease incidence in the field (1-10 scale)	Components		
		infection[a] frequency	latency[b] period	leaf[c] retention
Ibaarê	9	21	44	77
Mundo Novo	3	11	62	42
F_1	6	13	57	73
F_3 (d) (10 progenies)	4 – 9,5	5 – 23	45 – 62	61 – 85
Correlation coefficient with disease incidence in the field		0.38	0.68[*]	0.75[*]

(a) Infection frequency: Number of lesions per leaf.
(b) Latency period: Number of days between inoculation
    and sporulation of 50 per cent of the lesions that
    ultimately sporulate.
(c) Leaf retention: Number of days from inoculation to
    50 percent leaf drop.
(d) F_3: Width of variation. Field data from F_2 mother
    trees.

The Agaro Coffee Type

The Agaro coffee type consistently shows a lower disease
incidence in the field than commercial cultivars. This coffee
type carries the SH4 resistance gene, but the matching race
(v4v5) is widely present in the Campinas coffee rust population.

The components of resistance were analyzed with the race
v4v5 in both the laboratory and the greenhouse. The differences
between the Agaro type and commercial coffee types were only
minor (Table 2). In field and greenhouse inoculations, however,
early necrosis of lesions on certain Agaro plants was observed.
This early necrosis may be related to environmental conditions
as it cannot be demonstrated in leaf disc tests. Plants showing
this characteristic usually perform poorly in the field.

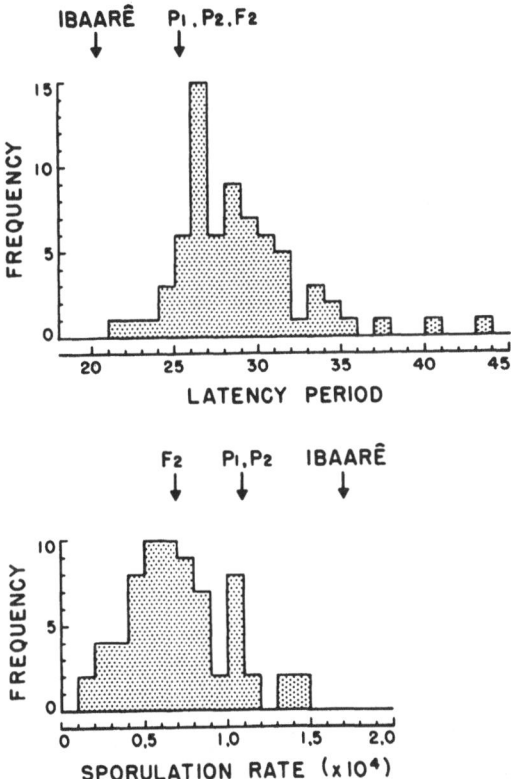

Fig. 1. Variation for latency period and sporulation
intensity in leaf disc tests of 70 F3 progenies
of the Agaro x Catuai cross, compared with both
parents and the highly susceptible control Ibaarê.
Abscissa: frequency of trees in each class.

        Seventy F3 populations of an Agaro x Catuai cross were
inoculated in the greenhouse and the laboratory in 1978. Substantial transgressive segregation for latency period, infection fre-
quancy and sporulation intensity was observed under both conditions
(Figure 1). Coefficients of correlation were significant between

the components of resistance observed in the greenhouse and laboratory experiments. A selection program has been started for increased levels of partial resistance, based on the most resistant F_3 populations.

Other Introductions of *C. arabica* from Ethiopia

Campinas has a collection of 200 coffee introductions that were collected by the FAO mission to Ethiopia in 1964 (Meyer et al., 1968). In general, the productivity of these introductions is very low; great variability for growth habit and adaptive ability exists.

Seventy three individual plant progenies grown from seeds were inoculated in the greenhouse and the leaf disc test was applied to the mother plants (Eskes and Toma-Braghini, 1980). Normally, the level of partial resistance of these progenies was equal or slightly higher than that of the commercial Brazilian cultivars. The variation for infection frequency was considerable, but only a slight variation for latency period was observed. The correlation between the responses in the greenhouse and in the leaf disc test was not significant. Rust severity in the field varied greatly among the introductions. This may be due to differences in resistance, and also to the different growth habits and the variation in yield potential. (See below).

Effect of Leaf Age, Light Intensity and Productivity on Partial Resistance of the Mundo Novo and Catuai Cultivars

Leaf age does not greatly affect resistance in commercial *C. arabica* cultivars. After inoculation, however, the infection frequency may be twice as high on old leaves than on the youngest fully grown leaves and the latency period may be a few days longer on the young leaves.

High light intensities before inoculation significantly increase the infection frequency of Mundo Novo coffee seedlings. However, high light intensities after inoculation were found to decrease the infection frequency (Table 4). The latter observation may be due to overheating of the leaves by direct sunlight which may kill the fungus in the leaf. A significant interaction was observed between the effects of light intensity before and after inoculation (Table 4).

In three experiments the effect of productivity on field susceptibility was demonstrated. In field observations on 14 lines of the Mundo Novo cultivar, a significant positive correlation between yield and disease incidence was found (Figure 2). In the Catuai cultivar, the epidemic of coffee leaf rust was followed on branches with and without berries within the same tree. Leaves on branches with berries had twice as many lesions as leaves on branches without berries. This effect was confirmed by leaf disc inoculations (Eskes and Zink de Souza, 1981).

RESISTANCE OF *C. Canephora* cv. KOUILLOU

Field Resistance

The Kouillou cultivar is commercially grown at relatively low altitudes in the State of Espirito Santo, Brazil. Although nearly every plant shows sporulating lesions, leaf rust is never severe. Considerable differences in disease severity were observed among individual plants of the collection at the Campinas

Table 4. The effect of light intensity, before and after inoculation, on the number of rust lesions per leaf of *C. arabica* cv. Mundo Novo seedlings.

Shade before inoculation (%)	Shade after inoculation (%)			Average
	76	57	34	
76	16	19	15	17
57	32	21	11	21
0	49	40	20	36
Average	32	27	15	

ANOVA

Source	df	ms	F	p
Repetition	6	97	1.7	n.s.
A: shade before	2	2050	35.6	0.005
Error (a)	12	58		
B: shade after	2	1614	24.1	0.005
A x B	4	371	5.6	0.005
Error (b)	36	67		

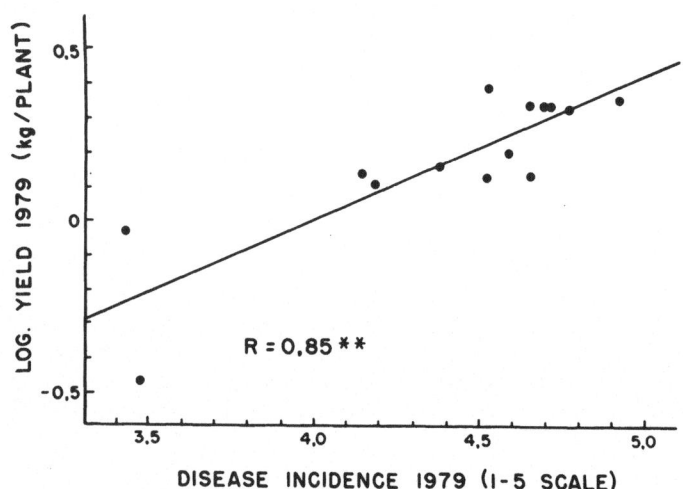

Fig. 2.   Relationship between yield and disease
incidence of 14 lines of Mundo Novo.

Research Station (Figure 3). As will be discussed later, some
immune or nearly immune plants were shown to possess race
specific resistance (Eskes et al., 1981b).

Repeated leaf disc inoculations carried out on 23 plants
correlated well with the field observations, indicating that the
differences in severity were caused by variability for resistance.
Considerable variation was observed for all components of partial
resistance, especially infection frequency and sporulation intensity.
Disease severity in the field correlated best with infection frequency and infection type (Table 1).

Inheritance of Resistance

Progenies of crosses among plants with different levels of
resistance were inoculated in the nursery (Eskes et al., 1981c).
From the segregation ratio for resistance in these progenies, it
was concluded that one nearly immune parent possessed monogenic resistance. (This resistance is already matched by a new

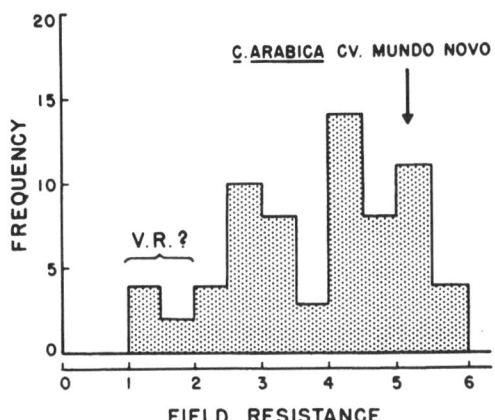

Fig. 3.   Variation for field resistance, measured over
a five-year period, of 68 plants of the Kouillou
cultivar of *C. canephora* . Classes 1 and 2
indicate immunity and resistance, respectively.
Classes 3 to 6 indicate increased incidence of
sporulating lesions.

race of *H. vastatrix*). Another parent, which showed a low infection
type on adult leaves but with a high type on old leaves, also appeared to have monogenic resistance. Two parents showed high
levels of probably polygenic resistance, which was mainly expressed by a low infection frequency, but, under certain conditions
also by a lower infection type or by a heterogeneous infection type.
This resistance was more pronounced on adult leaves than on old
leaves.

Effect of Leaf Age and Light Intensity

In leaf disc experiments a significant effect of leaf age on
the resistance of the Kouillou cultivar was found (Table 5). In
general, the adult leaves were more resistant than the very young
leaves or the old leaves (Eskes et al., 1981c). Depending on the
plant, this resistance was expressed by a lower infection frequency,
by a longer latency period and/or by a lower infection type. In
extreme cases, the adult leaves were immune, whereas sporulating

Table 5.   Effect of light intensity before inoculation and
           leaf age on the percentage of sporulating leaf
           discs of four plants of the *C. canephora* cv.
           Kouillou, two plants of Icatu and of *C. arabica*
           cv. Catuai.

Plant code	Average field resistance (1-6)	Leaf age		Light intensity	
		adult	old	shade	sun
Kouillou 66-3	2.3	7	54	14	47
Kouillou 70-11	2.3	19	63	36	46
Kouillou 70-14	2.8	6	53	21	38
Kouillou 68-15	5.3	72	50	43	79
Icatu H 4782-13-72	4.8	37	58	39	56
Icatu H 3851-2-689	4.8	44	49	29	64
Catuai H 2077-2-5-81	5.5	76	84	70	90

lesions usually developed on the old leaves. The effect of leaf age
decreased when the general resistance level of the plant was lower.

   The effect of light intensity was studied in greenhouse, nursery and laboratory experiments. Light intensity in the nursery
was high (66 per cent of the total radiation) and low in the greenhouse (15 per cent of the total radiation). Air temperatures in the
two environments were similar. The infection type of the Kouillou
plants was significantly higher in the nursery than in the greenhouse (Table 6). Interactions between environments and genotypes
were also significant, mainly because the increase in susceptibility
was greater for the more resistant genotypes. This influence of
light intensity was confirmed by leaf disc tests on shaded and un-
shaded leaves collected in the field. In this experiment, both the

Table 6. Infection types (0 – 9 scale) obtained by
          inoculation of Kouillou plants, under
          greenhouse and nursery conditions, and
          average field resistance score (1 – 6 scale).

Plant genotype	Greenhouse (low light intensity)	Nursery (high light intensity)	Average field resistance
66–1	2.8	6.0	2.1
66–3	5.6	6.6	3.1
70–14	7.2	7.4	3.1
68–10	9.0	9.0	4.5
67–12	6.2	8.2	4.6
68–15	7.6	8.8	5.3
Mundo Novo	8.0	8.8	5.2

ANOVA (Greenhouse/Nursery):

Source	df	ms	F
A: Genotypes	6	23.83	36.1**
B: Environment	1	27.66	41.9**
A x B	6	2.96	4.5**
Error	56	0.66	

components of partial resistance and the infection type were
affected by light intensity (Table 5).

RESISTANCE OF THE ICATU POPULATION

Introduction

The Icatu population originates from a cross between a tetraploid *C. canephora* (4n = 44) and *C. arabica* (2n = 44). The present
population results from several backcrosses with *C. arabica*
(Monaco, 1977). Progenies of the most promising plants were
selected and are advanced to obtain homogeneous productive lines
with resistance to coffee leaf rust.

Complete vertical resistance expressed by hypersensitive
reactions was found in about 60 per cent of all plants. The susceptible plants of the population showed great variability for disease severity in the field. Plants with very high levels of incom-
plate resistance were found but plants were also present that are
more susceptible than the Mundo Novo cultivar.

Expression of Incomplete Resistance

Most Icatu plants with intermediate levels of resistance
showed heterogeneous reaction types, after artificial inoculations,
which were characterized by the simultaneous occurrence of non-
sporulating and sporulating lesions on the same plant and even on
the same leaf. Greenhouse inoculations of 26 grafted Icatu plants
showed that the percentage of lesions producing spores was better
correlated with field incidence (r = 0.68) than with latency period
(r = 0.60) or with the total number of sporulating lesions (r = 0.58).

The above heterogeneous reactions of Icatu plants were probably not due to variation in the fungus, as reinoculation with spores
from the best sporulating lesions did not result in increased susceptibility. Thus, the heterogeneous reaction seems rather a type
of incomplete resistance, which can be affected by light intensity
and leaf age. These effects were confirmed in leaf disc tests
(Table 5).

THE CATIMOR POPULATION

This population is derived from a cross between Caturra
and the Hybrid of Timor, which is a natural hybrid between *C.
canephora* and *C. arabica* . It contains several vertical resistance
genes (Rodrigues, 1975), incomplete resistance is also frequently
observed. Seedlings of susceptible plants showed little variation
for resistance in greenhouse tests. Seedlings of intermediately
resistant or resistant plants showed considerable variation for
infection type. Heterogeneous reactions were common. Seedlings

of a few plants showed hypersensitive reactions on the young leaves, whereas the old leaves were normally susceptible. Seedlings from other plants showed heterogeneous reactions on leaves of all ages. In one descendance, a 3 R: 5 MR: 3 S segregation was obtained, suggesting monogenic inheritance of the resistance, with incomplete dominance (unpublished observations).

These results suggest that the heterogeneous reaction type in the Catimor population may at least partly be due to major genes that give incomplete protection.

INCOMPLETE VERTICAL RESISTANCE

The existence of incomplete vertical resistance was demonstrated by inoculations with compatible and incompatible races. In *C. arabica* it was found that the heterozygote SH_4 sh_4 gives only partial protection to incompatible races under high light intensities: in inoculations with race II (V_5) in the nursery, an F_2 population segregating for SH_4 showed many intermediate (heterogeneous) resistance types and only 25 per cent of the seedlings were completely resistant. In comparable experiments in the greenhouse (lower light intensity), the same population showed the expected 1:3 ratio.

As indicated earlier, in the Icatu population, plants were often found that showed a heterogeneous reaction type after inoculation with race II (V_5). Based on these reactions, plants were classified as R, MR, MS and S. In 1978, a new rust race (isolate 2) was detected that overcame simultaneously the resistance of certain MR and MS plants indicating the presence of incomplete specific resistance (Table 7).

In the Kouillou population, two new races were discovered in 1980. In the leaf disc test, both these races caused more disease on certain MR, MS and S plants (Table 8) indicating that in this coffee population incomplete specific resistance was also present.

INCOMPLETE VERTICAL PATHOGENICITY

Three cases of incomplete vertical pathogenicity were detected. A rust race was found that partly matched the resistance of the SH_3 gene of *C. arabica*. Entire plants or leaf discs of plants carrying the SH_3 gene showed heterogeneous or high infection types after inoculation with this race in the laboratory (temperature $22 \pm 2°C$, low light intensity). However, the same plants only developed low infection types when inoculated with this race in the nursery or greenhouse. In the field, this race is found on SH_3 sh_3 plants with some sporulating lesions.

Table 7.   Infection types (0 – 9 scale) obtained in leaf disc
           experiments of Icatu and Catimor plants inoculated
           with two field isolates of coffee leaf rust and race
           II. The field isolates were derived from spores
           collected from lesions on resistant hosts and
           inoculated on resistant hosts in leaf disc tests.
           The spores of the resulting lesions were used in
           this experiment.

Pathotypes	Mundo Novo	Icatu plants				Catimor plants	
		a	b	c	d	a	b
race II	9	1	4	6	7	2	9
isolate 2	8	9	9	9	9	5	6
isolate 3	9	6	6	9	9	5	5

Table 8.   Relative resistance of 7 Kouillou plants as
           measured by the percentage of sporulating
           leaf discs. Three field isolates of coffee
           leaf rust were compared with race II.
Entries: relative resistance = $\frac{T}{C}$, where T is the
           result of the tester and C the result of the
           susceptible control Mundo Novo (Turkensteen, 1973).

Pathotype	Kouillou genotype						
	66–13	67–12	67–1	69–7	69–15	66–3	68–7
race II	1.0a	0.9a	0.6b	0.3a	0.1b	0.1a	0.5b
isolate 11	1.0a	0.5b	0.1c	-0.5b	-0.7c	0.0a	0.7ab
isolate 10	0.0b	1.0a	1.0a	0.5a	0.7a	0.1a	0.9a
isolate 12	0.8a	0.9a	0.8b	0.5a	0.7a	0.1a	0.9a

In each column, entries marked with the same letter did not
differ significantly at $P \leq 0.05$.

Indications for incomplete vertical pathogenicity were also observed in relation to resistance genes of the Kouillou and Icatu populations. Intermediate pathogenicity of isolate 12 on the Kouillou plant 66-12 is shown in Table 3. Intermediate pathogenicity of isolate 3 on Icatu plants a and b is shown in Table 7.

SIDE EFFECTS OF INCREASED VIRULENCE

Plants without the resistance factor that was matched by the new rust races detected on Icatu and Kouillou were inoculated with those new races and the common race II. The new rust races showed a significantly lower pathogenicity than race II on the Kouillou plants 67-12, 67-1, 68-7 and 69-15 (Table 8). This effect of the extra virulence was not observed on other plants like 66-3. Also, the Icatu isolate 2 consistently showed a lower pathogenicity than race II on Mundo Novo. This lower pathogenicity was expressed by a lower sporulation intensity, a longer latency period and often by a heterogeneous reaction type. The lower pathogenicity of isolate 2 was even more pronounced towards certain plants of the Catimor group (Table 7).

DISCUSSION AND CONCLUSIONS

Methodology

The leaf disc inoculation test is a very valuable tool for determining quantitative or qualitative differences in resistance of coffee plants to *H. vastatrix*.

The proposed 0 to 9 assessment scale for the infection type includes a range for heterogeneous reaction types, which seems realistic for the coffee - *H. vastatrix* relationship.

Resistance of *C. arabica*

In comparison with the more susceptible coffee type Ibaarê, the Brazilian coffee cultivars possess a certain level of partial resistance that is expressed by a shorter leaf retention period, a longer latency period and a lower infection frequency. The latter two components of resistance seem less effective under high light intensities and relatively high temperatures and when the productivity of the coffee trees is high.

Increased levels of partial resistance were found in F_3 progenies of an Agaro x Catuai cross. Further studies are needed to determine whether this resistance is sufficient under

field conditions and in plants with high productivity. In the Agaro
population, a factor conditioning early necrosis of the lesions was
also detected. This factor appeared to be more prominent in plants
with a low productivity, which makes it doubtful if this type of
resistance can be incorporated into a high yielding coffee cultivar.

Resistance of *C. canephora* cv. Kouillou

The incomplete resistance of this cultivar is highly variable
in its expression; resistance is influenced by leaf age, light intensity and season. The level of field resistance of many plants is
related to the infection type. The most important components of
resistance are infection frequency and sporulation intensity.

No clear distinction between partial resistance, *sensu*
Parlevliet (1979), and other types of incomplete resistance was
possible. Many plants normally show high infection types, but,
under certain conditions, low infection types can also be observed
on these plants. Some plants with high levels of incomplete resistance that appear to be inherited polygenically have been identified.
Those plants have been used in crosses with *C. arabica* as a start
for a breeding programme based on the resistance of the Kouillou
cultivar.

Resistance in Derivatives of *C. canephora*

The Icatu and Catimor populations showed a similar resistance expression as Kouillou. In all three populations, heterogeneous reaction types were frequent. Incomplete vertical resistance was demonstrated in Icatu and Kouillou. Certain types of
vertical resistance in this population seem to operate at different
levels of effectiveness, depending on leaf age, light intensity and,
possibly, also the genetic background. Therefore, no clear distinction exists between this resistance and partial resistance,
sensu Parlevliet (1979).

Host-Pathogen Interactions

Testing of the new races detected in the Icatu and Kouillou
populations together with race II, showed a decreased pathogenicity of the new races on certain coffee types but not on others.
These new races are probably derived from race II and differ from
it in only one gene for virulence. If this hypothesis is true, one
should conclude that this gene has negative side effects for pathogenicity to other coffee genotypes. This phenomenon would confer a
specific type of stabilizing selection, in which the negative effect
of the virulence genes can be detected only in relation to some

coffee types but not in relation to all of them. If this effect cannot
be overcome by subsequent mutations in the fungus, the implication
is that certain genotypic combinations for resistance to coffee leaf
rust may be durable, although they possess vertical resistance
genes that can be individually overcome by the fungus.

The Relation of the Results of Resistance Theories

The results showed that incomplete specific resistance to
coffee leaf rust is rather frequent and that *H. vastatrix* also
possesses races with intermediate pathogenicity to certain vertical resistance genes. This wide variation for incomplete vertical resistance and pathogenicity apparently also occurs in other
host-pathogen relationships (Parlevliet, 1979) and it may explain
a great part of the total variation observed for incomplete resistance. Therefore, durable resistance may not be obtained by just
selecting any genotype with incomplete resistance.

It is suggested that all types of resistance and pathogenicity,
including the incomplete types, should be studied simultaneously
in order to clarify the host-pathogen relationship. If only one type
of resistance, or pathogenicity, is taken into account the conclusions
of the research will probably be incomplete.

The heterogeneous reaction type of coffee leaf rust is
common, especially in *C. canephora* and its derivatives. This
reaction type does not seem to be related to fungal variation, but
is rather an expression of incomplete resistance or incomplete
pathogenicity. It can be affected by environment or by the physiological conditions of the host tissue. This explanation of the
heterogeneous reaction types may also be valid for other pathosystems.

The observed effects of environment and leaf age on the
coffee - *H. vastatrix* relationship resemble the complex relationship, observed in many other pathosystems, between host genotype,
pathogen genotype, development stage of the host, and environment.
Host x pathogen x environment interactions have been reported for
a number of rust diseases (Chandrashekar and Heather, 1981;
Eversmeyer et al., 1980; Lewellen et al., 1967; Milus and Line,
1980; Stubbs, 1980; Zadoks, 1961). Also, the effect of environment and development stage on resistance is widely recognized for
other pathosystems. It is suggested that these complex relationships constitute an important stabilizing factor in achieving the
balance between host and pathogen in natural pathosystems. Therefore, it may be wrong to derive theories on durable resistance by
considering the host-pathogen relationship only. The balance of the
system is probably based on very specific stabilizing forces, which
determine different ecological conditions for the degree of incom-

patibility of each host-pathogen relationship. The detection of
these forces may be difficult under uniform conditions of host,
pathogen and environment, which often prevail in the agricultural
pathosystem.

REFERENCES

Carvalho, A., Ferweda, F.P., Frahm-Lelivel, J.A., Medina, D.M.,
    Mendes, A.J.T., and Monaco, L.C., 1969, Coffee, *in:* "Outlines
    of Perennial Crop Breeding in the Tropics", F.P. Ferweda, ed.,
    Misc. Pap. Landbouwhogesch., Wageningen, 4:189-241.
Chandrashekar, M., and Heather, W.A., 1981, The effect of pre- and
    post-inoculation temperature on resistance in certain cultivars
    of poplar to races of *Melampsora larici-populina* Kleb,
    Euphytica, 30:113-120.
Chaves, G.M., Cruz Filho da, J., Carvalho de, M.G., Matsuoka, K.,
    Teixeira Coelho, D., and Shimoya, C., 1970, "A ferrugem do
    cafeeiro (*Hemileia vastatrix* Berk.& Br.)", Edição Especial
    SEIVA, Universidade Federal de Viçosa, MG., Brazil.
Chevalier, A., 1947, "Le caféiers du globe, Fascicule III" Paul
    Chevalier, ed., Paris.
Cramer, P.J.S., 1957, "A review of literature of coffee research in
    Indonesia", Inter-American Institute of Agricultural Sciences,
    Turrialba, Costa Rica, Misc. Publ. No. 15.
D'Oliveira, B., 1957, "As ferrugens do cafeeiro", *Revista do
    Café Português*, Separata No. 3.
Eskes, A.B., 1980, Ocorrência de un isolado de raça V_3V_5 de
    H. vastatrix pouco virulento em condições de laboratorio,
    in: "Abstracts 8th Brazilian Congress on Coffee Research,
    Campos de Jordão, S.P., Brazil, 25-28 November 1980".
Eskes, A.B., Toma-Braghini M., and Van de Weg, E., 1980, Grau de
    resistência a *H. vastatrix* observado em varios introduções
    de *C. arabica* provenientes de Etiopia, e em cruzamentos entre
    plantas do cultivar "Kouillou" de *C. canephora,* *in:* "Abstracts
    8th Brazilian Congress on Coffee Research, Campos de Jordao,
    S.P., Brazil, 25-28 November ,1980".
Eskes, A.B., and Zink de Souza, E., 1981a, Attaque de ferrugem em
    ramos com e sem produção de plantas do cultivar Catuai, *in:*
    "Abstracts 9th Brazilian Congress on Coffee Research, São
    Lourenço, M.G., Brazil, 24-27 October 1981".
Eskes, A.B., Toma-Braghini, M., and Carvalho, A., 1981b, Testes com
    raças novas de *H. vastatrix* diferenciadas em *C. canephora* cv.
    Kouillou e nas populações de Icatu e Catimor, *in:* "Abstracts
    9th Brazilian Congress on Coffee Research, São Lourenço, M.G.,
    Brazil, 24-27 October 1981".
Eskes, A.B., Toma-Braghini, M., and Hoogstraten, J.G.J., 1981c,
    Segregação para resistência a *H. vastatrix* em cruzamentos entre
    plantas de cultivar Kouillou de *C. canephora,* *in:* "Abstracts
    9th Brazilian Congress on Coffee Research, São Lourenço, M.G.,

Brazil, 24-27 October 1981".

Eskes, A.B., and Toma Braghini, M., 1981, Assessment methods for resistance to coffee leaf rust (*Hemileia vastatrix* Berk. & Br.), *FAO Plant Prot. Bull.*, in press.

Eversmeyer, M.G., Kramer, C.L., and Browder, L.E., 1980, Effect of temperature and host-parasite combination on latent period of *Puccinia recondita* in seedlings of wheat plants, *Phytopathology*, 70:938-941.

Lewellen, R.T., Sharp, E.L., and Hehn, E.R., 1967, Major and minor genes in wheat for resistance to *Puccinia striiformis* and their responses to temperature changes, *Can. J. Bot.*, 45:2155-2172

Locci, R., Minervini Ferrante, G., and Rodrigues, C.J., 1971, Studies by transmission and scanning electron microscopy on the *Hemileia vastatrix* - *Verticillium hemileiae* association, *Rivista di Patologia Vegetale*, Serie 4, Vol. 7:127-140.

Marques, D.V., and Bettencourt, A., 1979, Resistência à *Hemileia vastatrix* numa população de Icatu, *Garcia de Orta, Sér. Est. Agron.*, 6:19-24.

Meyer, F.G., Fernie, L.N., Narasimhaswamy, R.L., Monaco, L.C., and Greathead, D.J., 1968, "FAO Coffee Mission to Ethiopia". Food and Agriculture Organization of the United Nations, Rome, 200 p.

Milus, E.A., and Line, R.F., 1980, Characterization of resistance to leaf rust in Pacific Northeast wheats, *Phytopathology*, 70:167-172.

Monaco, L.C., 1977, Consequences of the introduction of coffee leaf rust into Brazil, *Ann. N.Y. Acad. Sci.*, *287; 57-71*.

Nelson, R.R., (ed.) 1973, "Breeding plants for disease resistance. Concepts and applications", Pennsylvania State University Press.

Parlevliet, J.E., 1979, Components of resistance that reduce the rate of epidemic development, *Ann. Rev. Phytopathol.*, 17:203-222

Rodrigues, C.J., Jr., Bettencourt, A.J., and Rijo, L., 1975, Races of the pathogen and resistance to coffee rust, *Ann. Rev. Phytopathol.*, 14:49-70.

Stubbs, R.W., 1980, Environmental resistance of wheat to yellow rust (*Puccinia striiformis* Westend. f. sp. *tritici*), *Fifth European and Mediterranean Cereal Rust Conference, Bari and Rome, Italy, Proceedings*: 77-81.

Sylvain, P.G., 1955, Some observations on *Coffea arabica* L. in Ethiopia, *Turrialba*, 5:38-53.

Turkensteen, L.J., 1973, "Partial resistance of tomatoes against *Phytophthora infestans*, the late blight fungus", Doctoral Thesis, Agricultural Research Reports 810, Wageningen.

Van der Graaff, N.A., 1981, Selection of Arabica coffee types resistant to coffee berry disease in Ethiopia,

Visveswara, S., 1974, Periodicity of *Hemileia* in Arabica selection
    S. 795, *Indian Coffee*, 38:49-50.
Zadoks, J.C., 1961, Yellow rust on wheat. Studies in epidemiology
    and physiologic specialization,

DISCUSSION

DINOOR: You have shown a positive correlation between disease level
and yield. You may have meant to show that plants with a potentially
higher yield are more susceptible to rust. This will mean that
higher levels of resistance are needed for high-yielding selections.

ESKES: Correct. This poses a problem when selecting for incomplete
resistance. Firstly, we need observations on the field plants over
a period of several years, including years with high yield. Secondly,
it may be difficult to combine high yield with high levels of incomplete resistance. Thirdly, if we need to select for high levels
of resistance, we may easily select incomplete vertical resistance
instead of polygenic resistance. We try to overcome this problem by
studying the inheritance of the resistance before selecting a certain
coffee genotype.

VAN DER GRAAFF: In your work you found interactions of resistance
with leaf age and with environment. These are due to scaling, but
ranking of the material does not change. Thus, if you choose your
scale carefully, your tests will still be sufficient to measure
differences between genotypes.

ESKES: Probably no ideal screening method for incomplete resistance
to coffee leaf rust exists. Certain types of resistance only show up
under certain experimental conditions. We would need to imitate
field conditions, which again change from day to day and from year to
year. However, if the genetic variability for resistance is great,
we are able to detect this resistance under most of our experimental
conditions.

JOHNSON: I believe *C. arabica* is a tetraploid and *C. canephora* a
diploid. You showed a cross between them from which it was deduced
that there were both major and minor genes controlling resistance.
However, in segregating generations from such a cross one would
expect varying numbers of chromosomes and a major effect might be
due to several genes on a single unpaired chromosome being transmitted.

ESKES: We are talking about the Icatu population. This population
consists basically of tetraploid plants, but some degree of aneuplody
is still found, so your suggestion is indeed valid. However, the
major genes to which I referred are resistance factors that are
expressed by a hypersensitive reaction. Two races are already known

that match some of the Icatu resistance factors. Therefore, I
suppose that these factors could be called major genes. When I
referred to minor genes, I meant to say that also a large
variation for quantitative resistance exists in the Icatu population.
So far, we have identified some plants which carry genes for
incomplete vertical resistance.

RESISTANCE TO COFFEE BERRY DISEASE IN ETHIOPIA

N.A. Van der Graaff and R. Pieters

Food and Agriculture Organization
of the United Nations
Rome, Italy

SUMMARY

A review is given of a program to obtain Arabica coffee types that are resistant to CBD. In various parts of Ethiopia, coffee trees were selected (mother trees) that showed a low level of CBD in areas where the disease was heavy. These trees were re-appraised for their level of resistance through repeated field observations, inoculations in the field, and inoculations of seedlings. The mother trees had consistently less disease than the population of non-selected trees in the same area. In field inoculations, the group of mother trees also showed considerably less disease than other trees. Differences in resistance were found among mother trees by means of field observations, field inoculations, and seedling tests. The correlations among these responses are given. Mother trees were multiplied and large progenies were planted from the trees that passed selection thresholds. From these progenies, further information on resistance was obtained through visual disease estimates, counts of berries, and inoculations of detached berries. The correlations among observations and tests are given. The nature of resistance to CBD was also studied and it was concluded that resistance is most likely horizontal. Progenies differed in resistance to vascular wilt. In multilocation trials, coffee types differed in attack by leaf rust, leaf blight and blotch leaf miner. Disease and pest severity was related to provenance of the mother trees. The progenies were not susceptible to brown eye spot. Distribution of seed from the best coffee types begun in early 1978. In 1980, 15×10^6 seeds were distributed.

INTRODUCTION

Arabica coffee is growing on some 400,000 hectares in Ethiopia. Coffee is the main export product of the country. Main coffee growing areas are located in the south and south-western areas of the country; a relatively small and geographically isolated location exists in the east (Figure 1).

Ethiopia is the centre of origin and the centre of domestication of *Coffea arabica*. The history of the crop in the country is practically unknown; it was probably domesticated in the south and south-west of the country in an area that was outside the traditional Ethiopian civilization.

Coffee in Ethiopia occurs between 1200 and 2100 metres. Rainfall in the coffee areas varies between 1000 and 2000 mm per annum; there is a marked dry season. The coffee occurs under four different systems (Institute of Agricultural Research, 1971):

(a) Forest coffee (60 percent), which is sometimes referred to as 'wild' coffee, but is exploited for many years: self sown seedlings have been transplanted to give an irregular, but dominant understorey in the forest, which itself is secondary. The forest is mostly thinned.
(b) Small holder coffee (37 percent): plots of varying sizes around dwellings.
(c) Semi-plantation coffee in the forest: seedlings raised in nurseries and planted, more or less regularly, in thinned forest.
(d) Plantation coffee: plantations established on previously cleared land; the seedlings raised in nurseries and regularly planted; shade trees often planted.

Most of the Ethiopian coffee is a low input – low output crop. Due to this, production prices have always been among the lowest in the world. Ethiopian *C. arabica* is a traditional crop, which is in balance with its indigenous parasites (Van der Graaff, *Chapter 23* in this volume).

In the late 60's, Coffee Berry Disease was introduced in Ethiopia (Mulinge, 1973). Coffee Berry Disease is an anthracnose of green coffee berries, and is caused by *Colletotrichum coffeanum* Noack sensu Hindorf. It forms black lesions on green coffee cherries; under unfavorable conditions, scab like lesions develop.

Coffee Berry Disease spread rapidly over south-western Ethiopia and by 1975 most of the area was probably infested. In 1978, the disease was also found in the eastern area of Harerge. Average losses due to disease amount to some 20 to 25 percent of the total crop (Van der Graaff, 1981). The losses on individual farms vary considerably; in high rainfall, high altitude areas, losses are up to 100 percent.

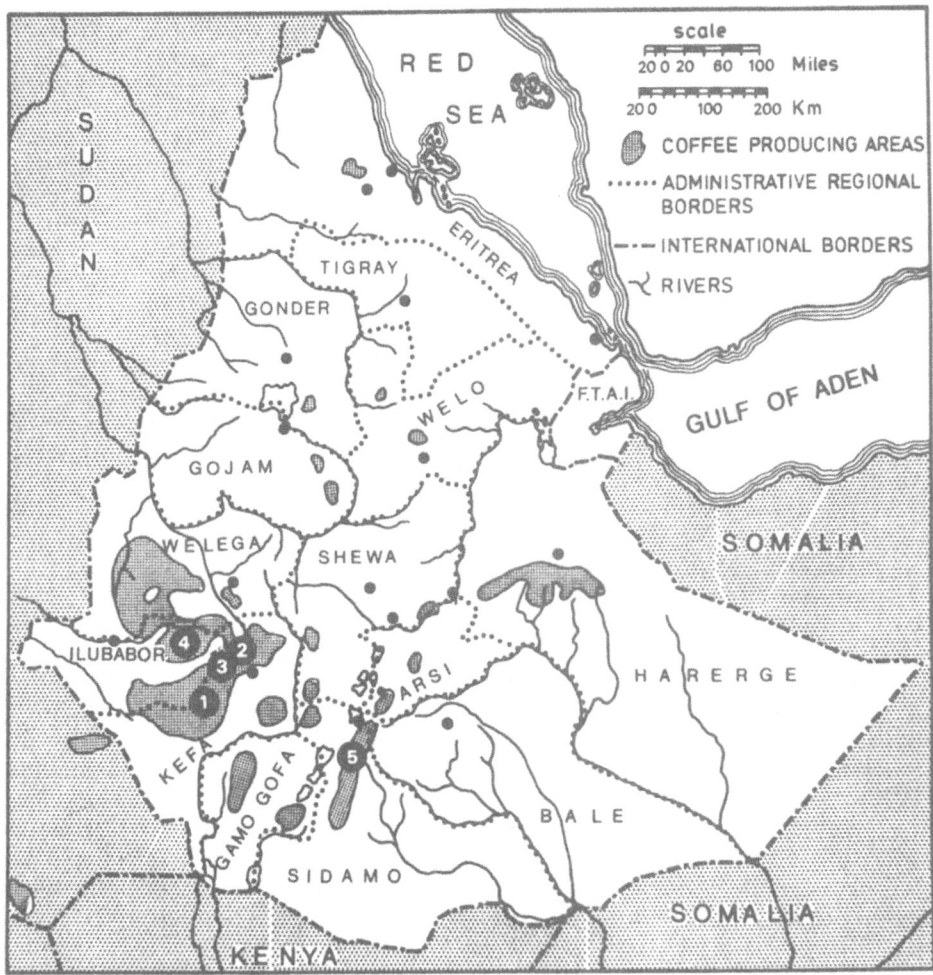

Fig. 1.   Major coffee producing areas in Ethiopia. Locations
where considerable numbers of mother trees were
selected are marked. Map according to Van der Graaff
(1981), changed. 1: Washi, Wush wush; 2: Agaro,
Jachi; 3: Gera; 4: Metu; 5: Sidamo.

Control of Coffee Berry Disease is possible by fungicides,
however, costs of application are high while yields are low and
coffee is difficult to spray in Ethiopia. Resistance to CBD was
recognized to be the only suitable control method and therefore
a program was begun to select resistant material. This paper
describes the execution of this selection program. A full account
can be found in Van der Graaff (1981).

THE CBD PROGRAM

The heterogeneity of the coffee population in Ethiopia offered possibilities to select for resistance. Differences in susceptibility were reported from Kenya as early as 1932 (McDonald) and from Zaire in 1946 (Hendrickx and Lefevre). In 1964, Firman reported a high level of resistance in the Rume Sudan coffee type; a program to breed for resistance was begun in Kenya in 1972 (Van der Vossen, 1973).

In Ethiopia, a resistance program was designed by Robinson (1974), FAO coffee pathologist in Ethiopia in 1973 and 1974, and was reviewed by Person (1974). As all grades of disease intensity were present in the coffee population, the working hypothesis was adopted that resistance was horizontal. Later, the program was amended to include other diseases and pests, however, its basic structure remained the same. The program consisted of:

(a) Selection of mother trees – A selection of 500 to 600 practically disease-free trees in areas with a high disease intensity;
(b) Planting of seedlings in nurseries – Immediately after selection, seed was collected to obtain a sufficient number of seedlings for the establishment of a 1000 tree progeny block;
(c) Observations and tests of mother trees – Mother trees were observed for a maximum of four years. Notes were taken on CBD, other diseases, pests and yield. If possible, quality samples were prepared. Tests on the mother trees were performed to determine their level of resistance;
(d) After an evaluation of the tests and observations, progenies of approved mother trees were planted in progeny blocks of up to 1000 trees on a farm in one of the areas where the CBD intensity was very high. Furthermore, progeny trials were established at other locations in the country;
(e) The disease resistance of progenies was re-evaluated through field observations and tests. Observations on progenies were made through a visual estimation of the percentage of diseased berries and regular berry counts. Regularly, tests were performed on detached berries;
(f) A separate program was begun to develop and apply tests for resistance to vascular wilt caused by *Gibberella xylarioides*
(g) In progeny trials, observations were made on the severity of other diseases, pests, yield and quality;
(h) A preliminary assessment was made to be able to distribute material. Based on data from the distributed material, the assessment will be updated at regular intervals.

The program started in 1973 and the first seed distribution begun in early 1979. Funding was obtained from the Ethiopian Government, UNDP and, through the Ethiopian authorities, from

EEC. A more detailed description is presented below:

Selection for Resistance

Individual trees without, or with a low level of, disease were selected from areas where disease intensity was high; a negative screening was made for low yield and for susceptibility to other diseases and pests. The main selection locations are shown in Figure 1. In total, 639 trees were selected in the period from 1973 to 1975.

Re-assessment of Resistance

Field observations were made for up to four years to determine the variability of the disease intensity of the mother trees. As a rule, a mother tree was discarded if more than one percent of its berries were diseased. In one area where many mother trees were selected, the population of mother trees was assessed during four years and was compared with non-selected trees (Table 1). Disease in the group of mother trees was consistently lower than in the group of non-selected trees. Within the group of mother trees, some trees had consistently more disease than others.

Table 1.   Mother trees, visual disease assessment. Differences between the group of mother trees selected in 1975 and non-selected trees at Gera, one of the selection areas. The percentage of diseased berries per tree was estimated. In the table, the class in which the median observation fell is indicated.

n: number of trees
M: percentage class of the median value

Year of observation	Selected trees		Unselected trees	
	n	M	n	M
1976	57	0.1-1	753	11-50
1977	55	0	560	51-90
1978	55	0.1-1	555	51-90

Testing of Resistance of Mother Trees

Mother trees were tested through inoculations in the field and on seedlings. In the inoculations in the field, branches were sprayed with a conidial suspension and a number of branches on the same tree were used as replications of the experiment. Branches were covered with plastic for 24 hours to ensure the presence of moisture needed for conidial germination and for penetration of the cuticle. The number of berries before inoculation and the number of healthy berries three weeks after inoculation were determined. Differences in responses between mother trees and non-selected trees are shown in Table 2. Kendall's rank correlation coefficients between field observations and field inoculations are given in Table 3 for Gera (mother trees) and Jima (a wide range of susceptibilities). Differences between trees in the field inoculation test may be partly due to microclimate, however, the differences between mother trees and non-selected trees are believed to indicate genetical differences.

Testing of resistance in young seedlings was based on Cook's inoculation test (Cook, 1973, a, b; Marakuru, 1976; Van der Vossen et al., 1976). The test was adapted to local conditions (Van der Graaff, 1978, 1981) after which seedlings from practically all mother trees were subjected to it. Seedlings of a mother

Table 2. Mother trees, field inoculation tests. Mean responses of three consecutive tests, made on a group of mother trees and randomly chosen non-selected trees in 1976. Data are presented for three selection areas.

n: number of tested trees

$\overline{x}$:   fraction of berries dropped or diseased (angularly transformed)

Location	Mother trees		Non-selected trees	
	n	$\overline{x}$	n	$\overline{x}$
Gera	77	30	22	61
Jachi (Agaro)	28	31	29	62
Wush wush	13	16	11	37

Table 3.   Mother trees (Gera) and coffee types ranging
           from susceptible to resistant (Jima). Kendall's
           rank correlation coefficients between field obser-
           vations and responses in field inoculations, and
           between field observations and responses in seed-
           ling tests. The highest disease intensity in the
           field in the indicated period and the average
           response of three consecutive field inoculations
           were used for the calculations.

Field observation	Field inoculation	Seedling test
Gera 1976–78	0.44**	0.27*
Jima 1975–76	0.63**	0.30*

tree were raised in a number of boxes containing up to 100 seedlings per box. Two coffee types were included to serve as a
reference. Seedlings were sprayed with a conidial suspension
at or just before the unfolding of the cotyledons; a re-inoculation
was performed after 48 hours. Boxes were kept closed 48 hours
before, in between, and 48 hours after the inoculations to maintain
a high relative humidity and, thus, to ensure infection. After three
weeks, individual seedlings were graded according to lesion size
and color on a five-class scale. Corrections were made according
to the inoculation date based on the results of the reference coffee
types. Highly significant differences were found among coffee
types, however, the correlation between field observations and
seedling tests was not very satisfactory (Table 3).

Propagation

    Selection thresholds were established through the comparison
of the results of field inoculations, seedling tests and field observations. Coffee agronomists from the coffee research station
propagated the selected and approved material. Up to 1000 progeny
trees were planted per approved mother tree. The progeny blocks
were planted at Gera, a farm established in an area where the
disease intensity was among the highest in the country. In the
period 1975 to 1978, 156 progenies were planted with a total of
approximately 120,000 progeny trees. Replicated progeny trials
were planted at Gera and at a number of other sites in the country.
At least one of the sites was thought to be highly conducive to leaf
rust.

Appraisal of the Resistance of the Progenies

The progenies were observed and tested to obtain information on the level of resistance and on the homogeneity of the resistance levels among trees within each progeny. The evaluation was made through:

(a)   Visual estimation at regular intervals of the CBD level of 100 trees of each progeny.
(b)   Berry countings at three week intervals on marked branches (one branch of each of 25 trees).
(c)   Tests on detached berries.

To ascertain the presence of the fungus, trees were sprayed with a conidial suspension one year before their first crop. This served to establish the fungus on the bark where it lives as a micro-epiphyte.

The percentage of diseased berries was estimated by visual observations. In many progenies, a small percentage of "off-type" trees could be identified with a high level of disease. In some progenies, these trees had other characters that marked them as "off-trees". These trees are most likely resulting from cross-pollination.

Berry counts were made at three weeks intervals from the start of berry expansion (some six to eight weeks after flowering) to 21 weeks after flowering, when most of the epidemic was over. It was observed that a considerable "physiological" drop occurred. This drop, which amounted to an average of 21 percent, varied considerably among progenies. After correction for the "physiological" drop, the percentage diseased and dropped berries due to CBD varied between zero and 36 percent. Counts made during a second year revealed that damage was higher in more susceptible progenies, but remained low in those with a high level of resistance while losses were almost complete in unselected material in the same area.

In detached berry tests, berries of progenies were arranged in boxes and inoculated. Boxes were then kept closed to ensure conditions highly conducive to CBD development. Each box contained 50 berries and represented one progeny; replications were made by using three or four boxes per progeny. The number of diseased berries was recorded until nine days after inoculation.

In preliminary experiments, it was determined that the susceptibility of small berries was high and variable, both in resistant and susceptible coffee types. In fully expanded, green berries, susceptibility was lower and less variable.

In the year the progenies had their first crop, detached berry tests were performed on fully expanded green berries at two to three weeks intervals. Susceptibility of the expanded green berries varied according to the testing date; there was a significant inter-action between date of testing and coffee progenies. Satisfactory correlations with field observations were obtained if the progeny averages calculated over all testing dates were compared with the field data. Rank correlations among the observations and test responses are presented in Table 4. The correlation coefficients were relatively low, however, it should be realized that the data were derived from pre-selected material in which high levels of susceptibility did not occur. For example, higher rank correlations were obtained in later years, when a range of susceptibilities were used as references in the testing programme. Criteria for distri-bution of selected material were determined based on visual field assessments, field observations and detached berry tests. "Off-types" were removed from resistant progenies.

Other Diseases and Pests

The system Arabica coffee-parasites was in balance before the introduction of Coffee Berry Disease. Resistance against diseases and pests was at an optimum level through generations of selection by farmers. New cultivars selected for resistance to CBD should also possess adequate resistance against other

Table 4.   Progenies at Gera in 1978. Kendall's rank correlation coefficients among disease inci-dence (number of trees with CBD), disease severity (mean percentage disease from visual estimates), percentage CBD determined through regular berry counts, and responses in detached berry tests averaged over a number of consecu-tive tests. Number of progenies per entry varied between 48 and 57. All correlations were highly significant ($p \leq 10^{-3}$). Data accord-ing to Van der Graaff (1981).

	DBT	Severity (berry counts)
Disease incidence	0.32	0.41
Disease severity (estimates)	0.36	0.42
Disease severity (berry counts)	0.35	–

diseases and against pests. Methods to obtain satisfactory resistance levels are described in the following sections:

(a) *Gibberella xylarioides* (vascular wilt). (Research on this
subject was mainly performed by R. Pieters). Differences in
susceptibility to vascular wilt were found in a coffee collection
at Jima research station (Van der Graaff and Pieters, 1978).
Differences were determined through the use of a grid in which
the occurrence of tree death's in groups of four trees of one
coffee type was compared with death in groups of four trees in
which each tree represented a different coffee type. Based on
these observations, two different tests were devised (Pieters and
Van der Graaff, 1980). A seedling test was used in which seedlings were inoculated by nicking the seedlings with a knife that
had been dipped in a conidial suspension of *Fusarium xylarioides*
The latent period (e.g., the time between inoculation and death
of the first seedling) and percentage of dead seedlings per test
were then determined. These data correlated well with field observations. A second test was devised in which the percentage of
conidial germination was determined on freshly exposed cambial
layers of twigs. Germination correlated well with field data.

In both tests, resistance appeared to be a quantitative character.
Differences in horizontal pathogenicity were discovered among
various isolates.

Based on field observations and test results, tentative selection
criteria were determined and all material was selected accordingly.

(b) Leaf rust *(Hemileia vastatrix)* , leaf blight and stem dieback
(Phoma tarda) , blotch leaf miner (predominantly *Leucoptera
caffeina*), brown eye spot (*Cercospora coffeicola*). In trials
conducted in various parts of the country, quantitative differences
in disease and pest intensity were observed among progenies. They
included differences in leaf rust, leaf blight and stem dieback,
and in the infestation level by a blotch leaf miner. It was possible
to prove that these differences were statistically significant. When
the coffee types were grouped according to the provenance of their
mother trees, differences were found to exist between provenances
(Table 5). These differences can certainly be related to climate,
Metu being wetter than Washi and Washi receiving more rain than
Agaro and surroundings. This trend can be extended to material
from Harerge province, an area with certainly much less rain
than any of the three indicated in the table. Coffee types from
that area are much more susceptible to leaf rust and leaf blight.
At present, the differences in leaf rust intensity are more system-
atically studied through leaf disk tests (Critchett, pers. comm.).

In the course of studies at Jima Research Station, it was
found that statistically significant differences in intensity of attack

Table 5.   Susceptibility to diseases and pests and the
           provenance of mother trees. Data from progeny
           trials at Metu (leaf rust), Gera (blight) and
           Agaro (leaf miner). The progenies were
           visually scored for the percentage of leaves
           showing symptoms. The data were grouped
           according to the provenance of mother trees
           of the progenies; in the table, the mean values
           per provenance are shown. In each row, data
           marked with the same letter did not differ signifi-
           cantly. Data according to Van der Graaff (1981).

Disease	Provenance		
	Metu	Washi	Agaro
Leaf rust	1.3 a	1.8 a	4.1 b
Blight	0.9 a	2.8 b	1.6 b
Leaf miner	16.8 a	24.8 b	29.1 b

by *Cercospora coffeicola* existed among coffee types from Harerge.
Coffee types from Western Ethiopia showed a very low level of
Cercospora both at Jima and at other locations in Western Ethiopia,
thus indicating a sufficient resistance level.

The observations in the progeny trials allowed material with
susceptibility to leaf rust, leaf blight and leaf miner to be discarded.

Yield

With respect to yield, the current low level of production per
unit area should be taken into account. At present, genotypic
differences are less limiting to Ethiopian coffee production than
agricultural practices, soil fertility, and ecological conditions.
Indications have already been obtained that improved cultural
practices can greatly increase the yield of traditional, unimproved
coffee. However, neither the long term agronomic effects of these
improvements nor their long term economic effects can be fore-
seen. Therefore, the distribution of genotypes with maximum
yield potential is not yet needed as it is unlikely that yield potential
will be a limiting factor in the near future. The gradual development and acceptance of better agricultural practices and the necessary higher inputs will only gradually increase the demand for the
genotypes that respond maximally to such inputs. At the present
stage, neither inputs or genotypes, not their interactions are

defined. It was, therefore, decided to consider for distribution progenies derived from mother trees yielding at average level or above average during the whole observation period. An extra year was available between seed distribution and planting of seedlings and thus it was theoretically possible to withdraw seedlings of progenies that yielded poorly in the first two years of bearing. Long-term observations on many locations are needed to study yield level and to determine interactions among yield, ecological conditions and agricultural practices. Only when such results have been obtained, it will be possible to identify the best geno-types.

Quality

Quality assessments were made on progenies before they were recommended for distribution. Quality was found to be vari-able: "the selections released on basis of CBD resistance falling mainly in the medium class and some lacking flavor" (White, 1980).

The Nature of the Resistance to CBD

Because coffee is a perennial crop, it is of the utmost import-ance to obtain an indication of the durability of its resistance. Without doubt, the chance for durability is much greater when resistance is horizontal than when it is vertical. These terms are used here to define the following situations:

Horizontal resistance is: quantitative – its expression depends on the conditions for disease development; polygenic or oligogenic – a rather continuous variation occurs in the host population between full susceptibility and complete resistance; non-specific–differential interactions between components of the host and of the pathogen populations are at a low level or absent.

Vertical resistance is: mostly qualitative – quantitative resistance does occur but it is rare; monogenic or oligogenic – variability in the host population is discontinuous; specific – interaction between elements of the host and of the pathogen populations is differential, being due to a gene–for–gene relationship.

The proof for the horizontal nature of resistance is elusive and full proof can probably never be given as its definition is nega-tive. Nevertheless, the resistance can be compared with its descriptors.

Quantitativeness: In all tests and observations made in the course of our study on CBD resistance, quantitative results were obtained. Practically all mother trees showed some disease in the field. For example, out of 55 trees selected in one location, 47 showed some disease in the period 1975–1978. Field inoculations

invariably produced disease, though lesions on the more resistant
trees often relapsed to inactive "scab" lesions. In seedling inoculation tests, a gradation of disease resistance was observed.
Quantitative results were always obtained in detached berry tests
on progenies. Where highly resistant mother trees were tested
in detached berry tests, the results were also quantitative.

   Specific differential interactions were studied in seedling and
in detached berry tests. In one case, specificity was suspected
and special detached berry tests and seedling tests were made to
elucidate this.   Significant differences were found among coffee
types and among inocula, but interactions were not significant. In
a detached berry test made with three isolates and 38 resistant
progenies of mother trees, differences were not observed among
isolates, but considerable differences existed among coffee types.
Interactions between isolates and coffee types were also significant.
In the latter experiment, differences remained quantitative and no
real inversions of resistance occurred. It is likely that the interactions were caused by other confounding factors like differences
in eveness between batches of berries and differences in the genetical homogeneity of the progenies from which the berries were collected. This will result in differences in standard error between
coffee types and may thus produce an apparent specific interaction
between host and pathogen. Comparable interactions were, for
example, found in experiments in which progenies were inoculated
with three inocula that differed in conidial concentration.

   Genetical studies to determine the number of resistance genes
are just starting in Ethiopia. In progenies with a satisfactory level
of resistance, some susceptible trees were always observed. Other
progenies showed a gradation in disease intensity. The variation
in disease intensity within a progeny will, apart from microclimato-
logical variation, depend on the homozygosity of the mother tree
and on the frequency of progeny trees derived from naturally cross-
pollinated seed. Due to the susceptibility of most of the coffee
population, cross pollination will result in an unacceptable level
of susceptibility.

   Summarizing, resistance is quantitative and little indication
of specificity has yet been found. Although some doubts remain,
it may be assumed resistance is horizontal.

Conclusions

   In Table 6, the selection procedure is indicated. While most
selection levels were arbitrarily established, the selection pattern
was relatively strict due to the large numbers involved. Seed
distribution started in 1978 with 200,000 seeds, in 1979 this was
increased to some 4.3 million and in 1980 to some 15 million seeds.
The Ministry of Coffee and Tea Development grows the seedlings

Table 6.   The steps in the selection and testing
            program for CBD resistance.

Selection criterium	Fraction discarded
Selection of mother trees:	$\pm$ 0.995
Discarded/lost/dead before seed was collected	0.31
Observation and testing:	
CBD	0.81
Leaf rust	0.49
Leaf blight	0.44
Vascular wilt	0.40
Blotch leaf miner	0.34
Total observation and testing	0.96

in nurseries and distributes them to farmers' cooperatives and
coffee state farms.

Much remains to be done. Long term research will be needed
on the nature and inheritance of CBD resistance and on resistance
to other diseases and pests. A further, long term, assessment of
all material will have to be made to identify the varieties best
adapted to the different ecological conditions. Finally, further
collection of CBD resistant germplasm is urgently needed. Although
many problems still remain and secondary problems are likely to
develop, the solution of the overwhelming problem of CBD in
Ethiopia appears to be a reality.

LITERATURE

Cook, R.T.A., 1973a, Work in progress in coffee research.
    Series III, Part IV. Detecting disease resistance in coffee
    plants, *Kenya Coffee*, 38:275-277.
Cook, R.T.A., 1973b, Screening coffee plants for CBD resistance,
    Coffee Res. Found. Kenya Ann. Rep., 1972/73:66-68.
Cook, R.T.A., 1975, Screening coffee plants for resistance to
    CBD. *Coffee Res. Found. Kenya Ann. Rep.*, 1973/74:64.
Firman, I.D., 1964, Screening of coffee for resistance to Coffee
    Berry Disease, *East Afr. Agric. For. J.*, 29:192-194.

Hendrickx, F.L., and Lefevre, P.C., 1946, Obsevations préliminaires
    sur la résistance de lignées de *Coffea arabica* a quelques
    ennemis, *Bull. Agric. Congo,* 37:783-800.
Institute of Agricultural Research; Jimma Research Station, 1972,
    Progress report for the period April 1971 to March 1972, Addis
    Ababa.
McDonald, J., 1932, Annual report of the Senior Mycologist for 1931,
    Kenya Dep. Agric. Ann. Rep. 1931:118-130.
Mulinge, S.K., 1973, Coffee berry disease in Ethiopia, *FAO Plant
    Prot. Bull.,* 21:86-86.
Murakaru, G.N.W., 1976, Influence of age of coffee seedlings on
    infection by *Colletotrichum coffeanum* (Noack) *(Glomerella
    cingulata* (Stonem.) Spauld.& Von Schrenk), *Kenya Coffee,*
    41:55-57.
Person, C., 1974, Consultancy report on the coffee berry disease
    program at Jimma Research Station, FAO, Rome.
Pieters, R., and Van der Graaff, N.A., 1980, Resistance to
    Gibberella xylarioides in *Coffea arabica:* evaluation of screen-
    ing methods and evidence for the horizontal nature of the
    resistance, *Neth. J. Plant Path.,* 86:37-43.
Robinson, R.A., 1974, Terminal Report of the FAO Coffee Pathologist
    to the Government of Ethiopia, FAO, Rome, AGO/74/443.
Van der Graaff, N.A., 1978, Selection for resistance to coffee berry
    disease in Arabica coffee in Ethiopia. Evaluation of selection
    methods, *Neth. J. Plant Path.,* 84:205-215.
Van der Graaff, N.A., 1981, Selection of Arabica coffee types
    resistant to coffee berry disease in Ethiopia, Med. Landbouw-
    hogesch., Wageningen, 81-11.
Van der Graaff, N.A., and Pieters, R., 1978, Resistance levels in
    Coffea arabica to *Gibberella xylarioides* and distribution
    pattern of the disease, *Neth. J. Plant Path.,* 84:117-120.
Van der Vossen, H.A.M., Cook, R.T.A., and Murakaru, G.M.W., 1976,
    Breeding for resistance to coffee berry disease caused by
    Colletotrichum coffeanum Noack *(sensu* Hindorf) in *Coffea
    arabica* L. I. Methods of preselection for resistance,
    Euphytica, 25:733-745.
White, R.G., 1980, Assignment report on coffee agronomy, FAO,
    Rome, AG:DP/ETH/78/004 Assignment Report.
Zadoks, J.C., 1979, Proposal to survey and improve the resistance
    of coffee in Ethiopia against diseases and pests, FAO Consult-
    ancy Report, typescript, 19 pp. + appendix.

DISCUSSION

WALLER: Van der Vossen, who is working in Kenya, has recently
claimed that CBD resistance in his material is due to major genes.
Your results appear to indicate a polygenetic system giving quantitative differences in resistance; would you like to comment on this?

VAN DER GRAAFF: Van der Vossen and Walyaro (*Euphytica*, 29:771-779) published a genetic analysis of a number of diallel crosses among coffee types with different levels of resistance. For their analysis, they used a seedling test. They concluded that some types possessed dominant resistance, some had resistance of intermediate types, others were susceptible, while a fourth category showed dominant suscept- ibility. In certain crosses, epistasis was observed. However, to determine whether the genotypic variation is additive or not, the non- genetical variation has to be additive over the whole scale used to measure resistance in the seedling test. This is certainly not the case in their data. In homogeneous material they tested, the mean grade varied from 4.2 to 11.9 (on a 1 to 12 scale) and S^2 varied between 0.70 and 7.91. In such data, five significantly different levels of S^2 can be recognized. This indicates scaling problems and this obviously throws serious doubts on their conclusions. Ranking of their results indicates that a scale may be found on which their results are additive. In their publication, they arrive at conclusions on the number of genes through (a) the distribution of grades of seedlings in the test; (b) the designation of part of the scale as a susceptible reaction. Concerning (a), their conclusions did not take into account the above-mentioned scaling problems. Concerning (b), there is no objective reason for this division. For example, in their experiments, seedlings of susceptible homozygous genotypes fall into grades 11 and 12; however, they also designate class 9 and 10 as a susceptible reaction.

PERSON: You raised many progenies from single trees. If there had been major genes for resistance, one would think that some of your selected parental trees would be heterozygous for a major gene, and the progeny would therefore show evidence of genetic segregation. Did you look for this, and, if so, did you find it?

VAN DER GRAAFF: We indeed looked for such segregation, but we never observed percentages of susceptible trees that would confirm a major gene segregation.

PERSON: Did the resistant trees exhibit any constant features such as, for example, an open canopy?

VAN DER GRAAFF: The progenies of mother trees differed among each other for such characters. Within progenies, such features were mostly constant.

PARLEVLIET: Do we need to worry about non-durability of resistance in the case of CBD and coffee? Why not simply use all resistance we might meet? (There are many pathosystems where this is valid).

VAN DER GRAAFF: You may be right, but facts are needed to support such an assumption. Therefore, the relevant parts of the pathosystem should be studied to obtain information on host-pathogen inter- actions, the mechanism of resistance and its inheritance.

ROBINSON: There are two further points which tend to support the horizontal nature of the resistance. The first is the susceptibility ranking of Kenya's cultivars in the 1920's; this ranking has not changed. The second is the fact that the CBD pathosystem is continuous and I do not believe that a vertical pathosystem can evolve in such a system.

VAN DER GRAAFF: The first point can be elucidated further by the data from the coffee improvement program in Zaire. Hendrickx and Lefevre reported in 1946 on the CBD resistance of the Arabica coffee types used in this program. Presently, the resistance ranking of this material appears unchanged and the most resistant material has remained so, although it is extensively grown. Concerning your comment that a vertical pathosystem cannot evolve, I agree in principle; however, as the pathosystem is new, certain adaptations of the pathogen may occur. Such adaptations may especially take place when, accidentally, large effects of single genes would be responsible for the resistance.

JOHNSON: If you have incomplete resistance to CBD, will the marketable product contain diseased berries and would they, if present, affect the marketable value of the crop?

VAN DER GRAAFF: A small percentage of the berries may be diseased, especially as late disease at the ripening stage cannot be completely prevented. However, this late disease also occurred before CBD was introduced and was caused by other, less pathogenic, *Colletotrichum* species (brown blight). We expect to bring CBD back to such a disease. Such late disease does not damage the beans but may cause slight problems if berries are to be "wet processed".

DINOOR: In your work, you seem to assume that resistance to CBD is horizontal. In a slide you showed (the results of berry tests from different coffee clones, inoculated by three inocula), there were clear differential interactions, indicative of quantitative vertical resistance. In screening for resistance, you are using a "black box" method, the pathogen being the unknown "black box". There are no details about the number of isolates involved, their origin, their characteristic pathogenicity, the number of growth cycles, etc. We do not know whether you selected for resistance against a *Colletotrichum* population representing one region or

various regions, or whether you selected for resistance against one
isolate. There is no way you can predict whether your selection
will stand up against epidemics in different parts of Ethiopia. It
may have been better to select separately against several isolates
and to combine the best resistances available. Such a selection may
then serve for different tactics of gene management.

VAN DER GRAAFF: In your comment, you presume that there are
differences in vertical pathogencicity of CBD towards *C. arabica*.
However, as I already stated, the differential interactions found are
most probably not indicative for quantitative vertical resistance.
The experiment in which we found differential interactions was no
exception; it was also possible to observe differential interactions
between date of testing and coffee type and between conidial concentration and coffee type (the latter experiment was made with a
single spore isolate). However, the increased test responses would
by no means come anywhere near the response levels of susceptible
cultivars. It should, therefore, be concluded that the interaction
is not an indication of incomplete vertical resistance; instead, it
is probably caused by: (a) a not fully satisfactory transformation;
(b) differences in genetic homogeneity among berry samples (the test
referred to was not on clonal material); (c) difference in physiological homogeneity among berry samples (b) and (c) produce differences in the standard error among berry samples and, thus, apparent
interactions. Concerning the testing procedure, in some stages, monoconidial isolates were used; in other stages, mixtures were used
from the location where the mother trees were selected, from the area
where the progenies were grown, and from our research station. In
some cases, incomplete vertical resistance was suspected; however,
in more detailed experimentation with the appropriate isolates this
could not be confirmed. The progenies were finally grown in trials
at a number of locations and no "breakdowns" of resistance were noted,
which would be expected when the race composition varied. This is in
concurrence with the durability of the (admittedly lower levels of)
resistance found in Zaire in the early 40's. The last part of your
question on the combination of resistance to various isolates and its
use in gene management is again a comment based on the assumption of
(incomplete) vertical resistance. If such were present, gene management would certainly not work, as, in contrast with diseases like
cereal rusts, allo-infection is of no importance in the CBD epidemic;
the fungus is always present on the bark as a micro-epiphyte.

DURABLE RESISTANCE IN POTATOES FOR DEVELOPING COUNTRIES

O.T. Page

Centro Internacional de la Papa (CIP)
Lima, Peru

INTRODUCTION

There is a great deal of general research information published on growing potatoes - an average of 1,820 papers and reports annually during the past five years according to Potato Abstracts· Although much of this information is concerned with diseases and pests, emphasis is skewed toward temperate zone diseases and pests. For example, less than two percent of the reports dealt with the potato tuber moth, root-knot nematodes and *Pseudomonas* wilt, which combined, are major production constraints of potatoes grown in a band encircling the world 30°N and 30°S of the equator.

While two-thirds of the world's population lives in developing countries, research on the major food crops of mankind is carried out primarily in the industrialized countries. It was for this reason that International Agricultural Research Centers such as CIP were founded to develop technologies to improve food production in the less developed countries. One of the principal mandates of the IARCs is to collect, classify and maintain all available germplasm relative to their specific crop interests. During the past six years CIP, through many difficult expeditions, has collected more than 13,000 accessions of the eight cultivated species of potatoes grown in South America. There are also about 150 species of wild tuber-bearing potatoes some of which have useful resistance attributes.

The collection of primitive cultivars and wild tuber-bearing potatoes is an important step in developing a management strategy for controlling diseases and pests. This is the germplasm resource for breeding for resistance to important fungal, viral and bacterial pathogens as well as insects and nematodes. Furthermore, the rapid erosion of primitive cultivars through the introduction of modern varieties, and the encroachment of man into the habitats of wild species has made the collection and maintenance of this valuable germplasm an emergency effort.

The breeding and selection of *Solanum tuberosum* cultivars grown in temperate potato producing areas of the world has been estimated to use less than 10 percent of the total genetic variability accumulated in the tuber-bearing *Solanums*. A study by Mendoza and Haynes (1974) of 80 North American potato cultivars has shown a very close relationship in the co-ancestry of the 10 leading varieties. When these cultivars are used in breeding the resulting offspring have a certain degree of inbreeding. Disease and pest resistance is largely incidental since selection is primarily for yield and cosmetic attributes.

POPULATION BREEDING

At CIP a population breeding approach, in which over 300,000 seedlings are screened each year, involves the application of continuous cycles of recurrent phenotypic selection in order to maintain wide genetic variability. This heterozygosity provides the basis for good yields and stability of performance over a range of tropical environments.

A population is characterized genetically by phenotypic frequencies reflecting gene frequencies for any given trait. If, for each generation only those plants selected for a certain attribute are mated, the gene frequency for that trait will increase as well as the frequency of desirable phenotypes. Progress, however, depends on the hereditary pattern of each trait. Selection for a character, like yield which is under the control of many genes as well as the influence of environment, will progress more slowly than selection for a character such as comprehensive resistance to potato virus Y (PVY) which is under a much simpler genetic control and apparently does not suffer environmentally-induced modifications. At present, CIP is emphasizing selection for comprehensive resistance (immunity) for PVY resistant types.

For example, 60 of 840 (7 percent) of diploid clones were found to be immune to PVY — that is the gene for PVY immunity was present in the initial population but at a relatively low frequency (Mendoza and Haynes, 1977). Given that the genetic control of immunity to PVY is simple, it was possible to increase the frequency of PVY immunity to 96 percent by assorted mating of immune genotypes through only four generations.

In breeding for root-knot nematode *(Meloidogyne incognita acrita)* resistance Mendoza and Jatala (1978) screened 1,020 cultivated diploid clones for resistanc. Only 28 resistant genotypes were

identified demonstrating that in the population the genes for
resistance were scattered and at a very low frequency.

Within this rather small sample of resistant plants bulk
crossing was made to concentrate genes for root-knot resistance.
Of 1,450 plants obtained, 52 were resistant indicating that considerable progress was achieved through one generation of selection.

These resistant individuals were then crossed with *Solanum
sparsipilum* a wild species with root-knot nematode resistance.
Also several susceptible diploid clones were utilized as parents
to gain information on the pattern of inheritance of resistance.

The interspecific hybrid showed 62 per cent of 1,058 plants
were resistant when resistant parents were mated while only 14.5
per cent resistant plants resulted from crosses of resistant parents
with susceptibles. Mating susceptibles resulted in 99 percent
susceptible progeny.

The results of breeding for PVY and root-knot nematode
resistance are but two examples which confirm that progress can be
achieved in disease and pest resistance by mating genotypes
selected by a stringent screening procedure. Through population
breeding it is possible to increase gene frequencies with the
concomitant increase in desirable phenotypes as a result of recurrent
selection.

An important aspect of population breeding in potatoes is the
possibility of recombining characters in such a way that a single
genotype may carry several desirable traits. For instance, in a
population being developed for adaptation to the lowland tropics
it may be necessary to combine heat tolerance with resistance to
late blight, bacterial wilt (*Pseudomonas solanacearum*), PVX and PVY.
It is evident that if selection for all these traits were applied
at the early stage of population development probably no single
individual would have survived since the resistance genes, although
present in the gene pool, would be widely isolated in subsamples.
Thus, it is necessary to select for individual resistance factors
along with required agronomic characters. Surviving clones of a
sample were crossed to survivors of another sample previously
screened for resistance to a disease. This approach increases the
recombination of two resistances such as blight and PVY resistance
in a background of improved adaptation and good yield. Clones with
combined blight and PVY resistance were then crossed with clones
having bacterial wilt and PVX resistance. Following this cyclic
breeding system a combination of characters was achieved.

It is important to note that in the screening scheme the
selection for resistance, either to disease or stress can be applied
to small seedlings. Only surviving genotypes are transferred to

the field to be selected for yield and other agronomic characters under appropriate conditions such as the lowland tropics.

Population breeding is a management strategy to produce long-term resistance to diseases and pests in order to reduce the risk of crop loss by small farmers in developing countries. However, there are also other practical routes on the way to sustainable crop yields which is the goal of durable resistance.

TRUE POTATO SEED

The potato has been cultivated as a principal food source for over a thousand years in the Andes. There the small farmer grows a mixture of clones suggesting that there may be reduced risk through a multi-line approach. In the wild, potatoes propagate themselves primarily by the production of true seed (TPS); each genetically unique TPS is essentially free of pathogens and pests. The use of TPS appears to be a practical way to produce sustainable yields. It is a disease and pest management strategy which although it has received no attention in developed countries offers many advantages.

During the past three years extensive research at CIP in agronomy, physiology and breeding and genetics has clearly demonstrated the practicality of the TPS approach. And so have the Chinese as observed during visits into remote areas of China by CIP scientists to look at potato crops grown from TPS on approximately 25,000 hectares. This is convincing evidence of the practicality of the use of TPS at farm level. TPS technology is appropriate for countries in southeast Asia where transplanting and intensive management characterizes small farm technology readily adaptable to growing and transplanting potato seedlings.

REFERENCES

"Potato Abstracts, 1976–1980" Commonwealth Agricultural Bureaux, Farnham Royal. Ed. M.S. Makal.

Mendoza, H.A. and Haynes, F.L., 1974, Genetic relationships among potato cultivars grown in the United States, *HortScience* 9: 328–330.

Mendoza, H.A. and Haynes, F.L., 1977, The breeding potential of a cultivated diploid potato population, *American Potato Journal*, 54:489–490.

Mendoza, H.A. and Jatala, P., in: "Planning Conference Report on Control of Nematodes", CIP, Lima (1978).

GENETIC RECOMBINATION AND MODE OF INHERITANCE OF PATHOGENIC

CHARACTERS BY *Phytophthora infestans* THROUGH SEXUAL REPRODUCTION

C. Laviola and M.E. Gallegly[*]

Istituto di Patologia Vegetale, Universita
Degli Studi, Palermo, Italy
*Dept. of Plant Pathology, West Virginia
University, Morgantown, U.S.A.

INTRODUCTION

By the late 1960's a good deal of knowledge had been accumulated on the resistance of *Solanum tuberosum* L. to *Phytophthora infestans* (Mont.) de Bary but very little was known about the mode of inheritance of pathogenicity in the fungus, its great variation and the origin of new physiological races being accounted for by mutation (Peterson & Mills, 1953; Toxopeus, 1956; Gallegly & Niederhauser, 1954). The discovery of sexuality in *P. infestans* (Niederhauser, 1956; Gough, 1957; Smoot *et al.*, 1958), provided a means to investigate the genetics of this pathogen but produced little immediate information.

The first report of genetic recombination in *P. infestans* was provided by Gough (reviewed in Smoot *et al.*, 1958) who established 2 single-oospore isolates genetically different from both parents. Two more single-oospore isolates, equally different from both parents, were obtained by Savage & Gallegly (1960). Romero & Erwin (1969) published the results of investigating the pathogenicity of 30 single-oospore isolates from 3 crosses. Among the progeny of one of these crosses, nine new physiologic races, different from both parents plus one parental combination, were recovered. With the exception of race 2 character, all the characters for pathogenicity segregated independently in a 1:1 ration, or very close to it, whereas A_1 and A_2 segregate in a 3:1 ration. This provided some account of recombination of characters in *P. infestans* from single-oospore isolates, but not a full understanding of the nuclear processes involved in the genetics of this fungs.

339

However, this has been studied further using 194 single-oospore isolates from germ-sporangia obtained through the single-zoospore technique developed by Laviola *et al.* (1978).

The aim of the present investigation was to study the pathogenicity as well as the compatibility type of the 194 single-zoospore isolates plus 29 single-oospore isolates, similarly established in a previous investigation (Laviola *et al.*, 1974).

MATERIALS AND METHODS

Isolates — The isolates of *P. infestans* used in this investigation were the 194 single-zoospore cultures and the 29 single-oospore cultures obtained from the crosses, in all compatible combinations, of isolate 63B, 60A, 473 and 455, whose characteristics are illustrated in Table 1. The F_1 isolates were derived as shown in Table 2.

Table 1.   Characteristics of the isolates of *P. infestans* used in this investigation.

Isolate	Compatibility type	Physiologic race	Origin Host	Origin Country
63B	A^1	3	Potato	Puerto Rico
60A	A^2	1,2	Potato	Mexico
473[a]	A^1	1,2,3,4	Potato	Mexico
445[a]	A^2	0	Potato	Mexico

[a] Isolate kindly supplied by Dr. C. Erwin, University of California, Riverside, California.

Table 2.   Number of single-oospore and single-zoospore isolates derived from the indicated crosses of *P. infestans*.

Cross	No. of single-oospore isolates	No. of single-zoospore isolates
63B x 60A	10	4
63B x 445	8	36
473 x 60A	3	14
473 x 445	8	140
Total	29	194

Physiological race identification - The physiologic race of
each F_1 isolate was identified according to the reaction of a series
of five differential hosts (Katahdin, r; Kennebec, R_1; 3RC-8, R_2;
Pentland Ace, R_3; 1563c(14), R_4) when inoculated with it. A minimum
of two pathogenicity tests were run with each isolate. When the
result of the second inoculation did not match that of the first
one, additional inoculations were made.

Compatibility type identification - For the compatibility type
identification each F isolate was double checked against the two A^1
(63B and 473) and the two A^2 (60A and 445) parental isolates. The
isolates which mated with 60A and 445 were identified as A^1, those
which mated with 63B and 473, as A^2.

RESULTS

Physiologic race identification - With the exception of
isolate 28 and isolate 159, which were lost during the course of the
investigation, all the F_1 isolates were tested for pathogenicity.
Most of them were pathogenic, and gave clear-cut reactions. Other
isolates required up to four separate inoculations to gain a clear
picture of their physiologic race. The pathogenicity of a few
isolates were never determined, even after five or six inoculations.

The pathogenicity character 4 occasionally appeared in four
single-oospore and thirteen single-zoospore isolates, even though
this character was not evident in either parent. This has been interpreted as spontaneous appearance in the sense of Gallegly &
Eichenmuller (1959), and as such was not taken into account in the
physiologic race identification.

Differential host 3RC-8 (R_2 genotype) reacted abnormally in a
number of pathogenicity tests. Relatively few lesions occurred, or
they were often small, slowly spreading and sporulated poorly.
Perhaps this characteristic partially accounted for the relatively
low frequency of pathogenicity character 2 among F_1 isolates. A
possible explanation for the 3RC-8 behavior could be the involvement
of a virus (mosaic) infection, evident in some plants, which may have
interferred with the fungal infection.

All of the possible physiologic races, except race 2 and race
2,4, were recovered from among the progeny of the four crosses studied (Table 3). In addition a few non-pathogenic isolates were
present among the progeny.

Compatibility type identification - Both A^1 and A^2 compatibility
types were recovered among both single-oospore and single-zoospore
F_1 isolates when isolate 445 was used as the A^2 parent, but only the
A^1 compatibility type was recovered when the A^2 parent was isolate
60A.

Table 3.   Genotypes recovered among the progeny of the four crosses
of *P. infestans.*

63B x 60A

    single-oospore   10, NP A^1

    single-zoospore   2, NP A^1; 1, 4 A^1

473 x 60A

    single-oospore   1, 2,3 A^1; 2, 1,2,4 A^1

    single-zoospore   2, 1,2 A^1; 2, 1,2,4 A^1; 2, 1,2,3,4 A^1

63B x 445

    single-oospore   1, NP A^1; 4, O A^1; 2, O A^2; 1, 4 A^1

    single-zoospore   2, NP A^1; 3, NP A^2; 5, O A^1; 2, O A^2; 2, 3 A^1;
                         1, 3 A^2; 2, 4 A^1; 1, 3,4 A^1; 1, 2,3,4 A^1

473 x 445

    single-oospore   1, NI A^1; 1, NP A^1; 1, O A^1; 1, O A^2; 1, O H;
                         1, 1 A^2; 1, 3 A^1; 1, 1,3,4 A^1

    single-zoospore   2, O A^1; 1, O A^2; 2, 1 A^1; 1, 1 A^2; 1, 3 A^2;
                         2, 4 A^1; 2, 1,2 A^1; 2, 1,3 A^1; 2, 1,3 A^2;
                         1, 1,3 H; 1, 1,4 A^1; 1, 2,3 A^1; 4, 3,4 A^1;
                         1, 3,4 A^2; 1, 1,2,3 A^1; 4, 1,3,4 A^1;
                         1, 1,3,4 A^2; 1, 2,3,4 H; 1, 1,2,3,4 A^1

NI: not identified as to pathogenic race; NP: non-pathogenic; H not
identified as to compatibility type because of abundant selfing.

<u>Frequency of appearance of pathogenicity and compatibility
type characters</u> - In instances where a number of single-zoospore
cultures were derived from one or two oospores, often two or more
genotypes (combination of pathogenicity and compatibility type
characters) were recovered, each being represented several times.
To study the frequency of appearance of compatibility types and
pathogenicity characters within the progeny of each oospore, only
a representative of each genotype was considered. Fifty-nine representatives were thus selected, three from cross 63B x 60A, 19
from the cross 63B x 445, six from the cross 473 x 60A, and 31 from
the cross 473 x 445. In addition, the 29 genotypes identified among
the single-oospore cultures have also been considered.

The ten single-oospore isolates established from cross 63B x 60A ($\underline{3}$ A^1 x $\underline{1,2}$ A^2) were non-pathogenic and all of A^1 type of compatibility. Among the three single-zoospore isolates from the same cross, two were non-pathogenic but one was race $\underline{4}$; all three of them were of A^1 type. Thus, the A^1:A^2 ratio among the progeny of this cross was 13:0. It is assumed that pathogenicity character $\underline{4}$ appeared spontaneously in the progeny since it was not present in either parent.

For the cross 63B x 445 ($\underline{3}$ A^1 x $\underline{0}$ A^2), out of eight isolates established from single-oospores, one was non-pathogenic, six were race $\underline{0}$ and one race $\underline{4}$. Two isolates of race $\underline{0}$ were of compatibility type A^2, whereas the other six isolates were of A^1 type. Among the 19 single-zoospore isolates, five were non-pathogenic, seven race $\underline{0}$, three race $\underline{3}$, two race $\underline{4}$, one race $\underline{3,4}$ and one race $\underline{2,3,4}$. The A^1:A^2 ratio among the single-oospore isolates was 6:2, and among the single-zoospore isolates was 13:6; the combined ratio was 19:8. Out of the 21 pathogenic isolates from this cross pathogenicity character $\underline{1}$ did not appear, character $\underline{2}$ appeared once, and the characters $\underline{3}$ and $\underline{4}$ each appeared five times; 13 of the 21 isolates were identified as race $\underline{0}$. The appearance of character $\underline{4}$ has been interpreted as spontaneous, whereas the appearance of the character $\underline{2}$ in the progeny, when not present in either parent, remains unexplained.

The frequency of appearance of pathogenic races and compatibility types among the progeny from cross 473 x 60A ($\underline{1,2,3,4}$ A^1 x $\underline{1,2}$ A^2) is as follows. Among the three isolates established from single-oospores, one was race $\underline{2,3}$ and two were race $\underline{1,2,4}$. Two of six single-zoospore isolates behaved as race $\underline{1,2}$, two as race $\underline{1,2,4}$, and two as race $\underline{1,2,3,4}$. All nine isolates were of compatibility type A^1. Thus, out of nine isolates, pathogenicity character $\underline{1}$ appeared eight times, character $\underline{2}$ nine times, character $\underline{3}$ three times and character $\underline{4}$ six times.

In the progeny from cross 473 x 445 ($\underline{1,2,3,4}$ A^1 x $\underline{0}$ A^2), out of the eight single-oospore isolates, one was not tested, one was non-pathogenic, three were race $\underline{0}$, one race $\underline{1}$, one race $\underline{3}$ and one race $\underline{1,3,4}$. The A^1:A^2 ratio among the single-oospore isolates was 5:2 (one was not identified). Among the 31 single-zoospore isolates, three were race $\underline{0}$, three race $\underline{1}$, two race $\underline{4}$, two race $\underline{1,2}$, five each of races $\underline{1,3}$; $\underline{3,4}$; $\underline{1,3,4}$ and one each of races $\underline{3}$; $\underline{1,4}$; $\underline{2,3}$; $\underline{1,2,3}$; $\underline{2,3,4}$; $\underline{1,2,3,4}$. Two single-zoospore isolates were not identified as to compatibility type, 22 were of type A^1 and seven of A^2 type. The combined ratio of A^1:A^2 for both single-oospore and single-zoospore isolates was 27:9. Among the 37 pathogenic isolates derived from this cross, pathogenicity character $\underline{1}$ appeared 20 times, character $\underline{2}$ six times, character $\underline{3}$ 22 times, and character $\underline{4}$ 16 times.

DISCUSSION AND CONCLUSION

The results obtained in this investigation clearly indicate that through the sexual process the pathogenicity characters segregate independently, apparently in a simple Mendelian manner, and that new physiologic races may arise by combination of pre-existing characters. They thus confirm and extend the results of Romero & Erwin (1969) on the occurrence of genetic recombination in *P. infestans*, already suggested from earlier studies (Gough, 1957; Smoot *et al.*, 1959; Savage & Gallegly, 1960). The results also show that segregation of pathogenic characters occurs upon oospore germination.

A more detailed account on these and other genetic implications will be given elsewhere. Here some of the implications of these findings and the role of the sexual stage in relation to epidemiology are discussed.

Genetic recombination in *P. infestans* through the sexual process is now definitely established. Where both compatibility types are present as occurs naturally in Mexico (Gallegly & Galindo, 1958), together with two or more physiologic races, there is the possibility that new and possibly more complex races may arise during a growing season by recombination of pre-existing pathogenicity characters. Under favourable environmental conditions, therefore, potato lines possessing even a high number of R-genes, but not protected by polygenic resistance, could be attacked and severely damaged. Consequently, it seems highly improbable that potato breeders could select, on the monogenic resistance basis, commercial varieties that would remain resistant for long.

The present knowledge on the origin of new physiologic races in *P. infestans* would favour, as a conclusion, the adoption of horizontal resistance. Indeed, potato breeders have long since switched to this type of resistance; but it could be suggested that in a potato breeding program both vertical and horizontal resistance should be considered. The combined incorporation of both types of resistance in a commercial variety would provide adequate protection, through horizontal resistance, against moderate attacks by any physiological race and simultaneously reduce selection pressure on the fungus. In such conditions the vertical resistance would be valid longer and so long as it remained effective, would provide complete protection against the physiologic races against which it is designed.

REFERENCES

Gallegly, M.E., and Eichenmuller, J.J., 1959, The spontaneous appearance of the potato race 4 character in cultures of *Phytophthora infestans*, Am. Potato J., 36:45-51.

Gallegly, M.E., and Galindo, A.J., 1958, Mating types and oospores of *Phytophthora infestans* in nature in Mexico, *Phytopathology* 48:274-277.

Gallegly, M.E., and Niederhauser, J.S., 1959, Genetic controls of host-parasite interactions in the *Phytophthora* late blight disease. *In*: Plant Pathology Problems and Progress 1908-1958; (Univ. Wisconsin Press, Madison) 168-182.

Gough, F.J., 1957, The sexual stage of *Phytophthora infestans*, the cause of potato and tomato late blight. Ph.D. dissertation, W. Virginia University, Morgantown.

Laviola, C., Gallegly, M.E., and Young, R.J., 1978, Genetics of *Phytophthora infestans*. I. Single-zoospore cultures from germ-sporangia and their significance in genetical studies, *Phytopath. medit.*, 17:39-44.

Niederhauser, J.S., 1956, The blight, the blighter and the blighted, *Trans. N.Y. Acad. Sci.*, 19:55-63.

Peterson, L.C., and Mills, W.R., 1953, Resistance of some American potato varieties to the late blight of potatoes, *Am. Potato J.*, 30: 65-70.

Romero, C.S., and Erwin, D.C., 1969, Variation in pathogenicity among single-oospore cultures of *Phytophthora infestans*, *Phytopathology*, 59:1310-1317.

Savage, E.J., and Gallegly, M.E., 1960, Problems with germination of oospores of *Phytophthora infestans*, *Phytopathology*, 50:573 (Abstr.).

Smoot, J.J., Gough, F.J., Lamey, H.A., Eichenmuller, J.J., and Gallegly, M.E., 1958, Production and germination of oospores of *Phytophthora infestans*, *Phytopathology*, 48:165-171.

Toxopeus, H.I., 1956, Reflections on the origin of new physiologic races of *Phytophthora infestans* and the breeding of resistance in potatoes, *Euphytica*, 5:221-237.

DURABLE RESISTANCE IN SELF-FERTILIZING ANNUALS

J.E. Parlevliet

Plant Breeding Department (I.v.P.)
Agricultural University
Wageningen, The Netherlands

SUMMARY

Durability of resistance varies greatly. It is easy to recognize good levels of resistance. The durability of such a resistance can hardly be assessed on small plots and in a short time span. But over the years experience has accumulated. Many resistances, both of a simple and of a more complex inheritance have shown to be durable. Several of these durable resistances appeared race-specific.

In those crop-pathosystems where introduced race-specific resistance genes break down rapidly, the cultivars often appear to carry reasonable levels of residual resistance. Breeders seem to accumulate other types of resistance along with the selection of race-specific resistance; there are no signs of a vertifolia effect.

When durable resistance is the aim the procedures to be followed depend on the crop-pathosystem.
1) When no problems of "loss of resistance" have been experienced despite extensive use of the resistance or when the pathogen concerned is not likely to be very adaptable the breeder should use any resistance including monogenic ones.
2) Partial, polygenic resistance, not confounded by major gene resistances is fairly easy to select for in the case of airborne and splash-borne diseases. Constant removal of the most susceptible phenotypes together with repeated recombination through crossing results in the accumulation of partial resistance. With soil-borne diseases the heritability is often so low that good selection techniques are not always available. Selection for high yields in

the presence of the pathogen can be useful in such cases.
3) Partial, polygenic resistance, however, is often confounded by
major gene resistances of short duration. In such cases the breeder
could remove not only the most susceptible phenotypes but also
those that show very little disease. This, together with repeated
recombination, may enhance the chances of accumulating partial
resistance.

There is, however, no approach that recognizes and selects
durable resistance with absolute certainty.

INTRODUCTION

The durability of resistance to parasites may vary greatly.
The problems of race-specificity and lack of durability occur
especially among self-fertilizing and clonal crops.

It is assumed that, before the advent of modern breeding
techniques, the agro-ecosystems were well balanced. Crops did not
show much disease; because they had reasonable levels of durable
resistance accumulated over large spans of time. This has changed
drastically. The rate of change in farming systems, in genetic
make up (plant breeding) and in the global interchange of crops
increased greatly. Such rapid changes in the agro-ecosystems are
inevitably accompanied by substantial imbalances. And the faster
one tries to remedy one imbalance the greater the chance that
other imbalances are overlooked. The breeding for disease resistance in self-fertilizing crops is a typical example of this. It
is easy to recognize and select genotypes that carry a high level
of resistance to a certain parasite. It is, however, impossible to
evaluate the durability of that resistance by testing it on small
plots and within a time span of a few years. A characteristic like
durability of resistance has such a low heritability that it can
only be recognized after vast testing in time and space (Johnson,
1978; 1979). No breeding program can accomodate this. However,
it is possible to increase the chance of selecting durable resistance when both the experiences from the recent past and the
available scientific information are used optimally.

DURABLE RESISTANCE

Johnson (1978; 1979) defined durable resistance as resistance
that lasts a long time when exposed to the parasite on a sufficiently large scale. Using this definition it is clear that
durable resistance is not rare. Durable resistance may be found
among simply and among complexly inherited resistances.

Examples of the former are the monogenic resistance in
cucumbers to *Cladosporium cucumerinum* and *Corynespora melonis* (25

years), in sorghum to *Periconia circinata* (25 years), in oats to
Helminthosporium victoriae (30 years) and in barley to *Puccinia
graminis* (T-gene, 40 years). Flax in the USA has been resistant
against *Fusarium oxysporum* f.sp. *lini* for some 40 to 50 years,
although initially a few resistance genes were overcome by
corresponding races. Especially against viruses resistance is often
durable like in cucumber to cucumber mosaic virus, in lettuce to
lettuce mosaic virus, in *Phaseolus vulgaris* to bean common mosaic
virus (gene I) and in potato to the viruses X, Y and A. See for
reviews Eenink (1976) and Parlevliet (1981). Meiners (1981),
reviewing the genetics of disease resistance in edible legumes,
concluded that most resistance is inherited in an oligogenic manner
and fairly often race-specific. However, in many cases the race-
specific resistance has not been of short duration.

   Of resistance with a more complex inheritance much less is
known. The partial resistance in barley to *Puccinia hordei*
(Parlevliet, 1981) and in potato to *Phytophthora infestans*
(Estrada and Turkesteen, 1979) appear polygenic and durable. These
partial resistances have been exposed to the pathogen on a vast
scale and over considerable periods and no signs of erosion have
yet been observed. The potato, although a clonally propagated crop,
illustrates this type of resistance very well. To the viruses
studied the potato cultivars vary in tolerance and in resistance
to infection. In the former there are less symptoms often because
of a reduced or retarded virus multiplication. The latter is
characterized by a smaller chance to become infected and diseased.
For both traits the potato clones vary greatly in a quantitative
way (Ross, 1977; Russell, 1978).

   The durability of resistance apparently varies widely. Race-
specificity, however, does not necessarily mean that the resistance is short-lived. Meiners (1981) concluded that race-specific
resistance in edible legumes is not always unstable. The large
variations in durability of race-specific resistances may be due
to: i) inherent differences in the genetic mechanisms underlying
the resistance as discussed by Parlevliet (1982). ii) to factors
in the environment that prevent the multiplication and spread of
the races of the pathogen virulent on the resistance genes used.
Such factors may be found in various crop protection measures or
being part of the agricultural system used. The race-specific
resistance of potatoes to the wart disease caused by *Synchytrium
endobioticum* has been very effective for more than 40 years
because of the accompanying phytosanitary methods.

   Irrespective of the causes plant breeders can and do use
the experience that some resistances have shown durability
while others broke down rapidly.

RESIDUAL RESISTANCE

In host-pathogen systems, where the pathogen is considered of some, but not of great importance the resistance accumulated in the crop seems sufficient to keep the pathogen at bay most of the time. In such cases plant breeders do not tend to select for resistance but rather to remove the most susceptible lines when the pathogen occurs. The barley-*Puccinia hordei* system is a good example. The average level of partial resistance is sufficient in Western Europe to keep leaf rust a minor disease. If, however, the barley acreage would be planted predominantly with very or extremely susceptible cultivars such as Midas or Akka, leaf rust would be a major pathogen in various regions of Western Europe.

In cases of severe disease pressure plant breeders started to select for resistance, and in this they became very efficient. Unfortunately the resistances selected so efficiently for were often effective for only very short periods. Barley-*Erysiphe graminis* f.sp. *hordei*, wheat-*Puccinia graminis* f.sp. *tritici*, wheat-*P. recondita* f.sp.*tritici*, wheat-*P. striiformis*, flax-*Melampsora lini*, lettuce-*Bremia lactucae*, tomato-*Fulvia fulva* and rice-*Pyricularia oryzae* are some of the examples. When the resistance breaks down, however, it does not fall back to that of the extremely susceptible cultivars. There appears to be some residual resistance. This can be illustrated with barley and its most important leaf pathogen, powdery mildew. European breeders have utilized race-specific resistance genes for more than 30 years. Parlevliet (unpublished) compared seven European cultivars, carrying different "broken" resistance genes, and their progenies obtained through repeated intercrossing, with lines taken randomly from the barley composite XXI produced in California. Table 1 shows that the European barley carried far less powdery mildew than the most susceptible composite lines. If these most affected (100%) composite lines represent the level of extreme susceptibility it is clear that the European barley cultivars and lines contain fair amounts of residual resistance.

From the lists of recommended cultivars published yearly by various countries it is possible to obtain some idea about the level of residual resistance. Often new cultivars enter these lists of recommended cultivars with a high value because of the presence of an effective race-specific resistance gene (Table 2). After some years this resistance gene becomes ineffective because of adaptation in the pathogen population. The resistance, however, never drops to levels similar to those of extremely susceptible cultivars or lines. Table 2 shows that the lowest values for barley to powdery mildew, the residual resistance, varies from 4 to 6 in the Dutch recommended cultivar lists and from 3 to 5 on the English lists (the lower values on the English lists compared with those on the Dutch lists relate to differences in the scoring and

Table 1. Percentage leaf area affected by powdery mildew
three weeks after heading of seven European barley
cultivars, of 200 of their progeny lines and of
200 barley composite XXI lines in 1977

Population	Leaf area affected (%)	
	Range	Mean
European cultivars	8 .- 23	15
European lines	7 - 35	15
Composite XXI	1 - 100	44

Table 2. Resistance values of spring barley to powdery mildew
according to the Dutch (D) and English (E) cultivar
lists. A 1 indicates extreme susceptibility, a 10
extreme resistance. On this scale Golden Promise
would probably score a 2 (D) or a 1 (E).

Cultivar	Resistance value at introduction		Lowest value for resistance		Years at lowest resistance	
	D	E	D	E	D	E
Union	8		6		10	
Volla	8		4[a]	3	1	1
Zephyr	> 6	> 6	4	3	9	8
Julia	7	5½	6	3	6	3
Mazurka	9	R[0]	5	5	7	6
Proctor		6½		3		14
Midas		> 6		5		10

a) after 8 years at a value of 6.

b) no figure was given, apparently not affected at all.

evaluation procedures). The most susceptible composite lines
(Table 1) would be considerably more susceptible than Golden
Promise. A one on either list might be flattering for such lines.
Similar residual resistances can be observed in other host-pathogen
systems were the resistance breeding is done through the incorporation of race-specific resistance. The values for residual
resistance in wheat to yellow rust and leaf rust on the Dutch
lists of recommended cultivars vary from 3 to 6. The most
susceptible cultivars, like Michigan amber for yellow rust, again
would score a value of 1.
It is not clear of what residual resistance consists of but it is
thought that partial resistance as described for barley to leaf
rust (Parlevliet, 1982) is a major component of it.

VERTIFOLIA EFFECT?

The concept of the Vertifolia effect (Van der Plank, 1968)
has been accepted fairly generally. This concept says that in the
presence of the major, race-specific resistance genes one cannot
select for race-non-specific resistance. The latter therefore
tends to get lost when the selection is for race-specific resistance. Van der Plank demonstrated this "Vertifolia effect" by
comparing six potato cultivars with and six without race-specific
(R) genes for resistance to *Phytophthora infestans* with races
virulent on these R genes. The cultivars carrying non-effective
R genes appeared to have considerably less field or partial
resistance than those without such R genes.

Van der Plank's reasoning seems logical. The observations,
however, contradict the universal occurrence of the Vertifolia
effect. If the Vertifolia effect exists the fair amounts of
residual resistance, even after prolonged periods of selection
for vertical resistance (VR), are difficult to explain. Even a
closer study of the potato-*Phytophthora infestans* relationship
does not reveal such a Vertifolia effect (Table 3). Of the 88
recommended potato cultivars almost half carried non-effective
R-genes. The average partial or field resistance of these
cultivars was somewhat higher than that of the cultivars without
R genes. According to the "Vertifolia effect" one expects a
considerable lower value for partial resistance.

The data of Table 3 and those concerning residual resistance
(Tables 1 and 2) suggest rather the contrary. If there is
selection for race-specific resistance there is simultaneously
some selection for partial resistance. How is this to be explained?
i) The VR genes for resistance do not always fully hide the
partial resistance. If studied in more detail the opposite appears
to be the case; partial resistance tending to be epistatic over
the race-specific hypersensitivity genes (Parlevliet, 1980). The
reduced infection frequency and the longer latent period,

Table 3.   Partial (field) resistance (1 extremely low,
            10 extremely high) of 88 potato cultivars to
            Phytophthora infestans according to the Dutch
            list of recommended cultivars of 1979.
            The cultivars are grouped according to their
            earliness (1 extremely late, 10 extremely early)
            and presence or absence of non-effective R-genes.

Range of earliness of cultivars	Partial resistance (mean and range)		
	with R-genes	without R-genes	range
$3\frac{1}{2}$ – 5	6.0 (15)[a]	6.3 (7)	4 – 8
$5\frac{1}{2}$ – $7\frac{1}{2}$	5.6 (19)	·4.8 (24)	3 – 7
8 – 10	5.0 (6)	4.3 (17)	3 – 6
all cultivars	5.5	5.0	3 – 8

[a] number of cultivars

components of partial resistance, can also be expressed when the
major genes are present (Clifford, 1974; Parlevliet, 1980) through
a reduced number of later appearing necrotic flecks. Plant breeders
therefore might tend to select lines carrying VR genes backed up
with partial resistance.
ii) The main reason, though, why the "Vertifolia effect" does not
generally occur, lies in the choice of the parents to be crossed.
The plant breeder, seeking resistance to a given pathogen, looks
for a good source of resistance. When he has found such a source,
often a cultivar or line introduced from somewhere else, he has to
cross it with a locally adapted good cultivar. In the process of
choosing this latter cultivar he tends to take one that is not all
that susceptible to the pathogen. Extremely susceptible cultivars
have little chance to enter such crossings. The parent with the VR-
gene too may carry variable levels of partial resistance. In fact
cultivars or lines without any partial resistance are rare
(Parlevliet and Kuiper, 1977; Niks and Parlevliet, 1978; Parlevliet
et al., 1980). The parents that constitute the basis of the
selection program therefore tend to carry moderate to fair levels
of partial resistance on top of the VR. The plant breeders already
do (consciously or semi-consciously) what is often advised; they
combine so called vertical and horizontal resistance. This, however,
does not seem to slow down the rate of "breaking down" of VR as
exemplified by the barley-powdery mildew and wheat-yellow rust
pathosystems in Western Europe. A more conscious approach toward
the selection of partial resistance, though, might increase the
still insufficient levels of residual resistance in such cases.

SELECTION FOR DURABLE RESISTANCE

When durable resistance is the goal in a breeding program it is necessary to realize that there is no guarantee that the resistance selected for is indeed durable. Only time and exposure on a large scale can give us the definite answer. But it is possible to increase the probability of durable resistance considerably by concentrating the breeding efforts on resistances that have lasted for a considerable time spab already and by avoiding types of resistance that have shown notoriously short duration.   Among the crop-pathosystems it is possible to discern three categories:
1) Those where simply inherited resistances have shown or can be expected to be durable.
2) Those where partial, polygenic resistance is present without the confounding effects of major VR genes.
3) Those where partial, polygenic resistance and major VR genes both occur.

1) <u>Simply inherited resistance</u>. In the section of "Durable resistance" a number of simply inherited resistances were mentioned, and there are more, that have shown durability even though such genes could be race-specific. For the edible legumes Meiners (1981) concluded that "Pathologists and breeders should continue to identify and use specific resistance wherever it can be found, because experience indicates that it is adequate for a majority of diseases of edible legumes". In some cases the lasting effect of VR genes is due to accompanying measures that reduces the chances of the formation and spread of corresponding races as with the potato-*Synchytrium endobioticum* pathosystem in Europe. If this is the case the exploitation of such resistance genes might lead to disappointment when done outside the agro-ecosystem where they proved to be durable.

New resistance genes are regularly found but information concerning their durability is lacking. The plant breeder might then look at the experience acquired with resistance genes to similar pathogens. If resistance to these pathogens tends to be durable one certainly should try the new resistance gene. As a rough guide one can state that the resistances that are so alarmingly short lived tend to be resistances to biotrophic or biotroph-like pathogens except viruses.

Simply inherited resistances are of great value in plant breeding. They are easy to manipulate and often easily recognized. One should use them unless there are good reasons to assume a short life span of the resistance concerned.

2) <u>Partial, polygenic resistance in the absence of major</u>
<u>VR-genes</u>. Some examples are the partial resistance in barley to
P. hordei (Parlevliet, 1982), in wheat to *Leptosphaeria nodorum*
(Brönniman, 1975; Scott and Benedikz, 1977; Rosielle and Brown,
1980), in oats to *Puccinia coronta* (Luke et al, 1975; Simons,
1966; Politowski and Browning, 1978), in potato to various viruses
(Ross, 1977; Russell, 1978), and in pyrethrum to root knot nematodes (Parlevliet, 1976). In some cases the term tolerance has
been used erroneously (wheat-*L. nodorum*, oats-*P. coronata* and
pyrethrum-*Meloidogyne hapla*).

Selection for this type of resistance depends on the pathosystem. If the heritability of the partial resistance is fairly
high (barley-*P. hordei*, wheat-*L. nodorum*, oats-*P. coronata*) the
selection is not difficult; one selects the more resistent phenotypes. Parlevliet et al. (1980) showed that the partial resistance
of barley to *P. hordei* can be selected for efficiently at any stage
of the breeding program, i.e. in the seedling stage, in the
individual adult plant stage in the greenhouse as well as in the
field and in the small plot stage. Even the removal of the most
susceptible phenotypes is effective. Of the European and the
Composite XXI barley populations discussed in the section on "Residual
resistance" 5000 individual plants were selected in 1976. In each
population the 30 percent of plants with the highest level of powdery
mildew were removed. The remainder showed a level of powdery mildew
of 12 percent (European population) and 37 percent (Composite XXI)
when tested in 1977. The unselected control populations (Table 1)
had 15 and 44 percent leaf area affected respectively.

If the heritability of partial resistance is lower one cannot
select effectively on a single plant basis anymore. Resistance to
infection of potato cultivars to virus diseases has to be selected
for in the field on fairly large plots because one can only identify cultivar differences by differences in the percentage of
plants infected (showing virus symptoms). The potato cultivars
Clivia, Grata and Prima for instance show a high, intermediate and
low percentage of plants infected respectively when exposed to the
leaf roll virus. This resistance to infection is virus-specific;
with virus Y the same potato cultivars show an intermediate, a
high and an intermediate percentage respectively (Ross, 1977).
With soil pathogens the heritability of partial resistance is often
very low. Selection for it is very difficult. A practical approach
then is to select for high yields (if that is the main selection
criterion) in the presence of the pathogen as suggested and carried
out in pyrethrum for its main pest, the root knot nematode (Parlevliet, 1976). This approach can also be followed when the heritability
is higher (Brönnimann, 1975; Simons, 1966).

When selection for partial resistance is practised it is easy
to under- or over evaluate the level of resistance.

i) The assessment of the amount of disease or pathogen is normally done at one moment. Differences in development stage of the host cultivars at that moment may result in incorrect evaluations. The partial resistances of late genotypes are over, those of early genotypes under evaluated.

ii) Differences in the amount of inoculum applied. Too low dosages may result in escapes. In order to avoid such escapes the inoculum dosage is often taken too high. Under such conditions major gene resistance shows up very well, partial resistance, however, may largely disappear.

iii) Stage and time of assessment. Partial resistance is not always fully expressed. In the cases of cereal-rusts and cereal-powdery mildew the differences in partial resistance are largest just after heading and smallest in the seedling stage (Parlevliet, 1981). The plant breeder should therefore assess the amount of disease or pathogen at the right stage of the plant. Even at the right stage it is important to realize that relatively small differences in the time of assessment can be important. In the case of leaf pathogens in the field the best moment is in general when the most susceptible lines approach the maximum level of disease. At an earlier date the differences between lines have not yet reached the maximum values, at a later date they tend to disappear again.

iv) Interplot interference. Especially with wind-borne pathogens the level of partial resistance can be seriously under estimated when the cultivars or lines are compared in the usual way, i.e. in small, adjacent plots (Van der Plank, 1968; Parlevliet and Van Ommeren, 1975). Table 4 gives an impression of the degree in which the partial resistance of cultivars like Julia and Vada are under estimated compared with susceptible cultivars like Sultan and L98. The plots isolated from one another give a good assessment of the level of partial resistance as experienced by the farmer. Although the differences between lines are considerably reduced in adjacent plots compared with the real situation, the farmer's fields, the ranking order remains the same (Parlevliet and Van Ommeren, 1975; Parlevliet et al., 1980).

It should be emphasized that the expression of major gene resistance is far less sensitive to the aspects of assessment mentioned above than partial resistance. Strong selection for resistance therefore tends to select major gene resistance if present, and one is often not certain that they are not. The best way to accumulate polygenic resistance is by constantly removing the most susceptible phenotypes only and to recombine frequently. This leaves also ample room for the simultaneous improvement of other characteristics.

3) Partial, polygenic resistance in the presence of major VR-genes. When the aim is partial resistance everything that has been discussed above under 2) is valid here too. The selection for partial resistance, however, becomes considerably more difficult

Table 4. Number of *Puccinia hordei* urediosori per tiller on
five spring barley cultivars approximately four
weeks after heading in 1973 at three different field
plot situations.

	Field plots		
Cultivar	Isolated	Adjacent	
	3 x 4 m	2 x 2 m	$\frac{1}{4}$ x 1 m
L98	1000	500	2300
Sultan	750	250	1700
Volla	110	40	700
Julia	17	12	450
Vada	1	15	130
Range	1000x	42x	18x

because of the confounding effects of the major genes. Strong
selection for resistance unavoidably selects the major genes, but
even moderate selection for lines or plants that show some disease
of a susceptible infection type does not safeguard against the
selection of major genes especially not when a mixture of races is
present. Fairly often a mixture of races is used with the idea that
most of the major genes become ineffective. This is not so; they
become at best partially ineffective and Table 5 illustrates what
the results can be. When a series of host cultivars varying in
R genes are exposed in the field to a mixture of races most of the
R genes are only partially neutralized. Depending on the frequency
of the races that are virulent on the various cultivars one may get
large differences in amounts of disease (of a susceptible type).
These differences, though, do not reflect differences in partial
resistance, they merely reflect which major genes are still most
effective (to that mixture of races). Unfortunately it is often
not possible to exclude this situation because the naturally
occurring pathogen population is most likely a mixture. In that
case it is almost impossible not to avoid the selection of major
genes. One increases the chances to select partial resistance, by
removing only the most susceptible lines or plants together with
the most resistant ones.

If one can control the composition of the pathogen population
in the field by introducing the inoculum well in advance of the
natural inoculum one should use the race (from the regionally
occurring population) with the widest virulence spectrum. With only
one race predominantly present both the partial resistance and the

Table 5. Amount of disease one might observe if host genotypes
of a self-fertilizing crop are affected by a wind borne
leaf pathogen when the host genotypes carry different
race-specific resistance genes and the pathogen
population consists of various races.

Resistance genes host cultivars	Virulence genes pathogen races				accumulated amount of [b] disease
	V1 (40) [a]	V1V2 (40)	V1V3 (19)	V1V2V3 (1)	
R1	+ [c]	+	+	+	70%
R2		+		+	55%
R1R3			+	+	40%
R1R2R3				+	10%

a) Percentage in the initial inoculum.
b) Expressed as leaf area affected by infections of a susceptible
   type.
c) a + indicates that the host is susceptible for that race.

major gene resistance are expressed much better, the degree of
confounding being strongly reduced. VR genes with an incomplete
expression, however, are extremely difficult to distinguish from
partial resistance. The chance of recovering partial resistance
is best when the plant breeder removes in each selection phase the
most susceptible and the highly resistant (not or hardly affected)
lines or plants.

It must, however, be realized that resistance cannot be
classified unambiguously into two groups nor is there an easy way
to discern durable from non-durable resistances. There is no
simple characteristic by which one can always discern polygenic,
partial resistance from monogenic hypersensitive resistance
without fail. Only a sound knowledge of host-pathosystems in
general and of the host-pathosystem implicated in particular is a
good guarantee that one uses the resistance that is most
appropriate for the situation concerned.

REFERENCES

Brönniman, A., 1975, Beitrag zur Genetik der Toleranz auf *Septoria nodorum* Berk. bei Weizen (*Triticum aestivum*), *Z. Pflanzen züchtung*, 75:138-160.

Clifford, B.C., 1974, Relation between compatible and incompatible infection sites of *Puccinia hordei* on barley, *Trans. Brit. Myc. Soc.*, 63:215-220.

Eenink, A.H., 1976, Genetics of host-parasite relationships and uniform and differential resistance, *Neth. J. Pl. Path.*, 82: 133-145.

Estrada, N. & Turkesteen, L., 1979, Breeding for resistance to late blight at CIP. *Report of the planning conference on the control of important fungal diseases of potatoes*, CIP, Lima, Peru, 1978:57-63.

Johnson, R., 1978, Practical breeding for durable resistance to rust diseases in self-pollinating cereals, *Euphytica*, 27: 529-540.

Johnson, R., 1979, The concept of durable resistance, *Phytopathology*, 69:198-199.

Luke, H.H., Barrett, R.D. & Pfahler, P.L., 1975, Inheritance of horizontal resistance to crown rust in oats, *Phytopathology*, 65:631-632.

Meiners, J.P., 1981, Genetics of disease resistance in edible legumes, *Ann. Rev. Phytopathol.*, 19:189-209.

Niks, R.E. & Parlevliet, J.E., 1978, Variation for partial resistance to *Puccinia hordei* in the barley composite XXI, *Cereal Rusts Bull.*, 6 (2):3-10.

Parlevliet, J.E., 1976, Breeding pyrethrum in Kenya, *Pyrethrum post*, 13:48-54.

Parlevliet, J.E., 1980, Minor genes for partial resistance epistatic to the Pa 7 gene for hypersensitivity in the barley-*Puccinia hordei* relationship, *Proc. 5th Eur. & Medit. Cereal Rusts Conf.*, Bari,:53-57.

Parlevliet, J.E., 1981, Disease resistance in plants and its consequences for plant breeding, in "Plant Breeding II", K.J. Frey, ed., The Iowa State Univ.Press, Ames, Iowa, 309-364.

Parlevliet, J.E., 1982, Models explaining the specificity and durability of host resistance derived from the observations on the barley-*Puccinia hordei* system, *Proc. Conf. on Durable resistance in crops*, Bari 1981, 57-80

Parlevliet, J.E. & Kuiper, H.J., 1977, Resistance of some barley cultivars to leaf rust, *Puccinia hordei*; polygenic resistance hidden by monogenic hypersensitivity, *Neth. J. Pl. Path.*, 83: 85-89.

Parlevliet, J.E., Lindhout, W.H., Ommeren, A. van & Kuiper, H.J., 1980, Level of partial resistance to leaf rust, *Puccinia hordei*, in West-European barley and how to select for it, *Euphytica*, 29:1-8.

Parlevliet, J.E. & Ommeren, A. van, 1975, Partial resistance of
    barley to leaf rust, *Puccinia hordei*. II. Relationship between
    field trials, micro plot tests and latent period, *Euphytica*,
    24:293-303.
Politowski, K. & Browning, J.A., 1978, Tolerance and resistance to
    plant disease. An epidemiological study, *Phytopathology*, 68:
    1177-1185.
Rosielle, A.A. & Brown, A.G.P., 1980, Selection for resistance to
    Septoria nodorum in wheat, *Euphytica*, 29:337-346.
Ross, H., 1977, Methods for breeding virus resistant potatoes.
    *Rep. Planning Conf. on Development in the control of potato
    virus diseases*. C.I.P., Lima:93-114.
Russell, G.E., 1978, "Plant Breeding for pest and disease
    resistance", Butterworths, London and Boston.
Scott, P.R. & Benedikz, P.W., 1977, Field techniques for assessing
    the reaction of winter wheat cultivars to *Septoria nodorum*,
    Ann. Appl. Biol., 85:345-358.
Simons, M.D., 1966, Relative tolerance of oat varieties to the
    crown rust fungus, *Phytopathology*, 56:36-40.
Van der Plank, J.E., 1968, "Disease resistance in plants", Academic
    Press, New York and London.

DISCUSSION

KRANZ: I believe that, in the tropics, the essential aspect is to
maintain diversity and that acceptance of concepts leading to a
narrowing of the genetic base is one of the chief mistakes adopted
from developed countries. A second important aspect may be the
feasibility of breeding techniques, which should be amenable to
local facilities.

PARLEVLIET: Breeding should, indeed, be on-site under conditions
similar to the agro-system in which the varieties resulting from the
breeding program are to be used. A good breeder will, of course,
ensure sufficient variation in his breeding program.

DE PONTI: I have a comment and a question. Firstly, the comment:
Breeding and distribution policies of some of the international
institutes have greatly increased disease and pest potential through:
a) the distribution of single varieties over millions of hectares,
thereby creating super-monocultures; b) the development and
distribution of F_1 hybrid varieties, which are extremely uniform,
instead of open-pollinated varieties, which possess inter-varietal
diversity. Secondly, my question. You advocated the use of a
single-race inoculum to detect race non-specific resistance. However, in this conference, it has also been argued for to select
preferably in the field (where mostly race-mixtures are present).
How do you think this problem can be solved?

PARLEVLIET: I agree, in principle, with your comment. Concerning your question in a number of cases, the variable natural pathogen population has to be accepted in the selection procedure. However, when control is possible, one should use a single race.

SHARP: What caused the durability of the resistance of the cultivars Felix and Manella, and what happened with the offspring of these cultivars?

SILFHOUT: The variety Manella probably has a combination of three to four genes. The selection was a very lucky one; at that moment, races possessed virulence to some of the resistance genes but not to the two to three gene combinations. Thus, this is an example of an accidental pyramiding of resistance genes.

PARLEVLIET: I agree that in both varieties the resistance was probably due to a lucky combination of genes conferring VR. The use of these varieties as parents in breeding programs resulted in the break up of the combination into single genes, to which matching pathogen races were present.

VAN DER GRAAFF: Minor gene effects for disease resistance have a low heritability. Through better, quantitative, scoring methods, it may be possible to increase heritability and the effectiveness of the selection. Presently, many scoring methods are meant to measure qualitative differences.

PARLELIET: This is not a general rule. Minor genes against foliar pathogens, such as rusts and powdery mildews, have a reasonably high h^2. In fact, I found it easy to select for plants with a higher number of minor genes. With other pathogens, especially soil pathogens, the h^2 of minor-gene resistance may be low. In these cases, better assessment methods may certainly be worthwhile considering.

ESKES: In tropical areas, monogenic resistance often has a shorter effective lifetime than in temperate zonex. This may be related to the lack of a dead-season, a higher inoculum pressure, the use of few phytosanitary control methods, and, due to the fact that many crops are new and relatively unadapted to tropical countries, a relatively low level of partial resistance. Thus, in tropical areas other breeding methods may be needed than those used in temperate zones to obtain durability of resistance.

PARLEVLIET: In principle, I agree that other breeding methods may be needed. Concerning your arguments, I assume that in annual crops in the tropics "dead season" type "bottle-necks" occur for pathogen populations. Also, I doubt whether "break-down" of resistance occurs much more rapidly in the tropics; for instance, we do not know how often resistances break down during the actual breeding process in temperate zones.

ZADOKS: Your analysis would gain in depth if you would separate your data into those in which regulatory interference has strongly contributed to the durability of resistance (many potato diseases) and those in which regulatory interference has not occurred (many cereal diseases).
Secondly, it seems to me that many monogenic resistances have been durable, especially in those cases where the infection cycle of the pathogen is of a long duration. Finally, it has often been said that we, as a group, are in a hurry to produce resistant cultivars, when we are working for or in the tropics. In my opinion, any hurry is counter-productive with respect to dura-bility of resistance. The breeder of annual crops in the tropics has to take his time and he must take care not to exaggerate disease pressure for the purpose of working fast.

PARLEVIET: I agree with the comment in the first part of your question. Concerning the second part of your quesiton, I believe that the elusive type of resistance occurs especially vis-a-vis specialized biotropic-like parasites. Concerning the speed of the breeding process, one cannot speed up the selection of "adapted" varieties (including durable resistance) without losing part of the adaptability complex.

ROBINSON: In one of the discussions, Buddenhagen mentioned that wheat and barley breeding methods produced a "boom and bust" cycle in rice vis-a vis rice blast. The approach of Buddenhagen would produce "stop-gap" cultivars, which are rapidly replaced with a steady progression of increasingly superior cultivars.

PARLEVLIET: This approach is all right, provided we indeed can discern the more durable forms of resistance from the less durable ones, but this is not often the case. One can increase the chance of durability by exerting less selection pressure for disease resistance; this was one of the main points of my paper: select against susceptibility rather than for resistance.

INTERNATIONAL TESTING AS A MEANS TO IMPLEMENT DURABLE RESISTANCE:

THE EXPERIENCE OF CIMMYT WITH WHEAT RUSTS

Enrique Torres and Sanjaya Rajaram

Pathologist and Breeder, respectively
Wheat Program, CIMMYT
Apdo. 6-641, Delegacion Cuauhtemoc
06600 Mexico, D.F., Mexico

INTRODUCTION

Populations of plant pathogens change rapidly through the selective increase of those isolates capable of overcoming the resistance genes in commercial cultivars. Durable disease control is needed to avoid disruptions in food supply. The task of obtaining durable resistance must blend a breeding component of finding and combining genetic mechanisms and an epidemiological design of cultivar management.

Breeding should make use of 1) broad based resistance, a concept that implies effective combinations of resistance genes, and 2) dilatory resistance (Browning, Simons and Torres, 1977) which reduces the rate of disease development, as in slow rusting. Cultivars with proven lasting resistance to local pathogen populations are reliable choices as sources of durable resistance (Johnson, 1978).

While resistance remains transitory, strategies for disease containment should be designed to prolong the effectiveness of resistance genes. To establish these strategies some characteristics of the pathogen population must be known, such as virulence structure, distribution of biotypes, and spreading pathways for the inoculum.

Multilocation testing, the procedure of evaluating germplasm in a number of locations, provides a means to assess the spectrum of resistance available in various cultivars. CIMMYT, the International Maize and Wheat Improvement Center, has adopted multilocation testing to group cultivars of similar range and pattern of resistance, and to define areas of similar spectra of virulence.

CLASSIFYING CULTIVARS FOR EXTENT AND INTENSITY OF RESISTANCE

The wheat program of CIMMYT releases international and regional nurseries that contain the most promising bread wheat, durum wheat, triticale and barley cultivars in the base program in Mexico, or in a region.   Thus, the International Bread Wheat Screening Nursery (IBWSN) is assembled with the best advanced lines in the bread wheat program.   The Regional Disease and Insect Screening Nursery (RDISN) and the Latin American Disease Observation Nursery (VEOLA) consist of superior cultivars from the Eastern Hemisphere and Latin America, respectively.

Cooperators assess rust levels in these nurseries as described by Statler and Nolte (1979): severity of infection is measured in the Cobb scale (0-100) and host reaction is determined within the range of infection types R, MR, MS, and S.   To evaluate results, infection types are given numerical values of 0.2, 0.4, 0.8, and 1.0 respectively.   Each datum is transformed from Cobb scale to a coefficient of infection (CI) by multiplying the numerical value of its infection type by its severity.

Cultivars with CI below 10 are regarded as resistant.   Cultivars with CI up to 30 are considered to have moderate resistance that can be useful in containing epidemics.   Cultivars with CI higher than 30 are classified as susceptible.   Multilocation rust assessments of a diversified wheat germplasm should yield at least four classes of cultivar response across sites: (1) cultivars resistant at all sites, (2) cultivars susceptible at all or most sites, (3) cultivars resistant at most sites and susceptible at few sites, and (4) cultivars resistant to moderately resistant at all sites.   Tables 1 and 2 give examples of these four classes of cultivar response for leaf and stem rust, respectively.

Cultivars in class 1 perform as resistant across sites and have a low average CI (ACI).   The resistance genes of these cultivars appear to be effective against pathogen populations with a wide range of virulence.   Hence, cultivars in this class are likely candidates as sources of durable resistance.   Class 2 is made up of cultivars that are susceptible at all or most sites.   Their resistance genes can be overcome by most pathogen populations.   Therefore, these cultivars should be grown only under low disease occurrence.

Class 3 consists of cultivars that have effective resistance except at a few places, and national or regional programs may release cultivars in this class when resistance is effective locally. However, the resistance base in these cultivars is relatively narrow and resistance is likely to be transitory.   Thus, cultivars in this class will be particularly vulnerable to virulence shifts and must be protected by management practices such as growing as diverse an array of cultivars as feasible.   Cultivar diversity by itself, however,

Table 1.   Multilocation data on leaf rust infection (average coefficient of infection and Cobb scale) of five wheat cultivars
that illustrate four classes of cultivar response across
sites
Source: CIMMYT's 13th International Bread Wheat Screening
Nursery (IBWSN)[a]

Response class	Cultivar	ACI[b]	Guana- juato Mexico	Parana Brazil	Buenos Aires Argentina	Itapua Paraguay
1	Bobwhite"S"	0.9	0	0	0	TM
2	Pavon"S"	21.2	60S	60S	80S	70S
3	Pvn"S"(Pato(R)- Cal/7C x Bb-Cno)	2.9	40S	TR	0	TMSS
	Maya 74"S"-Mon"S"	5.1	TR	TR	0	60S
4	Trush"S"	3.5	TR	20MS	10S	TS

a   13th IBWSN consists of 530 entries
b   Average coefficient of infection over 40 sites

does not assure genetic diversity, because different cultivars may
carry similar resistance genes.   International nurseries contribute
to elucidate genetic similarities among cultivars in class 3.   Relationships among cultivars can be determined from their response
across locations.   In Table 1, cultivars Pvn"S"(Pato(R)-Cal/7C x Bb-
Cno) and Maya 74"S"-Mon"S" are susceptible to different leaf rust
populations, namely those of Guanajuato and Itapua respectively, and
thus appear to have different leaf rust resistance genes.   They may
therefore be used in a disease control program based upon genetic
diversity, as elements of a cultivar deployment strategy.   In Table
2, cultivars Hahn"S" and Pj62-Fury x Wop"S" do not appear to have a
differential interaction with stem rust populations.   Based on the
fragmentary data presented in Table 2 it would not be justifiable
to use these two cultivars in a similar deployment strategy.

Cultivars in class 4 are moderately resistant across sites.
The occurrence of R host reaction for cultivar Trush"S" at Guanajuato
(Table 1) and for cultivar Burgus 2.Sort 12.13 x Kal-Bb at New South
Wales (Table 2) suggest the presence of genes for specific resistance
in these cultivars.   However, when reactions become MS and S as in
all other locations in Tables 1 and 2 for these cultivars, disease
severity is kept moderate.   This effect may be due to low frequency
of isolates virulent on these cultivars.   Nevertheless, regular
occurrence of moderate disease levels on a given cultivar suggests
the effect of some mechanism of dilatory resistance.   Class 4 cultivars are also regarded as _possible_ sources of durable resistance.

The ACIs for classes 3 and 4 are usually very similar, and are
not useful for cultivar classification.   Proper grouping of cultivars

Table 2.   Multilocation data on stem rust infection (average coefficient of infection and Cobb scale) of five wheat cultivars
that illustrate four classes of cultivar response across
sites
Source: CIMMYT's 13th International Bread Wheat Screening
Nursery (IBWSN)[a]

Response class	Cultivar	ACI[b]	Cairo Egypt	Rift Kenya	Izmir Turkey	New South Wales Australia	Santa Cruz Bolivia
1	Kvz-K4500.L.A.4	0.3	0	0	0	0	0
2	Anza	24.7	10S	30S	80S	30S	10MS
3	Hahn"S"	4.7	0	0	60S	0	-
	Pj62-Fury x Wop"S"	20.5	0	10S	100S	5MR	60MS
4	Burgus 2.Sort 12.13 x Kal-Bb	5.3	5MSS	5S	20MS	TR	TS

a   13th IBWSN consists of 530 entries
b   Average coefficient of infection over 23 sites

requires a complete scanning of data across locations. Variance for
class 3 cultivars would be larger due to high rust severity in one
or more locations, relative to that of class 4 cultivars that are
defined as being steadily resistant to moderate resistant.

In conclusion, cultivars in classes 1 and 4 are candidates to
have broad based resistance that is likely to be durable as well,
but certainly requires the crucial test of time.

CLASSIFYING SITES FOR THE RANGE OF VIRULENCE IN RUST POPULATIONS

CIMMYT issues two trap nurseries each year: the Regional Disease
Trap Nursery (RDTN) and the Latin American Rust Nursery (ELAR). Trap
nurseries consist of single-gene differential cultivars for the three
wheat rusts, current and past commercial cultivars, and promising
advanced lines. These nurseries monitor the virulence structure of
rust populations and detect new virulence patterns that may pose a
threat to advanced lines and/or commercial cultivars.

Data from trap nurseries and regional screening nurseries help
to pinpoint epidemiological units, based upon boundaries that may
have been suggested by data from international screening nurseries.
Analysis of rust infection data from the RDTN has allowed the
Eastern Hemisphere to be divided into four epidemiological units
(Saari, E.E., personal communication, 1978) 1) the Indian subconti-
nent, 2) Middle East and Eastern Europe, 3) the Maghrebian and
Iberian countries, and 4) East Africa and the Yemens. ELAR data
suggest the existence of three major epidemiological units in South
America, the Andean unit that stretches along Colombia, Ecuador,

Peru and into western Bolivia; the Chilean unit; and the Eastern Southern Cone unit that may cover large parts of Argentina and Brazil. Other epidemiological units of a smaller size have also been suggested, such as the Abapo-Izozog district in south eastern Bolivia. The relationships among all these units should be clarified as more data become available (Dubin, H.J., unpublished data, Quito, 1980).

Pathways for inoculum movement can be charted from changes in the structure of pathogen populations within an area. There appears to be relatively free movement of inoculum within each of the regions or units described above. For the Eastern Hemisphere there are indications of occasional interchanges of inoculum between region 2 which is centrally located, and peripheral areas of regions 1, 3, and 4 (Saari, E.E., personal communication, 1978). Similar phenomenon may occur between the Andean and Chilean units (Kohli, M.M., personal communication, Santiago de Chile, 1981).

SUMMARY AND CONCLUSIONS

We have shown how CIMMYT international nurseries are used to identify cultivars with low or moderate disease incidence worldwide. Such cultivars should be further tested as sources of durable resistance. Regarding gene management, nursery data facilitate the grouping of cultivars into clusters of similar genetic makeup, as well as the definition of epidemiological units and pathways. These geopathological data are necessary for strategic geographic placement of varieties (CIMMYT, 1980), and for efficient warning systems against new threatening pathogen species or biotypes. All these components of breeding and gene management add to durability of disease.

REFERENCES

Browning, J.A., Simons, M.D., and Torres, E. 1977. Managing host genes: epidemiologic and genetic concepts, *in:* "Plant Disease -- an advanced treatise", Vol. 1, J.G. Horsfall, and E.B. Cowling, ed., Academic, New York.

CIMMYT, 1980, CIMMYT looks ahead, El Batan, Mexico.

Johnson, R., Practical breeding for durable resistance to rust diseases in self-pollinating cereals, 1978, *Euphytica* 27: 529-540.

Statler, G., and Nolte, P., 1979, Wheat leaf rust in North Dakota in 1977 and 1978, *Plant Dis. Rep.*, 63: 336-340.

BREEDING FOR DISEASE RESISTANCE IN WHEAT, THE ZAMBIAN EXPERIENCE

W.A.J. de Milliano

Temporarily: Laboratory of Phytopathology
Binnenhaven 9
6709 PD WAGENINGEN
The Netherlands

INTRODUCTION

In 1976 the Horizontal Resistance Breeding Program for Wheat
(HRBP) was started in Zambia, by the Food and Agriculture Organization (FAO) of the United Nations, as part of the International
Program for Horizontal Resistance (IPHR). This program was based on
theories developed by Vanderplank (1963, 1968, 1975) and Robinson
(1976). Dr. L. Chiarappa, Mr. R.A. Robinson, Dr. W.C. James and
Mr. C. Keller (FAO) were advisors to the program between 1976 and
1978. In 1979 the program became part of the Dutch Technical cooperation with Zambia, and Professor Dr. J.C. Zadoks, Dr. J.E. Parlevliet
and Dr. A. Darwinkel became advisors to the program. In 1978 advice
on specific problems was also given by Dr. I. Buddenhagen, of the
International Institute for Tropical Agriculture (IITA) in Nigeria.
Ir. W.A.J. de Milliano was head of the project up until July 1981
when he was succeeded by Ir. P. Groot. Apart from the salaries of
two expatriats, the project has been mainly financed by the Research
Section of the Department of Agriculture in Zambia.

The aim of the program was to develop new wheat cultivars with
horizontal (durable) resistance to all important local pests and
diseases, and which were also good yielding. Cultivars were to be
developed for the warm and wet rainy season (mid-November to mid-
April) and the cool part of the dry season, referred to as the
irrigation season (mid-April to September). The priority was the
development of rain-fed wheat. However, the project was set up in
an area which was not suitable for growing wheat during the rainy
season but, instead, was suitable for irrigated wheat. Thus more
attention was given to irrigated rather than rain-fed wheat.

At present Zambia is producing 10 per cent of its requirements for wheat (Hurd, 1981). Zambia imports annually about 100,000 mt at the expense of much needed foreign currency, and the demand for wheat is still increasing.

In 1976 when the project began, wheat was not a completely new crop to Zambia, but many problems concerning the cultivation of wheat needed to be solved such as crop husbandry, epidemiology of pests and diseases, and finding suitable locations for wheat growing. Only a few cultivars recently introduced from Zimbabwe, and also from the International Institute for the Improvement of Wheat and Maize (CIMMYT), were grown commercially, mainly in the irrigation season.

MATERIALS AND METHODS

Location

The first work in the development of new lines was done in the Southern Province on the National Research Station (NIRS), located at 15° 46' S latitude, 27° 55' E longitude, at 987 m above sea level (SL). At that time, NIRS had the best infrastructure in Zambia for wheat breeding. There were several members of staff with expertise in wheat cultivation and a number of laborers including 2 for the HRBP; a large collection of wheat cultivars; equipment for land preparation; an office and a laboratory for HRBP and also housing; and because of the irrigation facilities, at least two generations of wheat could be grown per year. However, the NIRS had the dis-advantage, that it was located in an area which was not suitable for the cultivation of rain-fed wheat. The temperatures were too high, the mean monthly temperatures in December to March being between 23 and 25 $^{\circ}$C. The rainfall was irregular with dry spells of 3 weeks duration or even longer. The rainy season was often too short for the production of normal raid-fed wheat.

Since 1979 further selection and yield trials have been carried out on HRBP breeding material (lines and segregating material) in potential wheat growing areas. Selections were made at Golden Valley Farm (GV) in the Central Province, located at 14° 52' S, 28° 28' E at about 1,150 m above SL, during both the rainy and the irrigation season. At Mbala in the Northern Province, located at 8° 51' S, 31° 20' E at about 1,670 m above SL, selections were made during the rainy season only. Table 1 gives the number of locations, in brackets, used for each breeding stage in the rainy and the irrigation season in 1980 and 1981.

Cultivars

Initially 22 cultivars were selected as parents for the breeding program. In 1977 a further 12 cultivars were added, which were slight-ly better adapted to the conditions of the rainy season. The selection

criteria were:
- Seedling susceptibility to stem rust (*Puccinia graminis* Pers. f.sp. *tritici*) and leaf rust (*P. recondita* Rob. ex Desm. f.sp. *tritici*), with infection types that sporulated and showed no necrosis;
- Cultivars from several centers of origin, e.g. Zimbabwe, Kenya, Mexico (CIMMYT), Pakistan, Australia and North Africa;
- Cultivars with a diversity of characteristics including marker genes, e.g. short and long straw, early and late flowering and ripening;
- A minimum yield of 4,000 kg grain per ha with absence of rust during the irrigation season.

The parent cultivars had varying levels of rust severity in the adult plant stage. Several cultivars had a severe reduction in yield when stem rust was present at high enough levels. All cultivars were susceptible to *Helminthosporium sativum* Pamm., King and Bakke. The cultivars which were selected could be divided into three categories: commercial cultivars which have been used in Zambia, Zambesi I, Tokwe and Umniati; cultivars which have been tested in Zambia for several years such as Shashi, Bubye, Chenab 70, Mexipak, Super X and Turpin 7; and cultivars which have been tested only once or twice such as African Mayo, Giza 156, Emu 'S', Condor and SA42. Most cultivars had only been tested during the irrigation season.

Table 1.   Number of entries of wheat in breeding stages of the Horizontal Resistance Breeding Program in Zambia.

Breeding stage	Rainy season		Irrigation season	
	1980	1981	1980	1981
Segregation material + head selections	200 (3)[a]	3150 (2)	1500 (1)	1600 (1)
First yield test	78 (3)	133 (3)	126 (1)	44 (2)
Second yield test	1 (2)	45 (3)	28 (1)	72 (2)
National yield test	2 (2)[b]	9 (3)[b]	9 (4)[b]	14 (4)[b]
Field test	0 (0)	0 (0)	6 (4)[b]	9 (4)[b]
Multiplications	0 (0)	0 (0)	2 (2)	23 (1)

[a] Figures in the brackets indicate the number of locations at which entries were grown.
[b] In addition to the locations already mentioned, entries were tested at other locations.

Note: Between 1976 and 1979, polycrosses were made using Ethrel.

Early in 1976, information was available about the susceptibility of the cultivars to leaf rust but not to stem rust or
Helminthosporium. Susceptibility to both stem and leaf rust was
tested during the 1976 irrigation season, and during the 1976-77
rainy season the susceptibility to all three diseases was tested.
There was no inoculation with specific races, and no samples were
sent for identification of the race(s) present. By selecting for
susceptibility, it was hoped to eliminate vertical resistance, and
thus to increase the chances to obtain durable resistance.

Crossing and selection

Between 1976 and 1979, polycrosses were made using Ethrel
(2-chloroethyl phosphonic acid) which was applied with a knapsack
sprayer. During the first year, each plot received three applications
of Ethrel (1,500 ppm at about 1,000 l liquid per ha per application).
In the following years only one or two applications were made per
plot. 2,000 ppm (100 ml Ethrel per 20 l water) was applied at 1,900 l
liquid per plot with 1 application, at 1,200 l liquid per plot per
application with 2 applications per season. The development stages
at the time of Ethrel application, according to the Decimal Code of
Zadoks et al. (1974), were 39, 45 and 55, respectively, 39 being the
most important stage at which to spray. Ethrel was sprayed in every
alternate column of the plots (3 m^2, 5 or 6 drills, 18 or 20 cm
apart), perpendicular to the prevailing wind direction.

For the first polycross, all parent cultivars were paired,
assuming homozygosity, and the pairs were randomized. Groups of 4 x 8
plots were planned in blocks with a total area of 12 m x 12 m = 144 m^2;
2 irrigation pipes covered 12 m. In order to synchronize flowering,
planting was staggered, the late flowering cultivars being planted
before the early flowering cultivars. The crossing block was sur-
rounded with a strip 6 m wide of wheat cultivars susceptible to the
rusts. The surround was planted three weeks before the breeding
material, so that the wheat in the surround had passed the anthesis
stage when crossing began.

The second polycross, without selfing in between crossings, was
carried out during the 1977 irrigation season. The breeding block
was laid out in the same way as for the first polycross in 1976, and
only seed of wheat treated with Ethrel was planted (24 g per 3 m^2
plot).

In the second polycross, with one selfing in between crossing,
seed from each plot treated with Ethrel (24 g) was planted during
the 1976-77 rainy season. Natural selection was severe because of
drought and high temperatures, and only 100 to 150 heads per m^2
remained to be harvested. In the following irrigation season all
seed was planted in strips of 15 m long. Wherever possible seed
originating from one mother cultivar was planted in one strip. Drills

of wheat were sown alternatively, 2 treated with Ethrel and 3 untreated, and 3 treated with Ethrel and 2 untreated.

Artificial selection was done according to: single seed, head,
plant, hill and drill (row), plot, parent-related, sifting (selection
for grain size and to some extent grain filling), grain weight per
head in relation to the stand (number of heads per m^2).

Yield trials were carried out at several locations with lines
and segregating material. The first yield trials with G4 lines
(fourth generation after the last polycross) had 2 or 4 replicates
per planting per location. The second yield trials had 6 replicates
per planting per location and at least at one location there were
several plantings. After a minimum of two yield trials, a line could
be submitted for admission to the national yield trial (4 replicates
per planting per location; several locations).

Management

Land was prepared according to local practices at NIRS; usually
with a tractor and implements such as a plough, cultivator, disc and
spike harrow. Occasionally the land was prepared by means of oxen,
plough and spike harrow, or by hoe. Seeding was done by hand. In the
breeding blocks, wheat was the only agricultural crop used in order
to build up soil-borne diseases. In most of the other fields, wheat
was followed by another crop such as sunhemp, horticultural crops,
maize and soybean. Fertilizers were applied by hand, just before
planting and at tillering. Fertilizer levels were high for the first
three generations (up to G4), in order to create optimum conditions
for the development of crop diseases. Later generations were exposed
to low, as well as high, fertilizer levels. During the irrigation
season, sufficient water for optimum growth was applied up to G4,
but later generations were occasionally submitted to moisture stresses.
Supplementary or more frequent irrigation was applied in order to
favor the development of rust and/or to compensate for drought.
During the rainy season there were often late plantings in order to
avoid problems with weeds. As a result of the late planting selection
for drought tolerances occurred, and there were also low levels of
severity of *Helminthosporium*.

During the period 1976-78 weeding was done by hoeing, and in
the following years 2,4-D was also applied about 3 to 4 weeks after
planting in order to control broad-leaf weeds. The rate of application was 1.5 to 2.0 l per ha.

During the first three years insecticides were not applied except for Dieldrin 2%, which was dusted in order to prevent harvester
termites (*Macrotermes* sp.) from attacking the crop. Seed for yield
trials was treated against storage pests and fungi with Captasan-M.
Since 1979, Dieldrin WP has been applied before the planting of
yield trials.

On the field trials at NIRS, stem rust and leaf rust were introduced by means of spreader plants of one cultivar from another
location, or from a neighbouring field with wheat. Rusts were introduced artificially when natural infections failed to appear in
sufficient quantities during tillering.

The wheat was harvested by hand or with a mini-plot combine
harvester, and began when all wheat was ripe. Bundles were threshed
with a locally assembled thresher, or in a sack which was repeatedly
beaten with a wooden stick. Single heads were threshed with a head
thresher, or by hand.

Development stages were scored according to the Decimal Code of
Zadoks et al. (1974), and disease severity according to the scales
developed by James (1971). Observations were made per plot, and
recorded in duplicate. One copy was sent for computer analysis, and
the other copy was kept by the breeder.

RESULTS

Crossing with Ethrel

Cross-pollination occurred after application of Ethrel. The
percentage of cross-pollination varied with the season and the
cultivar, for example during one season the variation between
cultivars was as great as 0 - 99% with a mean of 40%. In addition
to its limited efficiency, Ethrel had several undesirable side effects. Wheat treated with Ethrel tended to have a higher level of
stem rust severity and higher numbers of aphids than untreated wheat.
The appearance and development of the crop was also significantly
influenced by the Ethrel and this made selection difficult. At NIRS
crossing proved to be very inefficient during the rainy season because of the short flowering period, which reduced the chance of
successful cross-pollination, and because of yield reductions in
the already low yields (less than 1,000 kg seed per ha without
Ethrel). During the irrigation season the conditions for crossing
were better. The yields after treatment with Ethrel were acceptable,
up to 3,000 kg seed per ha, but there were also yield reductions of
40% and higher.

Breeding stages

In Table 1, the number of entries for each breeding stage in the
period 1980-81, are given. There was some segregating material with
2, 3 or 4 polycrosses for further selfing and selection. There were
also a reasonable number of entries (mostly lines) in yield tests.
The best lines have been admitted to the national yield trial and
have been multiplied. Some of these lines have been tested under
farming conditions.

Breeding strategies

Several breeding strategies were tested and information is now available on 4 of these strategies:
early line selection in a mixture with;
- one polycross and grown at one location only (strategy 1)
- two polycrosses and grown at one location only (strategy 2)
- one polycross and grown at two locations (strategy 3)
- two polycrosses and grown at two locations (strategy 4)
With strategy 1), 2) and 4) a reasonable number of lines, 2 to 7 lines, were selected after several yield trials which had a relative yield of at least 80% of the maximum yielding entry. There were 4 local reference cultivars in each yield trial. Furthermore the lines had low levels of stem rust severity under disease pressure. After one poly-cross, early line selection in a mixture that had been grown at two locations, and yield tests, no lines remained.

In addition to producing good yields under pressure of rusts in a number of seasons, the best yielding lines of the HRBP had yields which were equal, or better, than the best foreign cultivars, that is about 5,000 to 6,500 kg grain per ha during the irrigation season, and between 1,000 and 2,400 kg per ha during the rainy season.

DISCUSSION AND CONCLUSIONS

With the use of a chemical, Ethrel, which made a cross-pollination possible in the self-pollinator wheat, the technique of crossing wheat became more simple. As a result large quantities of segregating material were produced. With local breeding and selection, starting with susceptible parent material, normal breeding results were obtained with limited resources.

As most of the work was carried out at NIRS, more attention was given to stem rust than to diseases which are important elsewhere but rarely occur at NIRS. The exclusion of vertical resistance to rusts could not be proved. The resistance of the newly bred lines to stem rust, and also possibly to leaf rust, may be durable because: the severities were low under disease pressure; there was no signif-icant correlation between severity of stem rust and relative yield over a period of at least three seasons; there were only sporadic appearances of the diseases; there was a small acreage of wheat; and the fields were scattered over a wide area.

After testing the yield, but also possible shortcomings, the new lines may be used in a new breeding cycle. More efficient male gametocides for wheat other than Ethrel have been found, and hope-fully they will soon be made available to developing countries for breeding purposes.

Small projects, such as the HRBP in Zambia, appear to be very suitable for the early stages of crop improvement programs, especially as the requirements can largely be met from local funds.

ACKNOWLEDGEMENTS

I am most grateful to Professor Dr. J.C. Zadoks for helpful advice and discussions. A grant for travel expenses was provided by the Agricultural University Fund, Wageningen.

REFERENCES

Hurd, E.A., 1981, Status of research and production in Zambia, "Proceedings of wheat workshop and inaugural biennial meeting of agricultural science association of Zambia", National Council for Scientific Research, Box CH 158, Lusaka, 19-20.

James, W.C., 1971, An illustrated series of assessment keys for plant diseases, their preparation and usage, *Can. Plant Dis. Surv.*, 51(2):39-65.

Robinson, R.A., 1976, "Plant Pathosystems", (Advanced Series in Agricultural Sciences 3), Springer Verlag, Berlin and New York.

Vanderplank, J.E., 1963, "Plant Diseases: Epidemics and Control", Academic Press, New York.

Vanderplank, J.E., 1968, "Disease Resistance in Plants", Academic Press, New York and London

Vanderplank, J.E., 1975, "Principles of Plant Infection", Academic Press, New York, San Francisco and London.

Zadoks, J.C., Chang, T.T. and Konzak, C.F., 1974, A Decimal code for growth stages of cereals, *Eucarpia Bull.*, 7:10.

BREEDING FOR DISEASE RESISTANCE IN WHEAT; THE BRAZILIAN EXPERIENCE

Martinus A. Beek

Food and Agriculture Organization of the United Nations
EMBRAPA
CNP Trigo, Passo Fundo, R.S., Brazil

ABSTRACT

A description is given of a program on horizontal resistance in wheat which was conducted at Passo Fundo, Brazil. The objective of the program was to determine the feasibility of accumulating useful levels of horizontal resistance to all locally important pests and diseases. The breeding and selection methods that were applied are described and results of trials to evaluate the program are presented. From these results, it is concluded that this new breeding approach is very suitable for the conditions at Passo Fundo, where many diseases have to be overcome and where stability of crop yield is the main interest.

INTRODUCTION

In 1975, an experimental wheat breeding program was initiated at Passo Fundo, R.S., Brazil. Its purpose was to determine the feasibility of accumulating useful levels of horizontal resistance to all locally important pests and diseases of wheat. Rio Grande do Sul is the principal wheat growing state of Brazil, but it suffers serious crop losses due to pests and diseases. Wheat yields are low; in 1978, the best yield in 30 years was produced but even then the state's average was only 1,200 kg/ha. The yields also fluctuate widely and vary from 340 kg/ha to about 1,000 kg/ha in spite of an increasing use of insecticides and fungicides. The regular application of these pesticides costs more than US$ 30 million each year.

The most important diseases are *Puccinia graminis, P.*

*recondita, Helminthosporium sativum, Septoria nodorum, S.
tritici, Erysiphe graminis, Gibberella zeae, Gäumannomyces
graminis,* barley yellow dwarf virus and soil borne mosaic virus.
Aphid damage varies from year to year, but can be as much as
total damage from all the diseases combined.

The experimental program was based on Van der Plank's
(1963, 1968) concept of horizontal resistance and Robinson's (1976)
suggestion that breeding for horizontal resistance should imitate
the behavior of maize in tropical Africa after the introduction of
Puccinia polysora. In addition to the overall objective stated above,
the program had two targets: (a) to create new wheat cultivars
that permitted the cultivation of stable-yielding wheat in Rio Grande
do Sul, without the use of foliar fungicides or insecticides, and
(b) to produce increased average yields of 3,000 kg/ha. In 1981,
these targets were closely approached, and all the current evidence indicates that they can be reached through a continuation
of the breeding program .

MATERIALS AND METHODS

The experimental program differed from traditional wheat
breeding in three fundamental aspects: (a) great care was taken
to avoid vertical resistance, (b) population breeding methods were
made possible through the use of a male gametocide, which permitted random polycrossing, and (c) there was simultaneous
screening for all desirable characters, including horizontal resistance to all locally important pests and diseases.

Horizontal resistance is the resistance that invariably
remains after vertical resistance has "broken down". It is
impossible to select for horizontal resistance if vertical resistance is present and operating. It follows, that all vertical resistance must either be absent or inoperative during the screening
process. As it is probably impossible to obtain wheat lines that
possess no vertical genes whatever, all vertical resistance must
be made inoperative by epidemiological methods. In Brazilian
wheats, the only vertical resistances were against *P. graminis,*
and *P. recondita.* One vertical pathotype (i.e. physiologic race)
of each of these pathogens was "designated" for this purpose of
the experiment. Eighteen Brazilian cultivars of spring wheat
were then chosen as parent lines on the basis of their susceptibility to the two designated pathotypes. The use of these two
pathotypes in all screening ensured that the vertical resistances
were inoperative in all parents and in all progenies in all subsequent screening generations. It should be added, that other
characters, such as local adaptation and synchronisation of
anthesis to prevent assortative mating, also contributed to the
choice of these eighteen parents. The resulting population was

called "Mixture 1". At a later stage, two superior populations with better yield potential were created and were called "Mixture 2" and "Mixture 3"; "Mixture 4" included European cultivars and was mainly created for research purposes.

The random polycross was achieved by the use of "Ethrel" (2-chloroethyl phosphoric acid) as a male gametocide. This chemical was applied as an aqueous solution of 2,000 ppm Ethrel a.i. and 150 ppm of giberellic acid 3 (GA$_3$) at a rate of 1,000 l/ha and produced an average male sterility of 80 per cent. The wheat mixture was sown in strips 1.80 m. wide. Alternate strips were sprayed with the male gametocide at stage 9 of the Feekes growth scale and became the female parents. The strips in between remained unsprayed and became the male parents. In most generations, the entire population was grown on 0.4 ha, and an estimated 5 million crosses were obtained in each crossing generation.

Two wheat generations were grown each year. A combined crossing and screening generation was grown at Passo Fundo during the winter, as this is the normal season for wheat. A summer generation was grown at Brasilia to multiply the selections of the screening generation. The multiplication generation was self-pollinated.

The screening generation was inoculated with the designated vertical pathotype of the two rusts already described. It was also inoculated with all of the other diseases listed above, as they often occurred at an abnormally low intensity. Inoculation was by an aqueous spore suspension applied by a motorised airblast, knapsack sprayer.

The first screening was a negative one and involved the removal of the poorer individuals prior to anthesis in order to eliminate undesirable pollen; about 30% of all plants were removed. This negative screening was conducted in both the male and female strips, as the male sterility of the latter was not absolute. All subsequent screenings were positive and were based on eye assessments of good health and good appearance. In order to eliminate problems of interplot and interplant interference, all assessments were relative rather than absolute; regardless of the amount of disease, only the least diseased plants were selected. Screening was conducted on a grid system, with the best two or three individuals per square metre being retained; selected plants were labelled with a plastic tag. A final screening was based on individual plants for the quality and health of the seeds, both of which were determined in the laboratory.

Additional breeding strategies have been established with emphasis on a high and low heritability factor for disease resistance. A line selection program was begun using the Single

Seed Descent method until the F_4 lines were planted in the field.
A single grain from each selected plant was grown to maturity and
a seed of each of the resulting plants was grown until seed could be
harvested, after which this process was repeated until F_4. When
hydroponic culture was used, three selfing generations could be
produced during the off-season. The F_4 lines were grown in the
field, and the best plants were selected from them after one generation of multiplication; line selection was applied (F_6). This
material was then ready for a new population with open cross-
pollination while seed from particularly promising varieties was
retained for multiplication and variety testing. This method might
be more efficient for the selection of characters having a low heritability. In addition to these two selection methods, a third selection method is being tested. As many diseases have a direct
influence on seed development, in this method the seed is selected
instead of individual plants. Seed from the Ethrel treated strips
is graded by means of (a) a blower, which separates the light seeds
from the heavy seeds, (b) a gravity table, (c) a dockage tester.
Finally, a natural selection program is being conducted for
scientific purposes only.

PRODUCTION OF NEW CULTIVARS

The most promising individuals in any screening generation
are separated from the population for selfing to form pure lines
and potential new cultivars. The traditional approach requires
screening during six generations of selfing to eliminate all undesirable segregants and, as this screening can only be conducted
during the wheat growing season, it requires one year for each
generation of selfing. Various attempts were made to reduce this
delay, partly owing to the limited duration of the experimental
program, but principally owing to the urgent need for new wheat
cultivars in Brazil.

The first possibility is the cultivation of mixed populations
in place of pure lines. The traditional arguments in favor of pure
lines carry relatively little weight in Brazil, where the milling
requirements are not stringent and plant breeders rights do not
exist. Populations with good yields and resistance exist in bulk
from the earlier screening work and could be rapidly multiplied
as stop-gap cultivars, should the need arise.

A second possibility is the multiplication of the best F_6 lines
as described under materials and methods.

A third possibility is the Single Seed Descent method.
Selected plants are directly processed to F_7 lines in the greenhouse by the quickest way possible and the best lines are introduced in variety tests.

A fourth possibility is the technique of using doubled haploids. This is now being attempted by anther culture. The frequency of haploid formation has increased steadily with the anther culture and now exceeds one green plantlet per thousand anthers. The main advantage is that doubled haploids are homozygous at all loci. These can be multiplied for immediate testing as potential new cultivars.

RESULTS

Evaluation of the Progress on Horizontal Resistance

In 1978, the most advanced population, called Mixture 1 generation 5, was tested. This population was obtained as follows:

Mixture of 18 wheat varieties	random open cross pollination	Passo Fundo 1976
Generation 2	random open cross pollination	Brasilia 1977
Generation 3	random open cross pollination + selection	Passo Fundo 1977
Generation 4	multiplication	Brasilia 1978
Generation 5		Passo Fundo 1978

An experiment was carried out on this Mixture 1 Generation 5 population to evaluate the progress gained from the horizontal resistance screening method. The experiment was performed with a factorial design, 4 replicates, plot size 4 x 9 metres, and a border of oats between plots to avoid interplot interference. The varieties used in the experiment were: (a) original mixture of 18 varieties, (b) Generation 5, (c) Nobre, (d) Jacuí. Nobre and Jacuí are important wheat varieties in Brazil. No chemical control was applied. The yield results of this experiment are shown in Table 1.

A second experiment was performed in 1979. This year was an extremely bad year, owing to adverse weather conditions: a severe frost occurred on the 19th of September when wheat had

Table 1.   Yields (kg/ha) from an advanced generation,
           the original mixture and two important local
           cultivars.

Generation 5	2939	a[*]
Original mixture	2545	b
Jacuí	2511	b
Nobre	2070	c

[*] Values marked with the same letters did not
    differ significantly at $p \leqq 0.05$.

just headed. In this experiment, the latest selection (Mixture 1,
Generation 7) yielded 15% more than the original mixture, although
very little value can be attached to yield figures. The latest selection showed a significantly lower level of disease for: (a) the total
area of the leaf diseased by powdery mildew, *Septoria* and
Helminthosporium as assessed on the third leaf at growth stage
10 and on the flag leaf at growth stage 11.1, (b) *Septoria* on the
ear at growth stage 11.1. The results are presented in Table 2.

In 1981, the horizontal resistance program had reached
the following stage:

(a)   From three different varietal mixtures, 6 different popu-
      lations have been created that are estimated to yield a
      minimum of 1,200 kg/ha in a very bad year and up to
      3,000 kg/ha in a good year, without the use of chemical
      protection. It is expected, by continuing the selection
      process, that the differences between the yield perform-
      ance in a bad and a good year will be reduced further.
(b)   A population is being developed with the objective of using
      the accumulated horizontal resistance of mixture 1 and the
      straw, tillering, root development and plant type of 7
      European wheat varieties. This population was begun by
      means of hand crosses. As this newly introduced material
      is highly susceptible, the new population has initially a
      relatively low level of horizontal resistance, and an appro-
      priate level of resistance can only be obtained in a long-term
      program, through many generations of recurrent selection
      and open cross pollination.
(c)   500 different F_6 lines are under selection in a trial with 3
      replicates. The best lines will be crossed again for the
      accumulation of higher levels of horizontal resistance, and
      the seed will also be used for variety tests in 1982. These
      lines originate in numbers from mixture 1 (200 lines),
      mixture 2 (200 lines) and mixture 3 (100 lines).

Table 2. Yield and various pest observations on the original mixture of 18 wheat varieties, the latest selection and a locally important wheat variety in 1979.

Observation	Stage	Leaf nr	Wheat type		
			Orig. mixture	Latest selection	Nobre
Yield (kg/ha)	–	–	753	862	760
1000 kernel weight	–	–	25.4	26.6	26.7
Specific weight	–	–	68.8	71.3	68.5
Aphids/tiller	10, early boot		9	9	11
Leaf rust[1]	10, mid-boot	1	0.3	0.1	0.4
" "	10, " "	2	1.4	0.6	0.7
" "	10, " "	3	2.3	1.3	1.5
Powdery mildew,[1]	10, mid-boot	1	0.1	0.0	0.0
H. sativum	10, " "	2	2.0	0.7	1.1
	10, " "	3	11.8	4.6	2.7
All diseases [1]	11.1 milky-ripe	1	47	14	49
	11.1 " "	2	95	91	97
	11.1 " "	3	99	99	99
Stem rust [2]	11.1 milky-ripe		2.5	1.7	0.0
Septoria [3]	11.1 milky-ripe		14.5	0.2	31.7

1: percentage of leaf; 2: percentage of stem; 3: percentage of ear.

(d)   4,500 head to row F7 lines have been produced by the Single
       Seed Descent method. Each line represents plants selected
       from mixture 1 in either 1977, 1978 or 1979.

CONCLUSIONS

The feasibility of the new breeding approach is considered
to have been convincingly demonstrated. The entire program
was conducted by a very small team, and it shows that the new
approach is ideal for developing countries where, given appropriate scientific assistance, it could easily be applied to wheat
and, with only some minor modifications, to other cereals, grain
legumes and other annual species.

LITERATURE

Robinson, R.A., 1976, "Plant Pathosystems", Springer-Verlag,
       Berlin. 184 p.
Van der Plank, J.E., 1963, "Plant Diseases: Epidemics and
       Control", Academic Press, New York. 349 p.
Van der Plank, J.E., 1968, "Disease Resistance in Plants",
       Academic Press, New York. 206 p.

EXPERIENCE OF USING DURABLE RESISTANCE IN THE USA

E. L. Sharp

Department of Plant Pathology
Montana State University
Bozeman, Montana USA   59717

SUMMARY

Durable resistance has been reported in a number of cultivars
for different crops in the USA and may be conditioned either
oligogenically or polygenically. In several cases resistance
has been durable even though resulting in hypersensitivity. In
other cases durable resistance has come about by inhibition of
one or more phases of the infection process. The infection type
may be susceptible but sporulation is reduced in both onset,
quantity or duration. Generally speaking, those disease organisms
with stabile populations and few physiologic races have been more
amenable to long lasting resistance. Diseases for which durable
resistance has been obtained includes those causing wilts, leaf
spots and rusts and occurring on such hosts as root crops, legumes
and cereals.

Research with stripe rust of wheat has shown that many wheat
cultivars contain minor genes that can be accumulated through
judicious crossings of parental materials to give additive
resistance of long duration. These resistance genes behaved as
recessives and were temperature sensitive. Most selected lines
conferred greatest resistance at relatively high temperature
regimes but some behaved in an opposite manner. By selection of
the most resistant types in each segregating generation following
crossing it was possible to start with normally appearing susceptible
parent plants and obtain useful resistance after several generations
of inoculation, selection and selfing. Since the starting materials
can be commercially acceptable cultivars the probability of
obtaining good agronomic types combined with useable resistance
is greatly increased. The method is applicable for crops of

different ploidy levels and preliminary results indicate that it
is applicable to several host-parasite systems.

INTRODUCTION

In evaluation of durable resistance and its efficacy in
protection of crop plants a number of factors must be considered.
First, how extensive has the crop cultivar been grown? Often a
cultivar may appear as having stable resistance in a relatively
small area or region but not have effective resistance to all
virulence types in all areas. From the practical standpoint,
however, the resistance may be very useful in limited areas of
cultivation. Furthermore, the degree of resistance may vary
somewhat from year to year due to such factors as inoculum dosage
and effects of environment on the various phases of the infection
process. In some years a specific cultivar may appear near
susceptible to a specific pathogen but show resistance at other
times. A new race may also parasitize a cultivar with a somewhat
higher infection type but be of generally low virulence or have
poor survival ability in competition with other virulence types.

MONOGENIC DURABLE RESISTANCE

There are a number of cultivars representing different crops
where an apparent durable resistance has been used in the USA for
control of plant diseases. These have included resistance
conferred on a monogenic and polygenic basis. In some cases the
durable resistance has involved hypersensitivity but more often
than not the durable resistance has resulted in curtailment of
development at some phase of the infection process. Pathogens
displaying stabile populations with few virulence types have
generally been most amenable for durable resistance but there are
also examples of durable resistance to pathogens having many
virulence types. Walker (1966) has reviewed several diseases of
vegetables where durable resistance has been developed via the
backcross method. The following diseases were controlled by single
dominant genes: cabbage yellows (*Fusarium oxysporum* f. *conglutinans*
Wr. Snyder and Hansen), Fusarium wilt of tomatoes (*Fusarium
oxysporum* f. *lycopersici* sacc., Snyder and Hansen), Verticillium
wilt of tomatoes (*Verticillium albo-atrum*, Reinke and Berth),
Fusarium wilt of peas (*Fusarium oxysporum* f. *pisi*, Linforde,
Snyder and Hansen), white rust of radishes (*Albugo candida* Lev.
Kunye), bacterial wilt of cucumbers (*Erwinia tracheiphila* E. F.
Sm. Holland) cucumber scab (*Cladosporium cucumerinum* Ell. & Arth),
cucumber mosaic and pink root of onion (*Pyrenochaeta terrestris*,
Hansen, Gorenz, J. C. Walker and Larsen).

Other examples of durable monogenic resistance occur in leaf
spot of maize (*Helminthosporium carbonum* Ull) (Ullstrup and Brunson,
1947), Helminthosporium blight of oats (*Helminthosporium victoriae*,

Meehan and Murphy, 1946) and stem rust of barley (*Puccinia graminis*
Pers.) (Moseman, 1971) and loose smut of wheat (*Ustilago tritici*
Pers. Rostr.) (Caldwell, 1968). Resistance in maize to leaf spot
is conditioned by the dominant gene Hm and is expressed as a
hypersensitive fleck. Resistance to *H. victoriae* (Meehan and
Murphy, 1946) is conditioned by a single recessive gene and
occurs in all oat varieties with the exception of Victoria oats and
some of it's derivatives carrying the dominant gene for susceptibility. When the victoria types were first used for resistance
to crown rust (*Puccinia coronata* Pers. Corda) victoria blight became
a serious problem and genes for resistance to crown rust were
closely linked with susceptibility to victoria blight. Wheeler
and Luke (1955) broke this linkage in exposing some 45 million
seeds to the toxin of *H. victoriae*. Of the 973 seedlings which
survived the toxin treatment and seedling inoculation, about one-
half were also resistant to crown rust. The T gene conditioning
resistance to stem rust in barley occurs in many spring barley
cultivars commercially grown in the U.S.A. Of more than 200
physiologic races of *P. graminis* evaluated on cultivars containing
this gene, only a few cultures were virulent. The virulent races
have not become prevalent in commercial fields. The wheat cultivars
Trumbull and Kawvale, wherein resistance to loose smut is conditioned
by single specific genes, have provided good protection from loose
smut for many years.

POLYGENIC DURABLE RESISTANCE

In cases where monogenic resistance was not effective for
long periods, durable resistance could often be obtained by polygenes for several plant diseases. Some examples are: Northern
maize leaf blight, maize rust, crown rust of oats, stem rust of
wheat, alfalfa rust and stripe rust of wheat. In Northern maize
leaf blight (*Helminthosporium carbonum* Pass.) it has been possible
to obtain long lasting resistance. It has been found that resistance could be easily accumulated by recurrent selection (Jenkins,
Roberts, and Findley, Jr., 1954). Luke, Barnett and Pfahler (1967)
have noted that the Red Rustproof cultivar of oats has escaped
significant damage from crown rust (*Puccinia coronata*) for more
than 20 years. There was no change due to changing physiological
race patterns. Rust developed as much as 14 days later than on
susceptible cultivars. Even though it appeared susceptible late
in the season to several races there was no appreciable effect
on yield. This aspect of slow disease development occurs with
several other host-pathogen combinations, is an important epidemiological aspect and will no doubt be covered in detail in other
papers at this meeting.

Hooker (1967) has discussed the durability of resistance of
maize to *Puccinia sorghi* Schw. The inbred lines which have been
used to produce commercial hybrids show medium to low intensities
of infection in the mature plant stage and only one widely used

inbred line expressed resistance as a seedling. Wide variations
in reaction types occurred in segregating populations derived from
65 different crosses with some segregants exceeding the range of
their parents. The parents and F_1 plants were uniform in reaction
type. Commercial maize hybrids have never developed rust in epiphy-
totic proportions. If susceptible lines of maize were used in
inbred lines it is probable that the disease could be very destructive.

Alfalfa rust (*Uromyces striatus* Schroet. var. *medicaginis*
Pass.) was reported effectively controlled via recurrent selection
(Hill, Sherwood and Dudley, 1963). Resistance to infection was
obtained even though selection criteria included hypersensitivity
along with rust-free plants and those lightly infected. Most
progress was noted in the 2nd to 5th cycles of selection.

Green and Campbell (1979) have recently reviewed the
historical development of resistance of wheat to stem rust. Durable
resistance has been produced for a number of wheat cultivars.
After the appearance of race 15B in 1950, the cultivar Selkirk
was released and has been outstanding for durability of resistance.
It possesses genes Sr_6, Sr_{23} and H 44-24 resistance genes (Sr_7b,
Sr_9d, $Sr17$). Analysis of races occurring from 1954 to 1969 showed
that no commonly occurring races combined virulence capable of
attacking the combination Sr_6 and Sr_9d. Two rare races combining
virulence on Sr_6 and Sr_9d did not survive in a natural rust
population. Wheat cultivars Manitou and Neepawa have also been
resistant to stem rust since they were released to farmers in 1965
and 1969. Redman, Regent and Renown have been resistant since 1957.
Canthatch retained resistance from 1960 to 1978 and its loss of
resistance seemed to be caused by a steady increase of virulence
that resulted in the loss of protection conferred by gene Sr_7a.
Several cultivars have remained resistant in rust nurseries since
first evaluated and these include Kenya Farmer, Kenya 117A, Kenya
35C, 2.B.2, R. L. 2520, Mida - McMurachy - Exchange II-47-26,
Frontan - K58 - Newthatch II-50-25, Stewart 63, Justin, Hercules,
Norteno 67, and Waldron. Green and Campbell (1979) stated that
the cultivars released in Canada since 1950 that had durable
resistance differed from those that became susceptible mainly in the
number and combinations of resistance genes they possessed. Cultivars with durable resistance had many resistance genes and their
resistance was genetically and physiologically complex involving
genes for hypersensitivity as well as for low receptivity, poor
sporulation, and long incubation period.

The yd2 gene for resistance to barley yellow dwarf virus
occurs in a number of barley cultivars and has furnished adequate
protection for a number of years. Its effectiveness is largely
dependent upon the genetic background in which it operates. With
different parental combinations the disease ratings on subsequent

progeny range from resistant to moderately susceptible.

An apparent durable resistance to stripe rust (*Puccinia striiformis* West.) as been noted in several wheat cultivars. In the northwestern U.S.A. the wheat cultivars Cheyenne, Rego, Nugaines and more recently Crest have been grown for years on extensive hectareages and have never suffered appreciable losses from stripe rust. Of these cultivars, Cheyenne and Nugaines are susceptible as seedlings but develop resistance to all prevalent virulence types as mature plants and at relatively higher temperatures in the field. Rego performs oppositely to many wheat cultivars and may be susceptible at relatively high temperatures and resistant at lower temperature regimes. In Rego, the response to temperature occurs in both the seedling and mature plant stages. Crest has been resistant to moderately resistant to all prevalent virulence types at all growth stages for more than 10 years. Since the type of resistance exemplified by Crest is receiving major attention in control of stripe rust in the USA, it is appropriate to review some research leading to its development.

DURABLE RESISTANCE IN STRIPE RUST IN WHEAT

During the late 1950's and early 1960's stripe rust epiphytotics frequently occurred in the northwestern USA. At the same time, large areas in the region were planted to a few wheat cultivars which were very susceptible to the disease. During the development of these cultivars, stripe rust was considered as a minor disease and they were never adequately evaluated to *P. striiformis*. In looking for resistant parental materials, the wheat plant introduction, P.I. 178383, was found to be highly resistant to all cultures of *P. striiformis* as well as resistant to all prevalent cultures of *Tilletia controversa* Kuhn (dwarf smut). Since these two diseases tended to occur in the same general areas, P.I. 178383 was used in several breeding programs as a non-recurrent parent. In Oregon, P.I. 178383 was crossed with the club wheat cultivar Omar and subsequently the cultivar Moro was developed. Originally only a necrotic fleck reaction occurred on Moro. After three years of large scale cultivation the resistance of Moro was overcome by a new virulence type and a completely susceptible reaction resulted (Beaver and Powelson, 1969). At about the same time Crest was developed in Montana using wheat cultivars P.I. 178383 and Westmont as parents. Later evaluations showed that Moro contained only one major dominant resistance gene while Crest contained the same major dominant gene plus several minor recessive genes for resistance. Research was undertaken to separate the resistance in P.I. 178383 into its many components (Sharp, 1968). In controlled environment studies, many distinct reaction types were noted in the F_2 progeny following crossing of P.I. 178383 with a susceptible cultivar, Itana. Plants stable for reaction type were obtained by selecting plants with specific reaction types at each generation for 8 generations of selfing inoculation and selection. After the selected plants were stabilized for reaction type they were test-crossed to a susceptible wheat cultivar. Table 1 shows that three detectable minor effect additive genes occur in P.I. 178383. All the minor genes were additive in action and were temperature sensitive with greater resistance expressed at a temperature of 15/24 C (dark/light) t-an at 2/18 C. The minor gene lines reacted similarly to all prevalent isolates of *P. striiformis* (Volin, 1971). Stubbs (1977) reported on the average coefficient of infection for these lines and several other wheat cultivars. He evaluated their reaction to several physiologic races over a period of several years and concluded that they displayed attributes of a general type of resistance. Once the minor gene lines were stabilized for reaction type they were crossed with a number of commercially acceptable wheat cultivars and advanced lines believed to also contain minor genes conditioning reaction to *P. striiformis* (Sharp, 1976). Transgressive segregation to greater resistance was obtained in many cases even though one parent appeared susceptible under normal environmental conditions. Figure 1 shows the reaction of four wheat lines selected from a cross of Lancer and 18/1. Lancer was susceptible and 18/1 contained one known detectable minor gene and was moderately resistant. The progeny lines from this cross were all more resistant than either parent when evaluated to seven physiologic races from both the USA and Europe. There were some differences in reaction between the lines of the cross but interestingly the individual lines behaved similarly with the different virulence types or physiologic races. A total of 49 wheat lines containing minor genes were evaluated with the different virulence groups. Figure 2 illustrates the general range of reaction types within this set using the Clement virulence type. Greater resistance was expressed at the higher temperature regime 15/24 C as compared to 2/18 C and the reaction types for mature plants in the field was generally similar to that obtained on seedling plants at 15/24 C. In order to obtain more information on the universality of minor genes in wheat cultivars normally appearing susceptible, a number of crosses were made between such candidates and resistance was sought in the subsequent segregating generations (Krupinsky and Sharp, 1979). As shown in Tables 2 and 3 such crosses also resulted in transgressive segregation for resistance. The parents and F_1 showed only susceptible reaction. In some cases evidence of resistance has appeared first in the F_3 or F_4 generation and the plants may still show segregation for different reaction types in the F_6 or F_7 generations. However, once resistant plants appear they can eventually be stabilized for useable resistance. It appears that a genetic background must be established before the resistance genes can be detected by conventional genetic procedures. Some advantages to using such resistance are that it can be readily followed in a breeding program and readily manipulated to further increase resistance by use of various

Table 1. Minor genes from wheat variety P.I. 178383 conditioning resistance to stripe rust (a)

No. of genes	Probable F_1 genotype	Profile TOC	F_2 Infection type classes				Ratio	P value
			(4,3)	(3-,2,1)	(1-,0)	(0-,00)		
1	aaBbCCDD	15/24	200	73			1:3	.50-.30
		2/18	268					
2	aaBbCcDD	15/24	151	100	11		1:6:9	.50-.30
		2/18	200	36			3:13	.20-.10
3	aaBbCcDd	15/24	35	24	11	8	7:15:23:19	.05-.02
		2/18	63	36	1		1:18:45	.30-.20

(a) Plants with minor genes crossed to plants with infection type 4.

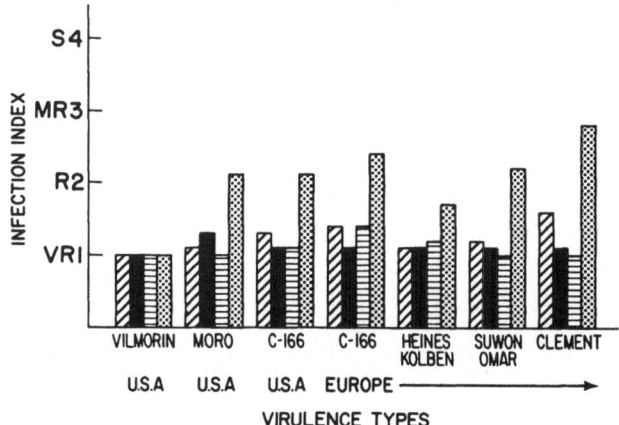

Figure 1.   Reaction of 4 wheat lines developed from Lancer X 18/1
            to 7 virulence types of *Puccinia striformis*.   Infection
            index: Lancer = 4, 18/1 = 3.

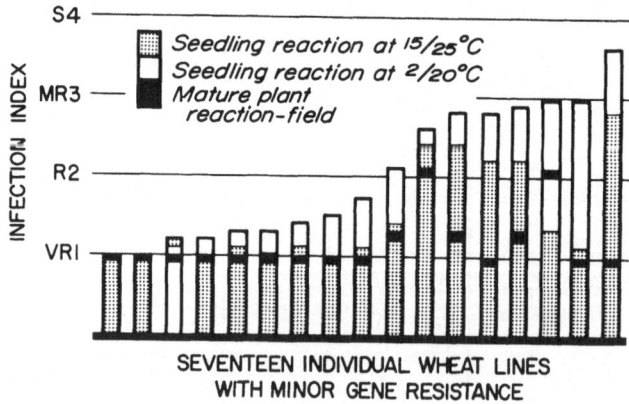

Figure 2.   Range of reaction of 17 minor gene wheat lines inoculated
            with the Clement virulence type of *Puccinia striformis*.
            Parental infection indices = 4 or 3.

temperature regimes. Since it is possible to obtain effective
resistance by crossing commercial wheat cultivars which normally
appear susceptible, undesirable traits often associated with
resistant parents of exotic or non-agronomic types can be avoided.
There is preliminary evidence that the principles may be applied
to other host-pathogen systems. Net blotch of barley (*Pyrenophora
teres* Died. Drechsl.) is one example of such a system (Bjarko, 1979).
The durability of such resistance appears promising and the methods
are worthy of further application and evaluation.

Table 2. Transgressive segregation for stripe rust resistance
          in segregating populations of winter wheat.

		Percent plants for each infection type				
Cross	Generation	(00,0-)	(0,1)	(1,2)	(3-)	(3,4)
Itana/Wanser	P_1					X
	P_2				X	
	F_1					X
	F_2				5	95
	F_3			7	22	72
	F_4		5	52	27	16
	F_5		25	72	3	

Table 3. Transgressive segregation for stripe rust resistance
          in segregating populations of spring wheat.

		Percent plants for each infection type				
Cross	Generation	(00,0-)	(0,1-)	(1,2)	(3-)	(3,4)
Shortana/Centana	P_1				X	
	P_2				X	
	F_1					X
	F_2				1	99
	F_3		2	2	9	88
	F_4	5	14	5	13	63
	F_5	19	77	3		
	F_6	37	62	1		

REFERENCES

Beaver, R.G., and Powelson, R.L., 1969, A new race of stripe rust pathogenic on the wheat variety Moro, C.I. 13740, *Plant Dis. Rep.*, 53:91.

Bjarko, M.E., 1979, Sources of and genetic action of resistance in barley to different virulence types of *Pyrenophora teres*, the causal organism of net blotch, M.S. Thesis, Montana State University, Bozeman. 81 p.

Caldwell, R.M., 1968, Breeding for general and/or specific plant disease resistance, Third International Wheat Genetics Symp., 263 p.

Green, G.J., and Campbell, A.B., 1979, Wheat cultivars resistant to *Puccinia graminis tritici* in western Canada: their development, performance, and economic value, *Can. J. Plant Path.*, 1:3

Hill, R.R., Sherwood, R.T., and Dudley, J.W., 1963, Effect of recurrent selection on resistance of alfalfa to two physiologic races of *Uromyces striatus medicaginis*, *Phytopathology*, 53:432

Hooker, A.L., 1967, The genetics and expression of resistance in plants to rusts of the genus *Puccinia*, *Ann. Rev. of Phytopath.*, 5:163.

Jenkins, M.T., Roberts, A.L., and Findley, W.R., Jr., 1954, Recurrent selection as a method for concentrating genes for resistance to *Helminthosporium turcicum* leaf blight in corn, *Agron.*, 46:89.

Krupinsky, J.M., and Sharp, E.L., 1979, Reselection for improved resistance of wheat to stripe rust, *Phytopathology*, 69:400.

Luke, H.H., Barnett, R.D., and Pfahler, P.L., 1975, Inheritance of horizontal resistance to crown rust in oats, *Phytopathology*, 69:400.

Moseman, J.G., 1971, Studies of inheritance of resistance in barley to pathogenic organisms, 1963-1969, *in*: Barley Genetics II:535.

Murphy, H.C. and Meehan, F., 1946, Reaction of oat varieties to a new species of *Helminthosporium*, *Phytopathology*, 36:407.

Sharp, E.L., 1968, Interaction of minor host genes and environment in conditioning resistance to stripe rust, 2nd European and Mediterranean Cereal Rusts Conference, Oeiras, Portugal, 158 p.

Sharp, E.L., 1976, Broad based resistance to stripe rust in wheat, 4th European and Mediterranean Cereal Rusts Conference, Interlaken, Switzerland, 159 p.

Stubbs, R.W., 1977, Observations on horizontal resistance to yellow rust *Puccinia striiformis* f. sp. *tritici*, *Cereal Rusts Bull.*, 5:27.

Ullstrup, A.J. and Brunson, A.M., 1947, Linkage relationships of a gene in corn determining susceptibility to a *Helminthosporium* leaf spot, *J. Am. Soc. Agron.*, 39:606.

Volin, R.B., 1971, Physiological race determination and environmental factors affecting the development of infection type in stripe

rust *Puccinia striiformis* West, Ph.D. Thesis, Montana
    State University, Bozeman. 90 p.
Walker, J.C., 1966, Pest resistance, *in:* Plant Breeding, Iowa
    State University Press, 219.
Wheeler, H.E., and Luke, H.H., 1955, Mass screening for disease
    resistant mutants in oats, *Science,* 122:1229.

DISCUSSION

PARLEVLIET: In the data (Figure 1) you showed some 6 or 7 races
reacting with 4 host genotypes. From these data I got the
impression that there are a few small host genotypes x pathogen
genotype interactions? Is that so?

SHARP: There were some small differences in interaction to the
different races within lines but the individual lines basically
reacted the same to the different races. In comparison to the
parents, all lines showed reduced infection type due to
accumulation of additive genes for resistance.

DINOOR: Is it known whether the wild emmer lines used in your
crosses are homozygous or not? If they are heterozygous you might
have extracted the yellow rust resistance polygenes only from the
wild emmer and not from the variety Lemhi for example. A parallel
selection from a selfing series of emmer could have clarified the
picture. It is also possible that genes derived from Lemhi are
only modifiers which activate resistance genes in wild emmer, while
they cannot act as resistance genes on their own.

SHARP: The wild emmers were received from Z.K. Amitae, Volcani
Institute, after he had selected for homozygous types. Occasionally
parent lines varied by one class of infection type (e.g. (4,3) →
(3-)) but this should not have materially affected the conclusions
made. Lemhi in some way contributed to the resistance of the progeny
subsequent to the cross of wild emmer lines and Lemhi.

MICKE: The resistance genes you accumulated in wheat following
crosses of susceptible parents appear to behave as recessives, re-
quiring two gene doses in the homozygous condition for expression.
This appears to be in contrast to the widespread opinion, that
useful genes for resistance are inherited in a dominant fashion,
where one gene dose is sufficient and the heterozygous type expresses resistance. Is there a difference between 'minor' and
'major' genes with regard to their inheritance?

SHARP: In stripe rust, both dominanat and recessive genes for resistance occur but in my experience those having minor effects
behaved as recessive. If heterozygous in terms of one gene the
plants were susceptible. At least 3 heterozygous minor genes

were required to detect a change in the phenotype so there were
dosage effects.

JOHNSON: Do you have any estimates of the number of genes involved
in these transgressive levels of resistance?

SHARP: By using different temperature regimes it was possible to
obtain evidence for 6 detectable minor genes in the crosses
resulting in some progeny being in a very resistant category.

JOHNSON: Is it correct to describe the gene action as additive when
there is no sign of it for two or three generations?

SHARP: I believe additions in the direction of resistance are often
made in the early generations, but since the loci are heterozygous
they are not detectable. With further selfing homozygous R genes
are obtained. Also a certain level of homozygous loci may be
required before resistant phenotypes can be detected, i.e. a
threshold value may be required.

JOHNSON: Many of the results were obtained on seedlings at various
temperatures. If you find lines with (00) reaction are they
completely resistant in the field?

SHARP: The reaction in the field is usually as good as or better
than reaction of the seedlings at 15-25ºC. However, plants with a
(00) reaction as seedlings will usually show a small amount of
sporulation as mature plants in the field (0 to 1- reaction).

DURABILITY OF BARLEY POWDERY MILDEW RESISTANCE GENES IN

DENMARK 1963-1980

J. Helms Jorgensen

Agricultural Research Department
Riso National Laboratory
Roskilde, Denmark

INTRODUCTION

Barley (*Hordeum vulgare* L.) is an important crop in Denmark. The spring barley area was 0.9 million ha in 1963 increasing to 1.6 million ha in 1980, which is 54 percent of the agricultural area. Within this period of time, spring barley varieties with major genes for resistance to the powdery mildew fungus (*Erysiphe graminis* DC. ex Merat f.sp. *hordei* Marchal) have become widely distributed, and the spectrum of varieties has changed continuously. Winter barley is grown on a very limited area, about 6,000 ha in 1980, and is always protected by fungicides.

The powdery mildew epidemic in the Danish spring barley crop dies out at harvest time due to the lack of overwintering hosts. Inoculum initiating the next year's epidemic consists of conidia transported by wind to Denmark from powdery mildew populations over-wintering on winter barley in neighbouring countries, primarily the northern provinces of West- and East Germany (FRG and GDR, respectively), and secondarily the eastern parts of the United Kingdom (UK). The northwestern provinces of Poland are of minor importance because of the small winter barley area, about 10,000 ha, compared to the 100,000's of ha in FRG and GDR. The severity of the epidemic in Denmark is a product of (1) the frequency of virulence genes in the initial inoculum (this is determined largely by the distribution of resistance genes in winter- and spring barley crops in FRG, GDR and UK), and (2) the distribution of resistance genes in the Danish spring barley crop.

The present communication continues a previous study covering the period from 1960 to 1976 (Jorgensen & Torp, 1978). It attempts

to record the durability of six powdery mildew resistance genes
utilized in Danish barley cultivation up to 1980. Data on the powdery mildew disease severity are from the Danish Agricultural Advisory
Service trials and from nurseries at the State Experimental Station,
Tystofte. The scores for disease severity on the resistant varieties are taken, when possible, as a percentage of the scores on susceptible varieties in the same trials/nurseries.

RESULTS

The data from 1977 to 1980 show that the disease severity on
varieties with genes $Ml-a12$ (Arabische resistance) and $Ml-a7$ (Lyallpur
resistance) is still high in Denmark in spite of the present limited
distribution of these varieties; but this is probably due to the
growing of varieties with these resistance genes in one or more of
the neighbouring countries. Further, varieties with gene $Ml-a9$
(Monte Cristo resistance) have become more diseased, but a corresponding distribution of varieties with this resistance cannot be
traced in neighbouring countries. Varieties with gene $Ml-a$ (Algerian
resistance) have also become more diseased and this may be explained
by the presence of this gene in a winter barley variety 'Valja' that
is rather widely grown in GDR. Resistance gene $Ml-a13$ (Rupee resistance) has remained effective, but it has been present only in a few
varieties with a very limited distribution, so that its durability
cannot be assessed yet. The other four of the above resistance
genes have had a durability of 2 to 7 years.

In contrast, resistance gene $Ml-(La)$ (*Hordeum laevigatum* resistance) has had a durability of about 12 years, remaining effective up
up to 1976 (cf. Jorgensen & Torp, 1978). Over the last few years
varieties with this gene have covered about 70 to 80 percent of the
Danish barley area, and up to 26 percent in the FRG, and probably
as a consequence, these varieties have now become more diseased in
Denmark.

DISCUSSION

The durability of gene $Ml-(La)$ cannot be explained either by
additional resistance genes present in the respective barley varieties (c.f. Torp *et al.*, 1978; Giese *et al.*, 1981), or by a greater
diversity in the gene background, i.e. the number of varieties per
unit barley area and time. One possible explanation lies in the
fact that gene $Ml-(La)$ confers an infection type 3n whereas the
other five genes confer type 0. This implies that $Ml-(La)$ virulent
(or aggressive) mutants (or recombinants) have to multiply in
competition with the avirulent (or less aggressive) part of the
pathogen population that occupies space on the barley leaves together with their infection type 3n colonies. In contrast, mutants with virulence corresponding to the other five resistance
genes can multiply without competition from the remaining part of

the pathogen population. Still another explanation may be that $Ml-(La)$ virulence (or aggressiveness) has a significant negative selective value when unnecessary and would decline unless strongly selected for by prolonged and widespread use of the corresponding resistance gene in the barley area. Data from the UK and Denmark favouring this explanation have appeared recently (M. Wolfe, this symposium).

REFERENCES

Jorgensen, J. Helms and Torp, J., 1978, The distribution of spring barley varieties with different powdery mildew resistance genes in Denmark from 1960 to 1976, *Kgl. Vet. og Landbohojsk. Arsskr.* 1978:27-44.

Giese, H., Jorgensen, J. Helms, Jense, H.P., and Jensen, J., 1981, Linkage relationships of ten powdery mildew resistance genes on barley chromosome 5, *Hereditas* 95:43-50.

Torp, J., Jensen, H.P., and Jorgensen, J. Helms, 1978, Powdery mildew resistance genes in 106 northwest European spring barley varieties, *Kgl. Vet. og Landbohojsk. Arsskr.*, 1978:75-102.

DISEASE RESISTANCE IN RICE

Ivan W. Buddenhagen

Departments of Agronomy & Range Science and
Plant Pathology, University of California
Davis, CA 95616

INTRODUCTION

Rice in different parts of the world is grown in many different agroecosystems and in different pathosystems. This means that the kinds and levels of resistance needed will differ among locations and thus, among rice improvement programs. Moreover, the rice ecology (dry or wet upland, irrigated, swamp, etc.) influences greatly the epidemiological potential of various common pests and diseases, beyond general climatic differences. The tendency for any new level of resistance to be short or long lasting is also dependent on these different factors, above and beyond the genetic plasticity of the pathogen and the type of resistance added to the host genotype.

It is revealing to contrast the California rice culture and rice improvement program with those of IRRI in the Philippines, IITA in West Africa and AICRIP in India. In contrast to the presence of many fungal and several different viral and bacterial diseases in tropical Asia and Africa, California rice has at present only one disease of any significance (stemrot: *Sclerotium oryzae*). Even this disease has low potential under the present agroecosystem since state yield is 7.4 T/ha on one-half million acres. Moreover, the pathogen is so far not known to be highly variable, being a low-grade necrotroph, with host resistances being only a gradation of severity levels. The basic Japonica rice genotype has escaped all but one of its Japanese ecosystem diseases and varietal development by default (validly) ignores essentially all the major rice diseases of the world.

THE DURABILITY OF RESISTANCE

Although "durable disease resistance" has been defined satisfactorily (Johnson, 1981) the operational steps required to develop
it in different kinds of crops to different kinds of pathogens are
far from clear. I consider it useful to examine the concept itself
in a broader context.

A cultivar that is widely grown for a long time has "ecosystem
balance." It will develop various levels of disease and insect
damage from local pests and pathogens, fluctuating with the environmental conditions and the dependency of the disease relationship on
these conditions. The aspect of ecosystem balance that is occupied
by the concept "durable resistance" is the effect over time of the
cultivar in influencing pathogen evolution. Pathogen evolution in
this sense is the selection effect of the cultivar on pathogen
population variability, including rare mutants or recombinants of
different virulence and the rapidity of their replacement of the
old population.

There are four major points influencing resistance durability:
1) potential magnitude of change to greater virulence; 2) Selective
effects of cultivar on that potential; 3) Rapidity of within season
replacement by new pathogen population (governed by type of pathogen
and the crop/pathogen system); and 4) off-season survival fitness.

By examining these four points for a given system we should be
able to predict the potential durability attainable, whether or not
we have the capability to develop it.

A fundamental point is the magnitude of difference between the
existing pathogen population virulence on the cultivar and that of
rare individuals so far undetected or undeveloped in it. If this
potential difference is small, then the resistance will be durable.

Selection methods often need to be improved so that this
potential magnitude of change is examined and the most virulent
strains are seen to be insufficiently virulent for major damage.
This is the desirable objective where one selects against race-
specific factors, i.e. for inoperation of, or susceptibility to
specific races, as recommended for stripe rust of wheat (Johnson
1981) and for blast of rice (Buddenhagen 1981a).

Relative durability or non-durability of resistance of a cultivar can also be viewed as the magnitude of cryptic error in the
selection process in varietal development. For, in a sense, all
resistance is durable. It never breaks down to the individual
pathogen that enabled resistance detection, if the test was valid
in terms of challenge and natural environmental amplitude for the
crop and the interaction. The resistance that "breaks down" was

never there in the first place. The "breakdown" is the expression of cryptic error not revealed in the original test due to sampling limitation - an insufficient range of the pathogen population (and environmental influence on the interaction) was sampled where the decision of "resistance" was made. The more completely the pathogen population and environmental influence is sampled, the lower will be the cryptic error. Resistance remaining after all the pathogen population and farm environments are sampled (obviously impossible) could only be overcome by new pathogen variability, through new mutation or recombination. So, how much non-durability is due to cryptic sampling error of existing pathogens and how much is due to later evolution of new forms? To increase durability, we need to examine: 1) how to reduce cryptic error in the challenge and selection process, and 2) what strategy to develop to reduce future pathogen evolution.

IS THERE A DURABLE RESISTANCE IN RICE?

Has durable disease and insect resistance been incorporated into new rice varieties and does it exist already in old traditional varieties? Hard data to answer these questions are almost impossible to obtain.

For blast (*Pyricularia oryzae*) there are good data from Japan (Ezuka 1979), IRRI in the Philippines (Ou 1979), CIAT in Colombia (P. R. Jennings, personal communication), and elsewhere that varieties previously considered resistant have become susceptible. There are abundant data that varietal reactions in blast seedling nursery tests vary greatly at different times and in different places (Ou, 1972). The variability is so great (Ou 1979) that one can question how breeding for specific race resistance can be expected to last or even to work at all at an operational level in an improvement program. Recently, a major blast epidemic developed in Korea on varieties that had spearheaded the green revolution ten years earlier (Crill 1979). However, good data from the tropics are rare for breakdown of blast resistance in new varieties, previously given a fair test for resistance.

If a uniform cultivar has been grown widely for a long time its tendency to become diseased should be stabilized at some level in each area, depending on the climate and the background pathosystem. If the agricultural system is stable also, then both the cultivar and its levels of resistance are "durable." If a new cultivar is introduced into the area there will be a period of time for adjustment of the old pest and pathogen populations to the new cultivar. Each will shift to an extent governed by the new genotype and by the potential magnitude for change existing in the pathosystem. A dominating part of this pathosystem effect will be the genetic potential in variability of the pathogen population for epidemiological competence and virulence.

Crops have major and minor diseases and some crops have more than others. Thus, there is no general rule for constant increase in pathogen virulence or in epidemiological competence and no general rule that hosts should become diseased or that major epidemics should occur. Much of our concern about "new diseases," new hosts for old pathogens (especially viruses), new strains of pathogens, is a reflection of a young documentary stage of a science, combined with "new encounter" situations and new modifications of agroeco-systems (Buddenhagen 1977). It is often difficult or impossible to untangle these various factors when a "new disease" occurs, or when a disease changes to greater importance.

For instance, blast on rice in Africa was unimportant (and hardly known) 50 years ago even though competent scientists examined rice culture. This must mean major epidemics were not occurring. Blast became important with the introduction of new varieties and expansion and intensification of rice culture, especially with greater use of nitrogen fertilizer. We do not know how much of this was due to new genotypes, expanded area, fertilizers or crop inten-sification or how large were shifts to greater virulence.

Crill recently wrote (1981) that "sheath blight was a minor disease of rice when IR8 was released in 1966. Today it probably causes more loss than any other fungus disease of rice..." and "there is little doubt that the sheath blight pathogen will continue to evolve by accumulating more genes for virulence, thus becoming a progressively more serious problem in rice production." So far as I am aware there is no evidence that the sheath blight pathogen populations on rice are any more virulent today than they were 30 or 15 or 5 years ago. That cultural practices have changed, making the disease more serious, however, is a fact.

The increasing importance of rice virus diseases in the tropics in the last 20 years is partly due to awareness, partly due to promo-tion of new and more vector and virus vulnerable genotypes, and partly due to crop-intensification affecting the ecology and epi-demiology of vector populations and the viruses they carry. That the populations of tungro virus (the major tropical Asian rice virus) have changed in either virulence or strains is not established. I have seen many traditional farmer cultivars hardly affected by tungro in eastern India alongside dozens of devastated new high-yielding dwarf experimental lines; a clear case of the injection of new intolerant varieties of high yield potential into an ancient system of virus/vector/old cultivar/agronomic balance. The old cultivars had a much higher level of tolerance and/or tolremicity (Buddenhagen 1981) to tungro; they have durable resistance.

The story of bacterial blight of rice in India followed a similar pattern. It was discovered after introduction of the new dwarf variety "TN1." Initially there was controversy over whether

it, like tungro, was a soil problem and not a disease and when it
was finally agreed to be a bacterial disease it was blamed on
introduction in the TN1 seed. In a way this was correct, it was
the introduction of the susceptible genotype of TN1 seed, not the
bacterium - that led to the epidemic. That variety was introduced
(along with more nitrogen fertilizer) into a balanced old cultivar/
bacterial/agronomic system in which the disease hardly showed. In
this case, however, disease expression is so involved with high
nitrogen nutrition that most old cultivars had little genetic
tolerance under the changed agronomy which alone revealed their
susceptibility. So, unlike for tungro, they were poor sources for
resistance breeding. I have advocated in India for bacterial blight,
a search for resistance (tolerance) among cultivars anciently grown
below large towns, where high nitrogen-charged waters (from sewage)
are used in irrigation. This should also work for blast disease
which is also closely tied to high nitrogen for maximum expression.

RESISTANCE BREEDING

     Breeding for disease and insect resistance in rice has been
reviewed by Khush (1977) and for insect resistance by Pathak and
Saxena (1980). (Additionally, many short notes appear in the
International Rice Research Newsletter, often on additional pests
and pathogens of more local importance which the interested reader
should check.) Of the 37 fungal diseases of rice (Ou 1972), Khush
discusses 4: (blast: *Pyricularia oryzae*; sheath blight *Corticium
sasakii*; brown spot *Helminthosporium oryzae* (= *Cochliobolus
myabeanus*); and, narrow brown leaf spot *Cerospora oryzae*. At the
time of preparation of that review (1976) Khush stated that no
breeding work had been done anywhere for resistance to sheath
blight, no active program on breeding for brown spot resistance
was underway and no rice breeding program was working on *Cercospora*
leaf spot.

     This leaves active resistance breeding programs only for blast,
with none covering the other 36 fungal diseases of man's most important food crop. This amazing situation, if true, must raise doubts
about the importance of these diseases, and of the availability of
disease loss data and the level of communication of such information
to breeders. Are all these other diseases of academic interest
only? Possibly the situation is not so serious, however. Khush
does add that past work has been done (successfully) on *Cercospora*
leaf spot in the United States, that advanced breeding lines are
evaluated for *Helminthosporium* in India and that IRRI was expanding
its efforts to screen and develop germplasm with higher levels of
resistance to sheath blight.

     Moreover, one might question what is meant by an "active breed-
ing program" for a given resistance. Breeders everywhere select
against whatever appears negative to productivity in their plots.

Many low-grade diseases which occur in such plots will be routinely selected against. If they occur consistently at low levels "extra susceptibility" is selected against automatically, maintaining them as minor pathogens. Maybe such a subtle approach is the most successful, since it is for these diseases where resistance seems most stable and durable, and breakdown of resistance is not recorded. Possibly if one developed for them a seedling screen that could detect hypersensitive resistance to massive allo-inoculum one could convert some of these minor diseases to major ones that, like blast, would quickly result in resistance breakdowns? It would be interesting to try.

Helminthosporium Leaf Spot

It is certainly an anomaly that the widespread airborne pathogen *Helminthosporium* which supposedly caused massive losses and the great Bengal famine of 1943 (Padmanabhan 1973) is not considered aggressive enough or important enough to merit inclusion in an "active resistance breeding program." I suspect that the Bengal rice devastation of 1943 was due instead to an epidemic of tungro virus and its vector, the green leafhopper, and that the leaf spotting of *Helminthosporium* was just the visible effect. It is probably correct to consider *Helminthosprium* leaf spot of rice as a manifestation of nutritional imbalance and thus resistance to it becomes resistance to the particular soil (or virus) problem inducing the imbalance. The predisposing factors will differ from place to place and thus routine <u>local</u> selection against brown spot symptoms will solve the problem <u>locally</u>. Such selections will not be resistant to *Helminthosporium per se,* however, and may show brown spot "susceptibility" in another environment without a fungal race difference being involved. Hence the conflicting data on resistance/susceptibility from screening nurseries and trials in different locations.

Cercospora Leaf Spot

The case of *Cercospora* leaf spot is also somewhat anomalous. Some US varieties were badly damaged in the 1930's and 40's and the route of attempting to find races was followed (Ryker 1943). "Race specific" resistance genes were detected and resistance was attributed to one or two major genes. New varieties were developed that were resistant and the disease has become insignificant. Why? Surely effort was not sufficient to pyramid resistance to the 16 races reported. Yet resistance has remained durable. In Asia and Africa, breeding lines vary in their reaction, often visible naturally, at some stage in their development and the more susceptible ones are discarded. (Although IR 20 is quite susceptible, this was apparently appreciated only after release and widespread planting). Thus, we have a fungal disease with easily dispersed conidia, a perfect stage, major vertical genes, and epidemiological potential on "highly susceptible" varieties and yet it remains a minor disease and (or since) resistance is durable. It would be interesting to examine in detail how challenge and selection were practiced in USA in the 1940's to determine how much (if any) concern was given to "races" in the selection process and how much underlying horizontal resistance was operating. An examination of Ryker's 1943 paper reveals that his races and subraces were not separated on the basis of hypersensitivity and that his classification of R, MR, S was based on duration of incubation period and size of spots--components of horizontal resistance. Since the disease affects mature plants, routine selection against high levels of disease in breeder's plots would be selecting for low r, and accumulation of genes for low r would be automatic. I have been informed (Fleet Lee, personal communication) that the early "race" work on *Cercospora* was dropped in the breeding and selection process in the USA. This is another example of where the use of the word "race" and simple categories of R, MR, S, etc. for interactions affecting disease potential obscure the important biological questions. I have discussed elsewhere (Buddenhagen 1981) how misleading the words "resistant" and "susceptible" can be. To understand what they mean in each case the methods employed to obtain these arbitrary reactions must be examined for their effect on the underlying biological phenomena affecting epidemiology and pathogen evolution.

Sheath Blight

Sheath blight is a typical *Rhizoctonia* induced disease and the literature on this pathogen (Parmeter 1970) is pertinent. As for any omnivorous necrotroph, resistance breeding is difficult. Environmental influences on disease development are strong, with effects of crop canopy and nitrogen status easily obscuring slight cultivar differences in "resistance" *per se*. Although variation of the pathogen has been recorded - and it is not necessarily a rice pathogen - one should expect little gene-for-gene coevolution and thus considerable stability of any minor gene resistance (tolremicity) that can be accumulated. The approach suggested by Khush (1977) of eliminating extra-susceptible breeding lines and using recurrent selection for accumulating minor genes seems good.

Blast

Blast appears to be the greatest enigma since "resistance" has broken down many times (especially in Japan), and fungal variability, even without a perfect stage, is great, anomalous, and controversial (Ou 1979, Ezuka 1979). The extensive work in Japan on identification and use of single genes for resistance has not lead to durable resistance. Earlier use of resistant Indica germplasm crossed with Japonicas was considered to provide more stable resistance for the Japonicas (Toriyama 1972), but the collapse of such varieties in Korea in 1978 indicates that this also, is questionable. IRRI has

made use of widespread testing in a seedling blast nursery to
identify "broad spectrum resistance" which has then been used as
"sources of resistance." The genetic nature of such "broad spectrum
resistance" is not well understood. Nevertheless, in lowland Asia
blast is seldom a problem today on varieties derived by this approach
(or by any other method). But, the "resistant" varieties from Asia
become severely blasted (with a few exceptions) when they are grown
under upland conditions in Africa (and probably in Asia). Blast
remains the most serious disease of upland rice and of irrigated
rice in higher elevations, and in Latin America generally.

    In spite of an extensive literature on blast, many key points
influencing resistance/susceptibility, epidemiology and fungal evolution are poorly understood. The mechanism of high fungal variation
is not known. The genetics of the host/pathogen interaction are
known only in the host, since the perfect stage has not been obtained
among rice isolates. The source of initial seasonal inoculum in an
area is usually not known, nor the role of alternative hosts in
pathogen ecology and evolution. The strong effect of variable plant
physiological status, as influenced by nitrogen (Matsuyama 1975),
water relations (drought) and aerobic/anaerobic root system conditions, in influencing blast susceptibility and epidemic development
is not understood.

    I consider the following to be a reasonable analysis of
published data and of my own studies and observations, and to be
important when considering breeding for resistance:

    1) There has been a long coevolution between rice (Morishima
et al. 1979) and the blast fungus in Asia and Africa, with a vertical gene system evolving on different rice species in each area,
presumably ancient wild pathosystems. Virtually nothing is known
about these systems (which still operate) nor whether different
vertical genes exist in the different species or even in Japonica
and Indica types of *O. sativa*. Nothing is known of the American
Oryza species and blast. The blast populations at a given time,
even in a localized area, are composed of mixtures of individuals
that contain different vertical genes. Japanese differentials
grown in Nigeria in breeding plots separated by a few hundred yards
revealed different "races." Earlier, Ou (1972) stressed that different lines were susceptible or resistant erratically in nurseries
grown at IRRI in different months. It is not possible to know how
much such conclusions are based, erroneously, on low frequencies of
a compatible type (a line with a few compatible lesions would be
given a low number for a row reaction and classed as "resistant").
Nevertheless, this type of information is an immediate alert for
a very variable pathogen population where any classification of
"resistance" based on hypersensitivity would be suspect on sampling
error alone.

2) Most of the vertical genes studied, govern in the varieties used, a hypersensitive block to disease development. Thus, the "resistance" is a "super-susceptibility" at cellular level, which is easily overcome by variability within the blast population, and results in "breakdown." The "breakdown" will then reveal some level of underlying horizontal resistance-epidemiological potential, fungal-host-environmental relationship. The hypersensitive vertical genes, when present in different varieties, may not give full hyper-sensitivity and thus may act instead to reduce r as they may do when "broken down" in the original variety.

3) As long as the vertical genes do not allow the disease cycle to operate (in a breeding nursery), the background resistance level of the genotype cannot be judged. i.e. a "breakdown" of the future cannot be judged for potential severity.

4) The blast populations are sufficiently variable and plastic that they can be expected to overcome any vertical gene and most combinations of vertical genes.

5) The blast fungus is sufficiently prolific and the spores sufficiently mobile so that a "breakdown" can expand from a single lesion to a region quickly enough to invalidate within a few years of widespread cultivation, a vertical gene apparently useful in breeders' plots.

6) Off-season survival and directional selection acting on survival probably temper rapidity of apparent appearance and spread of resistance breaking strains, but the extent of this is completely unknown.

7) The source for constant challenge to a hypersensitive immune variety has to come from some other variety or host in the region in order for a resistance breaking strain to develop after the first year of planting. If this does not occur, the breakdown must occur during the first year of planting and the delay in appearance is one of restriction on spread and survival. This point is important since, if it is the former, it means that a single variety planted on large areas would have a greater chance of durability than several varieties.

8) Blast does not normally occur in breeders' irrigated plots in the tropics, or if it does, it is erratic. Therefore, the breeder cannot select against overly susceptible plants or lines, and he usually relies instead on a separate seedling nursery test (often run by others) for information on blast "susceptibility."

9) The blast seedling nursery is manipulated to obtain maxi-mum physiological susceptibility (high N and aerobic upland condi-

tions), and maximum allo-inoculum generated on (usually) a single spreader variety and thus limited in fungal variability by the genotype of that spreader variety.

10) In the blast nursery, spore bombardment on candidate rows is enormous. The negative influence of a thousand hypersensitive reactions on a compatible infection attempt on a leaf is unknown, but possibly real. A row reaction is given regardless of segregation which occurs within the row. Low numbers are given for type of lesion and higher numbers for amount of disease. (But different types of lesions are usually present, at variable frequencies, on any leaf.) These numbers are grouped into R, I, or S categories. How these are used in making decisions depends on the breeder. Obviously, no one will save material that is "highly susceptible" often (rated 5-7). But what if in 10 nursery tests a line shows a 6 reaction in only one, all others being 0-1-2? This indicates either a "breakdown" of a vertical gene, or an error. The answer cannot be found in the literature. Khush states that 3 released IRRI varieties "have strong blast resistance," inherited from one common parent. The other 8 released varieties do not have "high levels of resistance" in their parentage and they classify as "susceptible" to "moderately resistant" in the blast nurseries. It is hard to know what any of this means. Probably only that a substantial part of the fungal challenge population in the nurseries carries vertical genes which can match the vertical genes in these varieties - and that different amounts of blast occur on them due to different levels of poorly detected horizontal resistance. Even Khush states that "IR5 and IR8 (the earliest of released IRRI varieties), classified as susceptible in the blast nursery, rarely get much blast where they are widely grown in lowland Asia. It appears that [they] have adequate levels of field tolerance that are not detected in the blast nursery." But the direction now is to insure that the advanced breeding lines "have good levels of resistance." If this means that they have complete hypersensitive vertical gene protection, one may question their vulnerability. Especially in the absence of knowledge of the number of such genes involved. It would be unfortunate if the route in the tropics now followed the earlier Japanese approach, with their history of "breakdowns."

11) The expression of blast that causes the greatest loss is neck and panicle blast, occurring between flowering and maturity. There is no guarantee that the seedling nursery measures the performance of a variety under an epidemic generated at this flowering stage. Correlation can be good or poor. As for many rice (and cereal) diseases, varietal differences from flowering to maturity stage are real and critical and it is here where the environment and genotype interact so strongly in influencing physiological conditions, maturity/senescence, pest damage - and yield loss.

12) I have detailed elsewhere (Buddenhagen 1981) an alternative
practical methodology that would enable the routine development of
lines having horizontal resistance to blast, which hopefully, would
be durable. The essence of this method is to avoid hypersensitive
resistance, avoid the seedling nursery blast test, utilize the
full-season relationship between candidate genotypes and blast
buildup to select for low r. The latter requires rice to be grown
under upland conditions and to challenge long breeding rows (F_3
onward) unidirectionally with a mixed population. The idea is to
manipulate the system so that the environment/plant physiological
status encourages blast development for a long period, especially
through the panicle stage, with self generated inoculum. I have
termed this aspect of resistance "tolremicity" - the ability of a
plant or cultivar to slow a self-generating epidemic. It is possible to breed for tolremicity in a logical fashion since it can
be measured or observed easily on thousands of breeding lines, if
field designs are made appropriate. Tolremicity may involve any
subcomponent of the requirements of epidemic development, and
recurrent selection for accumulation of genes for it can become
automatic. I believe much of successful breeding for resistance to
fungal diseases is where the natural environment in breeders' plots
has enabled automatic selection for tolremicity. I believe the
durability of tolremicity is based on the multicomponent possibilities of its maximum expression. Proof, however, will take time.
Tolremicity has general application since it can be analyzed for
any disease (or insect) system and modifications in breeding and
selection methods can be made to enable its selection.

13) Different rice ecologies and agroecosystems (and regional
pathosystems) require different levels of resistance or tolremicity
to blast. It should be obvious that using upland varieties in
breeding will enable generation of higher tolremicity since rice
under upland conditions is physiologically more vulnerable to blast.
Successful upland varieties are less susceptible to blast than successful lowland varieties, even from the same area. If there are
any "successful" varieties grown in the highlands of the tropics,
these also should have lower vulnerability. Likewise, old varieties
grown under the droughty conditons of "rainfed rice" in India and
similar ecologies should have evolved resistance, but the erratic
nature of vulnerability in this ecology may have encouraged vertical
gene selection (and pyramiding of such genes). I suspect it is such
environments that generated the famous blast resistance source of
Tetep and similar varieties. The Japanese have used upland
Japonicas as a source of resistance for years. The mistake they
made, in my opinion, was to try to extract from them, one or
several vertical genes, using a methodology that would not detect
"tolremicity."

14) Many of the major high yielding varieties (HYV's) can be
blasted, given appropriate conditions. It is uncertain if their

resistance has "broken down" or if they were known to be susceptible
at some level, even before release, without such knowledge causing
concern. Varieties such as Jaya, IR8, Ratna, Pusa 2-21, etc. have
different vulnerabilities to blast--from low to high. Pusa 2-21 was
selected in a blast escape area. It was susceptible when released,
but it is still grown. When I travelled extensively throughout
India in 1971-73 I observed blast severities on different varieties
in farms to be always in the same order. I questioned the meaning
of the striking differential reaction in blast nurseries compared to
the constant ranking for damage of the major HYV's in different
regions. This still means to me that each variety has a certain
vulnerability, that in a practical sense vertical gene protection is
not operating and that the amount of blast is a result of different
levels of tolremicity--effective similarly everywhere in the pathosystem. This has to mean that stabilizing selection is operating on
a regional basis, that these varieties have not "broken down"
because they always were "broken down." Whether or not erosion of
their tolremicity will occur, time may tell, but believable data
probably will not exist. It is, however, a conceptual mistake to
say that the resistance of Pusa 2-21 or IR 8 has "broken down."

15) At CIAT in Colombia, more concern has recently been given
to the strategy and tactics of varietal improvement in relation to
blast (Weereratne et al. 1981, Ahn 1981). This is partly because
of the collapse of several fairly recently released varieties
previously thought to be "resistant," which has led to questioning
of past approaches. The environment (and pathosystem?) in parts
of Latin America is highly conducive to blast development but the
original breeding plots at CIAT were in an inadequate blast area.
One new approach involves "pyramiding" vertical genes for resistance
by backcrossing to susceptible HYV's following crosses and inter-
crosses to different "broad spectrum resistant" parents. Although
this strategy seems logical to me, the methods used do not enable
pyramiding to be done in a logical fashion, but only by chance.
Another approach is to modify the blast nursery and the breeders'
plots to attempt better judgment of the lines' tolremicity. Yet
another is to mutate "broad spectrum resistant talls" and durable
upland talls to dwarf stature so as to require less (or no) backcrossing to susceptible HYV's. Also, through Rockefeller Foundation
support to Pennsylvania State University, potentially very useful
research on epidemiology in relation to breeding is developing,
linked with CIAT.

16) IRAT (Institut Researche Agronomique Tropicale) in the
Ivory coast and Malagasy has been pursuing research on horizontal
resistance to blast for some years (Bidaux 1978, Bidaux & Notteghem
1979, Notteghem 1979) as has IITA (International Institute of
Tropical Agriculture) in Nigeria since 1975, (Buddenhagen 1981).
Upland cultivars exist from landraces with durable resistance under

the most blast conducive conditions and new varieties appear to
have similar properties. None of these varieties is useful directly
in irrigated conditions and thus there has been a standoff in utili-
zing such material (and approaches) in the more important irrigated
systems. Since there has been little communication between these
African organizations and the major rice improvement programs of
the world, especially in Asia, it remains unknown if either the
approaches or materials would be useful generally.

17) In Brazil an 8-million acre upland rice culture exists
chiefly with two varieties that have moderate levels of horizontal
resistance. These two bred varieties, which differ considerably
from modern Asian dwarfs, are derived from old introductions by
the Portuguese. Some of the African material from Zaire is very
similar, presumably due to a similar source with Brazilian-African
trade in colonial times (unknown but possibly upland southeast
Asia). Conditions in Brazilian breeders' plots enable judgment on
what is really tolremicity and although more recently a seedling
blast nursery has been added, hypersensitive resistance is not
selected. An important need is greater resistance but there is no
evidence for (or against) "breakdowns" and the major present varieties have been grown widely for many years. According to Tanaka
(unpublished personal correspondence) the blast populations on the
two main varieties (IAC 47; IAC 25) are different and have different major vertical genes. The Brazilian upland culture would
be an ideal one for pursuing the goal of higher tolremicity in
the rice improvement programs, and for obtaining good data on
durability, and on the stability of vertical genes in the
pathogen population.

18) The recent history of blast in Colombia and Costa Rica
(P. R. Jennings, personal information) in relation to varietal
development and large acreage production is of considerable interest.
All old USA varieties introduced since 1950 were attacked in stress
areas in some years. The old variety Tapuripa was resistant in
Surinam (and remains resistant) and its dwarf derivatives have
remained resistant there, but are not widely grown elsewhere.   IR8
and some other IRRI varieties and the early CICA releases from CIAT
replaced the USA varieties in Colombia and elsewhere and they
quickly "broke down." But CICA 7, now in 5th year, has not "broken"
and this variety comes from a 3-way cross with only Colombia 1 as
a resistant parent. CICA 8, on much greater acreage now, comes from
a 3-way cross with Tetep as the only resistant parent. It had no
blast in the first year, but developed severe blast on a few drought-
stressed upland areas in the second year, which "recovered" before
harvest. In this same second year the great resistant source Tetep
(grown in small plots only) became diseased for the first time.
Isolates from Tetep and CICA 8 could reinfect but only very low r
values resulted.

The upland Brazilian varieties (very different genetically) remain generally healthy when grown in the Colombian llanos where even CICA 8 gets blasted. The concern of breeders is that this CICA 8 type "durability" is uncertain, is fortuitous and unpredictable and cannot be repeated in breeding except by accident. It is certainly true that it is not understood.

A situation similar to the CICA 8 story has occurred with blast in south USA. New varieties, Lebonnet and Labelle, resistant to the two common races became diseased by a new race, IC 17, in 1978 in one area in Arkansas and in several counties in Texas. By 1980 the diseases died out in Arkansas even though the varieties were not changed, and the area affected in Texas has not spread.

In my view there are two key points to be derived from these interesting stories: 1) Epidemic and damaging development of blast is a complex interaction of physiological plant status (physiological stress) and pathogen aggressiveness, operable only when microclimate is favorable for infection, spore production and dispersal. 2) Host genotype influences mostly the primary interaction (physiological status x pathogen aggressiveness) and this can be manipulated so that disease development occurs only under increasingly high stress - to the point of unimportance (genetically or culturally). The pathogen must acquire increasing aggressiveness as the host genotype is altered and this is extractable from the total pathogen population to some extent only, and at the expense of adaptability at lower resistance levels. Thus, low r is expressed due to a very narrow environmental/physiological range for disease expansion, and yearly buildup is reduced by negative directional selection in the off season.

In summary, the breeding of rice in temperate Asia in relation to blast has pursued methods that reveal major vertical hypersensitive genes, and it has resulted in breakdowns. In recent years, work in tropical Asia where lowland blast epidemiological potential is less, has utilized to a greater or lesser extent, upland seedling blast nurseries for decisions on "susceptibility." In general, if very superior lines were "susceptible" under these tests, they could still be selected, promoted and grown. Clear evidence for blast "breakdowns" in the Asian tropics do not exist. They do in Latin America but I question the degree of cryptic error in breeders' plots. A straightforward major vertical protection breeding approach has been pursued in Asia. These efforts leave gaps in understanding breeding and selection methodologies in relation to the wild or recently domesticated pathosystem, to the dynamics of pathogen variability, ecology and evolution, and to disease epidemiology. A comparison with the extensive research and thinking on wheat host/pathogen interaction (see reviews by Scott et al. 1980, and Johnson, 1981a) reveals this discrepancy. However, a start has been made to unravel the complexities and opportunities of the blast nursery system in relation to blast races (Ikehashi 1979). Tremendous opportunities exist for <u>useful</u> basic research at the ecological, evolutionary, epidemiological, host stress/ parasite, and genetic levels of the blast pathogen populations and of rice.

At the practical breeding level, methods to pursue and improve tolremic selection should be expanded. The extreme effect of the physiological status of the rice plant in influencing blast susceptibility and resistance needs to be appreciated. In a sense, blast is a physiological imbalance disease and a genetically uniform crop population ranges from susceptibility to resistance in a complex interaction with environmental conditions, pathogen pressure and variability. Breeders should understand that in a tightly coevolved system such as blast/rice <u>all</u> major vertical genes exist due to past natural selection that is temporary and localized in the wild and landrace pathosystem. Two hopes for durability of resistance lie in a) Pyramiding vertical genes in a logical and measurable manner, if many are not allelic; and 2) accumulating major and minor genes for resistance to reduce r, preferably in a recurrent selection system, and in an environment in which r is real and detectable. Both paths require creative research and more teamwork than now exists. Neither approach is being pursued vigorously with acceptable scientific rigor at present. This is not to say that stumbling progress is not, or will not, continue to be made. If, for unknown reasons, the limits of pathogen variability are breached by a selected cultivar, durability will have been accomplished.

Other Fungal Diseases

Some of the minor rice diseases are severe locally. To avoid introducing extra-susceptibility with new varieties it is important to select at several sites in the region where a new variety is to be promoted, and for the minor pathogens to be present at these sites. Vulnerability will be revealed and can be selected against. Most upland material developed in Asia from irrigated rices, and HYV paddy lines are severely affected by a husk disease complex under west African upland conditions. This disease may be caused by one or more of several fungi, including *Sarocladium* sp. (Synonym: *Acrocylindrium*), a newly discovered husk pathogen (Ngala and Buddenhagen, unpublished data). Grain weight is reduced and milling outturn is severely affected. Local old cultivars are hardly affected, as are many breeding lines derived partly from them. This problem and many others, including sheath rot, *Rhyncosporium* leaf and kernal smut and others can probably best be reduced (or maintained at a low level) by routinely selecting against them <u>in local ecosystems</u> favoring the problem. I would be reluctant to advise trying to detect races and perfect an artificial seedling or other test that obscures tolremicity and enhances pathogen variability.

It is of interest, considering their importance on other cereals, that rice can be attacked by two rusts (at least one obviously not coevolved), by two smuts and by downy mildew. Little attention is paid to any of these diseases on rice. They are isolated and unimportant. Let us not make them important by our methods of breeding and selecting new varieties! What better proof do we need that pathogens, even those in taxonomic groups with great disease potential on similar crops, do not readily evolve continuously toward higher virulence and greater epidemiological potential?

BACTERIAL DISEASES

The pathogen causing bacterial blight (*Xanthomonas oryzae*) is a rice-coevolved Asiatic organism (Aldrick et al., 1973). It is a systemic pathogen in the xylem. The evidence that its recent spotty appearance in Latin America (Ou 1979, Lozano 1977) is due to anything but recent introduction in seed combined with a possible confounding with a different bacterial disease from local grasses, is unconvincing. Further careful work is required to clarify the various alternatives. Bacterial blight has recently been found in Africa on a few experiment stations in the drier savanna of West Africa (Buddenhagen et al., 1979). In 1981 it was first detected in indigenous wild rice species in remote sahel-savanna locations, indicating indigenous evolution there separately from Asia (Buddenhagen, unpublished). Only in Asia, though, is it commonly present in farmers' fields, with a large endemic reservoir. Japanese workers classified their isolates into 3 groups (Toriayama 1972). We had extensive evidence of a range of virulence of many widely collected isolates on 8 "differentials" without strong race separation, and predicted the disease would become unimportant in the tropics in 5-10 years if breeders would refrain from pushing for major vertical genes (Buddenhagen and Reddy 1972). However, breeders and pathologists wish to find and prove that there are major genes and most any data can be squeezed to fit (Petpisit et al. 1977), especially if the pathogen is systemic, severity is subjective, and groups are arbitrary. Khush (1977) considers 3 centers for the 2 major tropical resistance genes, a Bengal center for gene Xa5 and a South Indian - Sri Lankan and Javan center for Xa4. Whether the different genes identified for "resistance" have any differing significance epidemiologically in the field and in different varieties (outside of Japan) and whether effective races have developed in the tropics as a result of deployment of resistant varieties is uncertain. Several IRRI and IRRI-derived varieties (such as IR 20, IR 22, Palman 579) classified as "resistant" to BLB were not developed for resistance to it but were selected for general performance in which low damage from BLB was a part. They are susceptible to BLB but they have high tolremicity. At present, where varieties containing Xa4 are widely grown in Indonesia, the disease is becoming more important, indicating a population shift

to greater aggressiveness in relation to the use of this resistant
dominant gene (H. E. Kauffman, personal communication). In tropical
India and neighboring countries probably a majority of the land area
is planted with varieties that would not be classified as "resistant"
based on major genes. What does this tell us about the disease?

In my opinion, a disease such as BLB, based on xylem invasion,
is an ideal case for breeding and selecting for tolerance and tolremicity (Buddenhagen 1981). With off season survival critical (and
possibly largely in seed) the pathogen's full ecological cycle is
important to consider in developing an appropriate methodology.

Bacterial Streak

Varieties differ in the amount of disease they will develop from
this stomatal-invading, parenchyma inhabiting bacterium (*Xanthomonas
translucens*). The disease, basically a minor one, is erratic in
occurrence, with initial inoculum probably usually from seed. The
disease occurs not only in Asia (Ou 1972), but erratically in
tropical west Africa where it is introduced and moved in seed into
new areas, and apparently present indigenously in wild rice in the
lake Tana area of Ethiopia (Buddenhagen, unpublished). "Races" have
been denied (Ou 1972) and reported (Shekawat et al., 1972) but their
reality or significance in the field is unknown. This is another
disease which apparently is kept in check by routinely selecting
against extra-susceptibility in breeding plots where the disease
occurs. Since the disease is erratic in occurrence, absent from
many areas, and easily transmitted in seed, varieties may exist and
be under development that are extra-susceptible and vulnerable.
This should be prevented by screening in endemic areas with uni-
directional spreader inoculum challenge; and by precluding introduction in seed into disease-free areas.

VIRUS DISEASES

Viruses that can inhabit rice are many, ranging from curio-
sities inoculated to rice in greenhouses and originating from other
hosts to those causing major internationally important field
diseases (Ling 1972, Shikata 1979). Rice viruses are still
imperfectly known, especially in the tropics (and China) where "new"
virus diseases are becoming apparent, (Ling 1978, Buddenhagen 1979).
Ragged stunt, very recently discovered in Asia, became devastating
in Indonesia on newly released vrieties in 1978. All but 4 viruses
on rice that are known are leafhopper or planthopper borne and their
geographic distributions are discrete. In some cases vectors range
beyond the range of the viruses, such as for *Nephotettix* and
Sogatodes in west Africa. The comparative identity of virus
diseases in different areas of the Asian tropics is still uncertain,
with inadequate characterization, isolation and serology combined
with a limited range of similar symptoms (Saito et al.; 1975,

Saito, 1977; Shikata 1979).

Tungro

The rice green revolution owes a large debt to Tungro virus because this disease, long considered in Indonesia to be a "soil sickness" and named "mentek," led the Dutch to breed for resistance to it as early as the 1930's (Van der Meulen 1975). A very success- ful 4-location field screening was conducted on a bulk from a cross made in 1934 of a resistant Bengal variety (Latisail) with a blast resistant Chinese variety (Tjina) which led to the selection of Peta, the mother of IR 8 and the variety whose genes are in many of the major tropical Indica HYV's. One would speculate that to be selected, released, and become widely planted, Peta (and other sisters) would have to have had considerable tolerance/tolremicity to both tungro and its vector, the green leafhopper. This proved to be so in tests at IRRI conducted 30 years later. Thus, many early HYV's had (and still have) sufficient tolerance/tolremicity to tungro, green leafhopper and blast, derived from this early Dutch approach to breeding and multilocation field selection for yield and performance to locally occurring unidentified severe problems that were later shown to be very widespread. Even though they were less severe or more erratic elsewhere, intensification and modification of rice culture has led to major epidemics, revealed by growing susceptible varieties. By accident, IR 8 got green leafhopper resistance from Peta (as well as resistance to the American rice delphacid *Sogatodes*) but not resistance to tungro virus. Yet Peta is resistant also to tungro as well as to hoja-blanca virus, vectored by *Sogatodes* in America. As so often in rice, varieties resistant to one disease or insect turn out to be resistant to others on distant continents. Good comparative work on the genetics and mechanisms of such phenomena are needed.

Tungro is the only virus of rice that is clearly coevolved with rice. However, recent work has thrown doubt on its cause, as a small bacilliform virus and a spherical particle have been seen in plants diseased with "Tungro" symptoms (Saito 1977). If this is true it throws all old tungro literature into doubt, especially strains and breeding for resistance. It is tropical Asian, endemic in north- eastern India-Bangladesh, where rice may have been first domesti- cated (Morishima et al., 1979) and present throughout southeast Asia where various wild rice species are indigenous. Since it is not vertically transmitted (through the egg) it has to survive through the off-season locally in perennial wild rice or alterna- tive perennial grasses, or reinvade from areas with overlapping rice cycles. The survival route is not established. This is important in considering durability of resistance, since if a weed screen is imposed for survival then any potential new virulence induced by a resistant rice variety could be blocked in the off-season by negative directional selection.

As mentioned earlier, very successful breeding for resistance
that has remained durable was conducted with no knowledge of the
disease in the 1930's. IRRI began breeding for resistance in 1965
with a mass screening cage challenge method. Varieties with 30% or
less infected seedlings were classed "resistant" (Ou 1972). This
approach measures part of tolremicity but confounds vector resistance and does not measure tolerance. A tungro epidemic at IRRI
and in the Philippines generally in 1971 revealed vast differences
in field resistance of many breeding lines and since then breeding
materials have been regularly subjected to field tungro pressures,
which has greatly aided avoiding overly susceptible material. In
India, field screening has been developed that reflects the
totality of plant tolerance and varietal tolremicity. With such
methods, combined with using the high levels of tolerance in
ancient cultivars in the endemic areas, tungro should have been
reduced to an insignificant disease, barring virus variability and
directional selection to greater virulence, but, it has not. Many
Indian HYV's are not résistant, reflecting a lack of concern of its
importance, and a willingness to gamble with occasional epidemics.
Recently, 91 HYV's released by state and national organizations
in India and by IRRI were evaluated for field resistance in a
realistic manner for 3 years. Only 7 were judged resistant, 27
as intermediate and 55 as susceptible (Anjaneyulu, personal
information).

There is little information on virus variability and strains
(or even if there are several viruses involved), and little information on genetics of resistance. Probably one major limitation has
been a lack of appreciation that tolerance is involved and that it
can be precisely measured without resort to lumping into "resistant"
and "susceptible" categories. Also, that tolremicity can also be
studied separately from tolerance. In any case, some varietal
resistance (that from Gam Pai) is not universally effective. Also,
reports of new strains in Indonesia require confirmation and precise
virus identification. Comparative studies are clearly required at
an international level and they might go far to precluding blow-ups
in the future. I see no reason why tungro vulnerability should
exist in any major tropical Asian variety. The same applies to
vulnerability to its vector, the green leafhopper, a pest in itself.
Tungro has probably appeared in southern Japan (Shikata 1979) but
even that is uncertain and the vulnerability of Japonica rices is
not known.

Grassy stunt

The durable resistance to this virus conferred by a single
dominant gene obtained from a single entry of the wild rice *O.*
nivara is a classic success (Khush 1977). This was a straightforward case of backcrossing with already good materials, for
incorporation of a resistance gene from a poor-type related

wild species (the only such example in rice). The disease has been
widespread in tropical Asia, but usually is of low incidence. The
range of the brown planthopper vector exceeds that of the virus,
which is strictly tropical Asian. The source of seasonal inoculum
in the absence of overlapping rice culture is unknown. The reasons
for low incidence and absence in temperate Asia are also unknown.
The resistance from *O. nivara* is probably immunity, but this has not
been studied. Field tolerance also exists within sativa cultivars,
which may be either plant tolerance or varietal tolremicity. New
sources of resistance have been found recently, (the type and
mechanism of resistance are unknown (Ou et al. 1976). This disease
is possibly due to a co-evolved rice pathogen.

Hoja Blanca

This American virus caused devastating epidemics on USA
varieties in the Caribbean area 1957-65, followed by direct vector
damage 1965-69. IR8 and the CICA varieties were then released and
the problem faded and remains unimportant (P. R. Jennings, personal
information). Also, new varieties were bred and released for
resistance in the U.S.A., and have remained resistant.

In all probability the hoja blanca virus is not originally a
rice-evolved virus since it is confined to America and occurs in
some areas where American *Oryza* species are not present. The
original host or hosts of the two planthopper vectors (*Sogatodes*
spp.) is also not clear from the limited published literature. Some
unknown grass must be the original virus/vector host but no information on this is available. It certainly would be pertinent to
know the susceptibility of the indigenous American *Oryza* species.
The long incubation period (30-36 days) restricts epidemic increase
and makes direct transmission to vector offspring a must for a good
epidemic.

The disease first became important in the 1950's in the
Caribbean region and it was found that the "local" Indica type
varieties and those from southern USA were susceptible (Lamey,
1969). Resistance was found in most Japonica varieties and in a
few Asian Indicas--including Peta and others that had been bred for
"mentek" (tungro) resistance by the Dutch. It would be interesting
to determine if all the Indicas from Asia, and the Japonicas, found
resistant then are also resistant to tungro, and to determine the
comparative inheritance of resistance to the two viruses.

Extensive field tests with different varieties in many Latin
American countries did not reveal any virus strain differences.
Virus resistance may be due to a single dominant gene. This resistance is probably plant tolerance operating also to provide
tolremicity. Resistance to the vector also exists (possibly
multigenic) and some HYV's from Asia (IR 8), bred in the complete

absence of both vector and virus, have resistance. The disease
became unimportant for a while where vector-resistant varieties
were grown. However, in Cuba IR 8 is susceptible to the vector,
indicating a new biotype, and spraying for insect control is
practiced, even on other varieties. There is now a resurgence of
hoja blanca in Ecuador, and also in Colombia, for reasons which
require investigation.

The Temperate Viruses From Japan

The major viruses on rice in Japan and Korea (stripe, dwarf,
and black-streaked dwarf) are probably not originally rice viruses,
since they and their leafhopper and planthopper vectors have a wide
graminaceous host range and rice is a fairly recent immigrant, at
least to Japan. These viruses must also exist in China, but little
information is available to the Western reader.

It is of considerable interest that the traditional lowland
Japonicas are susceptible to all 3 viruses and that many Indicas are
resistant; (upland Japonicas are also resistant to stripe). This
is just the opposite for hoja blanca and there apparently has been
no exposure and no natural selection for this resistance to exist,
thus the resistance is "accidental." Where investigated it appears
due to one or two major genes. Presumably they act to confer
tolerance and tolremicity but I cannot be certain from the literature. They cannot be considered to have co-evolved and thus are
probably constitutive genes, existing for an evolutionary reason
other than to confer resistance to the viruses against which they
are used.

Rice Yellow Mottle Virus

This beetle transmitted African virus was first described
from Kenya (Bakker 1974) and later it was found in Zanzibar,
Tanzania and west African countries (Raymundo and Buddenhagen 1976;
Buddenhagen, unpublished data). It can be mechanically transmitted
easily. HYV's from Asia and Colombia (closely related) are highly
susceptible. Many west African 'sativa' landraces are highly
resistant as are many *O. glaberrima* lines. Resistance, simply
inherited, is easily developed and simple inoculation and evaluation
procedures have been developed which measure tolerance (Sarkarung
and Buddenhagen, unpublished). It is not certain if this is a
rice virus, evolved on African rices, or whether the resistance of
African *O. sativa* landraces and *O. glaberrima* are fortuitous. It
may be a virus of African grasses. It became obvious with the
recent importation of susceptible Asian *O. sativa* Indica HYV's.

In summary, most so-called rice viruses are probably not co-
evolved with rice, with tungro (and possibly grassy stunt and
ragged stunt) as exceptions. Strain differences appear to be

minimal. Resistance is major gene and is either fortuitous and unrelated to the viruses, or is probably coevolved for tungro and in
either case, appears durable. In the case of tungro and RYMV,
resistance is to be sought in the endemic area, for the others (not
rice viruses) in germplasm quite different from the genotype and
region of the virus problem. Possibly this is not as illogical as
appears at first glance. For a non-coevolved virus to appear as
important, the local germplasm has to be susceptible. The local
germplasm in most areas has a common origin and a certain degree
of relatedness. Distant species, subspecies or cultivars have less
relatedness, with greater chance of different genes. Some of these
genes can be expected to confer "resistance." Viruses on rice
should not be a problem on rice anywhere and if they are it reveals
inadequate efforts or tactics in incorporating existing potential
resistance in *Oryza* into the mainstream of varietal development and
promotion. Several viruses that exist in the bush and can affect
rice (such as MSV in Africa) are now of minor importance, or are
even unknown. These can be kept minor by a realization of how such
viruses become important and carrying out a strategy of preventative
breeding locally for such potential problems.

INSECT RESISTANCE

I will not cover this area in detail. The reader is referred
to Khush (1977), and Pathak and Saxena (1980). However, a few
points are pertinent to this discussion.

The HYV technology alone has shifted the importance of the insect groups. In Asia, borers have declined in importance relatively
and planthoppers and leafhoppers have become very important. Some
reasons for this are postulated but the area is inadequately re-
searched. The rise of the hoppers is history in Asia but it is
happening and will continue in Africa and Latin America unless
tactics of varietal improvement and insecticide use can avoid
copying the Asian path. In Latin America, the *Sogatodes* planthopper
has been reduced to an unimportant problem on the HYV's (except in
Cuba) by the development and use of resistance (that may be multigenic). This route was initiated early on, probably due to the
blow-up of hoja blanca virus, vectored by *Sogatodes* on the old
varieties, about the time the advantages of HYV's were being
realized. The first major HYV from Asia (IR 8) was resistant,
fortuitously, or rather, as I think, due to the method of field
selection of its mother parent in Indonesia, for tolerance to many
things. In any event, the reduction of the virus and its vector
by the use of resistance was quick and very successful.

In Brazil, and in the Sahelian areas of Africa, where new rice
irrigation projects are developing, one may observe many hoppers
and other insects that are still at low populations, but obviously
ready to develop as major problems (and possibly transmit "new"

viruses from the bush) if the right varieties are grown for them.
Additionally in some sub-sahelian areas of Africa, the gall midge
Orseolia (Pachydiplosis) oryzae, is already a major problem. One
must ask why, since resistance from Asia to at least two biotypes
exists. Is there another biotype, or do those who promote new
varieties largely ignore such problems and thus promote susceptibles
because it has been minor under the old rice system, or do they
expect insecticides to solve the problem? It is in these same areas
where green leafhoppers (Nephotettix spp.) exist as potential
threats.

Likewise, in wetter areas in West Africa, intensive rice projects have resulted in increased importance of stem borers and other
"minor" insect pests. I know of one project in Nigeria where four
insecticidal sprays are now insufficient to control borers.

In most national programs one can find routine insecticide
trials. The hope of the entomologist is that spraying will make a
big difference in yield. In some programs, one may find insecticide
application on breeders plots. Commercial projects also want a
routine insecticide recipe. The role of resistance in maintaining
low populations of many different potential insect pests before they
become problems, and of consideration for avoiding the destruction
of biological enemies of pests by spraying is certainly given short
shrift, as major efforts are given to immediate yield in massive
coordinated trials of largely introduced lines and introduced
insecticides.

Great opportunity exists for developing and executing strategies and tactics that will result in types of resistance that will
maintain pest populations low in areas undergoing rapid change in
intensification of rice production. Such efforts need to be
integrated with efforts towards not upsetting, but rather,
encouraging natural biological control.

Most "resistance" to insects that breeders have concentrated
on in rice is the single major gene type. This resistance has been
durable in the case of the green leafhopper but not for the brown
planthopper. It has been regionally restricted in usefulness whether
or not durable.

The Bph1 gene for brown planthopper resistance was detected by
methods very similar to those that detect major vertical genes for
blast resistance, i.e. a + or - reaction. The original green leafhopper resistance, on the other hand, was developed in the field
and is a form of tolerance/tolremicity. Is there something to be
learned here? Also, no immunity has been found to stem borers and
progress in accumulating minor genes that appear to keep populations
low is being made.

In my view, whether one obtains a type of resistance that keeps populations low and that is not going to breakdown, or whether a different kind of resistance is obtained that is soon overcome is largely a result of the type of screening procedures that are used to detect "resistance." Additionally, of course, if major vertical genes have co-evolved, they will be findable (both in the host and in the pest or pathogen) in old endemic areas. In such areas, the vertical resistance gene should be expected to be transitory if deployed intensively. I would expect, however, that genes would also exist in such areas for tolerance and population suppression. The problem is to design methods to find them and use them and to suppress the overwhelming desire to rely on a major blocking gene and reaction.

Great opportunities exist for manipulating pest populations of the future by breeding for population suppression. Not only major pests of today but minor or localized pests, and even storage pests, can be manipulated to our advantage through careful analysis of methodology, evolution, and concepts underlying preventative suppression breeding.

REFERENCES

Ahn, S.W., 1981, The slow blasting resistance *in:* IRAT Blast Symposium, March, 1981, Montpellier, France (in press).

Aldrick, S.J., Buddenhagen, I.W., and Reddy, A.K., 1973, The occurrence of bacterial leaf blight in wild and cultivated rice in Australia, *Aust. J. Agric. Res.,* 24:219-227

Bakker, W., 1974, Characterization and ecological aspects of rice yellow mottle virus in Kenya, *Agricultural Research Report* 829, Wageningen, Amsterdam, 152 p.

Bidaux, J.M., 1978, Screening for horizontal resistance to rice blast *Pyricularia oryzae* in Africa. *In:* "Rice in Africa," I.W. Buddenhagen and G.J. Persley, eds., Academic Press, New York, 159-172.

Bidaux, J.M., and Notteghem, J.L., 1979, Nature et stabilité des des facteurs de virulence de *Pyricularia oryzae* en Cote d'Ivoire et a Madagascar, *L'Agron. Trop.,* 28:1135-1144.

Buddenhagen, I.W., and Reddy, A.K., 1972, The host, the environment, *Xanthomonas oryzae,* and the researcher. *In:* "Rice Breeding," International Rice Research Institute, Los Banos, Philippines, 289-295.

Buddenhagen, I.W., 1977, Resistance and vulnerability of tropical crops in relation to their evolution and breeding, *Ann. N.Y. Acad., Sci.,* 287:309-326.

Buddenhagen, I.W., 1979, Rice breeding for tropical African conditions and problems. Presented at the First Annual Research Conference, IITA, October 1979, Ibadan, Nigeria.

Buddenhagen, I.W., Vuong, H.H., and Ba, D.D., 1979, Bacterial leaf
    blight found in Africa, *Int. Rice Res. Newsl.*, 4(1):11.
Buddenhagen, I.W., 1981a, Practical breeding for yield stability
    and durable resistance to rice blast *in:* IRAT Blast Symposium
    March 1981, Montepellier, France (in press).
Buddenhagen, I.W., 1981b, Conceptual and practical considerations
    when breeding for tolerance or resistance. *In:* "Plant Disease
    Control," R.C. Staples and G.H. Toenniessen, eds., John
    Wiley and Sons, Inc., New York, 221-234.
Crill, P.J., Ham, Y.S., and Beachell, H.M., 1979, The rice blast
    disease in Korea and its control with race prediction and gene
    rotation *in:* "Evolution of the gene rotation concept for rice
    blast control, "International Rice Research Institute, Los
    Banos, Philippines.
Crill, P., 1981, Twenty years of plant pathology at the IRRI, *Plant
    Dis. Rep.*, 65:569-574.
Ezuka, A., 1979, Breeding for and genetics of blast resistance in
    Japan. *In:* "Proceedings of the Rice Blast Workshop,"
    International Rice Research Institute, Los Banos, Philippines,
    27-48.
Ikehashi, H., 1979, Implication of the international rice blast
    nursery data to the genetics of resistance, *IRRI Res. Paper
    Ser.* /40, International Rice Research Institute, Los Banos,
    Philippines.
Johnson, R., 1981a, Durable disease resistance *in:* "Strategies for
    the control of cereal diseases." J.F. Jenkyn and R.T. Plumb,
    eds., Blackwell, Oxford, 55-63.
Johnson, R., 1981b, Durable resistance: definition of genetic
    control, and attainment in plant breeding, *Phytopathology*
    71:567-568.
Khush, G.S., 1977, Disease and insect resistance in rice, *Adv.
    Agron.*, 29:265-341.
Lamey, H.A., 1969, Varietal resistance to hoja blanca. *In:* "The
    Virus Diseases of the Rice Plant," John Hopkins Press,
    Baltimore, MD., 293-311.
Ling, K.C., 1972, "Rice Virus Diseases," International Rice
    Research Institute, Los Banos, Philippines, 134 p.
Ling, K.C., Tiongco, E.R., Aguiero, V.M., and Cabauatan, P.Q.,
    1978, "Rice Ragged Stunt Disease in the Philippines, *Int.
    Rice Res. Newsl.*, 16:25 p.
Lozano, J.C., 1977, Identification of bacterial leaf blight in
    rice caused by *Xanthomonas oryzae* in America, *Int. Rice Res.
    Newsl.*, 2(4):4.
Matsuyama, N., 1975, The effect of ample nitrogen fertilizer on
    cellwall materials and its significance to rice blast disease,
    Ann. Phytopath. Soc. Japan, 41:56-61.
Morishima, H., Sano, Y., and Oka, H.I., 1979, Observations on wild
    and cultivated rices and companion weeds in the hilly areas
    of Nepal, India and Thailand. National Institute of Genetics,

Misima, Japan, 97 p.

Nishi, U., Kimura, T., and Maejima, I., 1975, Causal agent of waika disease of rice plants in Japan, *Ann. Phytopath. Soc. Japan,* 41:223-227.

Notteghem, J.L., 1979, Etude de quelques facteurs de la resistance horizontale du riz a la *Pyriculariose, L. Agron. Trop.,* 34(2):180-195.

Ou, S.H., 1972, "Rice Diseases," Commonw. Mycol. Inst., Kew, Surrey, England. 368 p.

Ou, S.H., Nuque, F.L., Ling, K.C., and Aguiero, V., 1976, Two possible new sources of resistance to grassy stunt virus disease of rice, *Int. Rice Res. Newsl.,* 1(1):10.

Ou, S.H., 1977, Possible presence of bacterial blight in Latin America, *Int. Rice Res. Newsl.,* 2(2):5-6.

Ou, S.H., 1979, Breeding rice for resistance to blast - A critical review, *In:* "Proceedings of the Rice Blast Workshop," International Rice Research Institute, Los Banos, Philippines, 81-137.

Ou, S.H., 1980, Pathogen variability and host resistance in rice blast disease, *Ann. Rev. Phytopathol.,* 18:167-187.

Padmanabhan, S.Y., 1973, The great Bengal famine, *Ann. Rev. Phytopathol.,* 11:11-26.

Parmeter, J.R., 1970, "*Rhizoctonia solani,* biology and pathology," Univ. of Calif. Press, Berkeley, CA., 255 p.

Pathak, M.D., and Saxena, R.C., 1980, Breeding approaches in rice, *In:* "Breeding plants resistant to insects," F.G. Maxwell and P.R. Jennings, eds., John Wiley and Sons, Inc., New York, 421-455.

Petpisit, V., Khush, G.S., and Kauffman, H.E., 1977, Inheritance of resistance to bacterial blight in rice, *Crop Sci.,* 17:551-554.

Raymundo, S.A., and Buddenhagen, I.W., 1976, A rice virus disease in West Africa, *Int. Rice Com. Newsl.,* 35-38.

Ryker, T.C., 1943, Physiologic specialization in *Cercospora oryzae, Phytopathology,* 33:70-74.

Saito, Y., 1977, Rice viruses with special reference to particle morphology and relationship with cells and tissues, *Rev. Plant Protec. Res.* 10:83-90.

Saito, Y., Roechon, M., Tantera, D.M., and Iwaki, M., 1975, Small bacilliform particles associated with penyakit habang (tungro-like) disease of rice in Indonesia, *Phytopathology,* 65:793-796.

Sakaguchi, S., Suwa, T., and Murata, N., 1968, Studies on the resistance to bacterial leaf blight, *Xanthomonas oryzae* in the cultivated and wild rices, *Bull. Natl. Agric. Sci., Ser. D.,* 18:1-29.

Scott, P.R., Johnson, R., Wolfe, M.S., Lowe, H.B., and Bennett, F.A., 1980, Host-specificity in cereal parasites in relation to their control, *In:* "Applied Biology, Vol. 5," T.H. Coaker, ed., Academic Press, London, 349-393.

Shekhawat, G.S., Srivastava, D.N., and Rao, Y.P., 1972, Host
    specialization in bacterial leaf-streak pathogen of rice
    (*Oryza sativa* L.), *Xanthomonas translucens* , *Indian J.
    Agric. Sci.*, 42(1):11-15.
Shikata, E., 1979, Rice viruses and MLO's and leafhopper vectors.
    In: "Leafhopper vectors and plant disease agents,"
    K. Maramorosch, and K.F. Harris, eds., Academic Press, Inc.,
    New York, 515-527.
Toriyama, K., 1972, Breeding for resistance to major rice diseases
    in Japan, *In:* "Rice Breeding," International Rice Research
    Institute, Los Banos, Philippines, 253-281.
Van der Muelen, J.G.J., 1951, Rice improvement by hybridization
    and results obtained, *Contr. Gen. Agric. Res. Stan.*, Bogor.
Weereratne, H., Martinez, C., and Jennings, P.R., 1981, Genetic
    strategies in breeding for resistance to rice blast *in:*
    IRAT Blast Symposium, March 1981, Montpellier, France.

DISCUSSION

KRANZ: I like your holistic approach, which is pragmatic and does
not require much research input. However, how many ecotypes do
you have to cater for? Certainly research, particularly by
epidemiologists, may help to reduce the number of apparently different
ecotypes to some essential ones.

BUDDENHAGEN: The different ways rice is grown in relation to
moisture regime largely dictate the need for different rice ecotypes.
Whether rice is flooded or is grown on aerobic soils has a major
influence on different disease and insect vulnerabilities. The level
of technological development also influences the type of rice one
would select. Within any particular ecology (i.e. good water control,
irrigation), there may be special areas where unbalanced soils
dictate the need for special tolerances, such as iron toxicity.

MICKE: Looking at your very promising pragmatic approach, I would
like to know your opinion as to where efforts of better seed
propagation, seed testing and seed certification, such as those
intended by the FAO seed program for developing countries, should
fit in.

BUDDENHAGEN: This would depend on the level of technology and
sophistication in a country. If rice varieties are developed, as
they should be, with sufficient resistance to seed-borne diseases,
farmers should be able to save their own seed without negative
effects. I recommend local breeding and local dissemination of seed
by whatever means are most efficient. I have seen several cases in
Africa where sophisticated seed programs have been tried with little

effect, because farmers already had their own varieties while the
main lack in the seed program was superior, locally-adapted
varieties with stable resistance. The problem was not seed
testing or seed certification, as such.

EVALUATION OF FABA BEAN CULTIVARS FOR SLOW RUSTING RESISTANCE

C.C. Bernier and R.L. Conner

Department of Plant Science
University of Manitoba
Canada

Uromyces viciae-fabae (Pers.) Schroet. infects faba beans
(*Vicia faba* L.), peas (*Pisum sativum* L.), lentils (*Lens culinaris*
Medic.) and several wild and cultivated spp. of *Vicia* and *Lathyrus*.
This rust is common on faba beans throughout the Mediterranean
region and can cause yield losses of 5-20 percent in Egypt. Since
the introduction of faba beans to western Canada in 1970, rust has
been reported in both Manitoba and Saskatchewan (Bernier, 1975).

In recent studies using seven faba bean inbred lines selected
for uniform reaction to either of two rust isolates, three different
dominant resistance genes were identified and seven rust races were
differentiated (Conner, 1981; Conner and Bernier, 1982b). These
genes are not expected to provide durable resistance because of
the high potential for production of virulent pathotypes on native
species of *Vicia* and *Lathyrus* (Conner and Bernier, 1982a). However, they might provide moderate durability when in multiline
cultivars.

In view of the often transitory nature of specific resistance,
faba beans were also evaluated for slow rusting (rate limiting)
resistance (Conner, 1981). Twenty-five open-pollinated accessions
showing moderate to high susceptibility were compared in small
plots over three years. Due to low seed supply, the number of
rows/plot and plants/row increased from one row with 12-15 plants
to three rows with 60-70 plants by the third year. A randomized
complete block design with five replicates was used. Two rows of
a highly susceptible line were planted at the ends of each plot
to act as spreader rows. These were inoculated with a mixture of
isolates in 1978 and race 1 and 2 at each site in 1979 and '80.
Starting two weeks after inoculation, the number of plants infected

and average percentage leaf area infected for each infected plant within a plot were recorded at weekly intervals for four to five weeks. Values of leaf area infected over time were summarized as area under the disease progress curve (AUDPC) (Wilcoxson *et al.*, 1975). Analysis of variance was carried out on these values and the means of each line were compared using Tukey's w- procedure.

Significant differences between lines were found in all three years for AUDPC values. Three lines consistently behaved as slow rusters over three years of testing and another six lines had values that shifted from low to intermediate (Figure 1). Final rust severity values were well correlated to AUDPC values (r = 0.69 - 0.86). Because of this association, final rust severity should be useful for selecting potential slow rusting lines in preliminary tests.

It would seem that durable resistance in a partially out-crossing crop such as faba bean might best be achieved by combining several slow rusting genotypes through a population improvement and recurrent selection program.

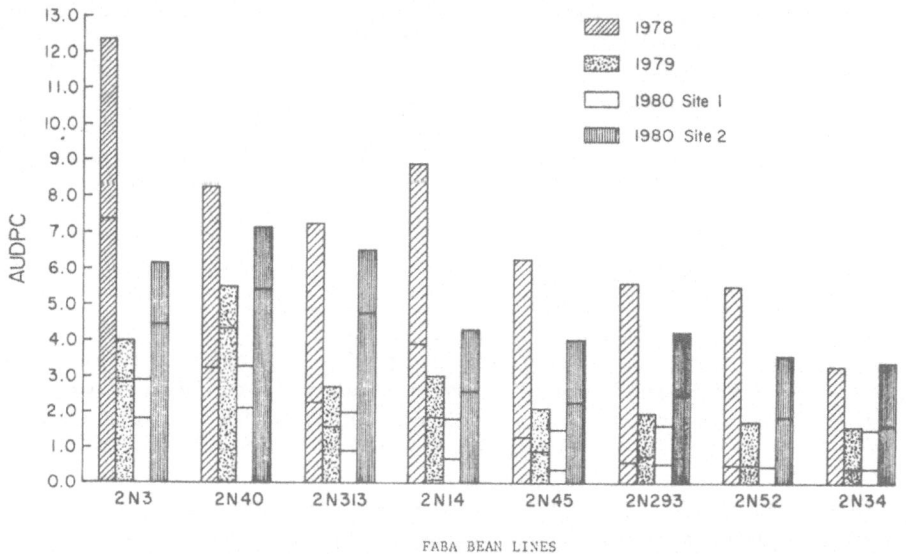

Fig. 1.   Area under the disease progress curve (AUDPC) for selected faba bean lines for three years.

REFERENCES

Bernier, C.C., 1975, Diseases of pulse crops and their control, In:
    Oilseed and pulse crops in Western Canada - A Symposium,
    J.T. Harapiak, ed. Western Co-operative Fertilizers Ltd.,
    Calgary, Alta, pp. 439-454.
Conner, R.L., 1981, Evaluation of resistance to rust (*Uromyces
    viciae-fabae*) in faba bean (*Vicia faba*) and pathogenic varia-
    bility in *U. viciae-fabae*, Ph.D. Thesis, University of
    Manitoba, February 1981, 118pp.
Conner, R.L., and Bernier, C.C., 1982a, Host range of *Uromyces
    viciae-fabae*, *Phytopathology* (in press).
Conner, R.L., and Bernier, C.C., 1982b, Race identification in
    Uromyces viciae-fabae, *Can. J. Plant Path*. (in press).
Wilcoxson, R.D., Skovmand, B., and Atif, A.H., 1975, Evaluation of
    wheat cultivars for ability to retard development of stem
    rust, *Ann. appl. Biol*., 80:275-281.

INTERACTIONS BETWEEN PEARL MILLET VARIETIES AND

Sclerospora graminicola ISOLATES

J. M. Waller[*] and Sarah L. Ball[**]

[*]Commonwealth Mycological Institute
Ferry Lane, Kew, Surrey, England
[**]Department of Agriculture and Horticulture
University of Reading
Reading, Berks., England

INTRODUCTION

Pearl or bullrush millet (*Pennisetum typhoides*) is a major cereal of semi-arid areas, being grown extensively in the drier areas of India and Africa (especially in the sahelian zone). It is an outbreeding plant originating in Africa where many different traditional landraces are grown, each particular to a certain area. Among the several diseases of this crop, downy mildew caused by *Sclerospora graminicola* (Sacc.) Schroet., is particularly damaging. The disease is systemic, infecting very young seedlings or developing tiller shoots. The pathogen is carried in the apical meristem so that all organs produced from an infected apical meristem become diseased. Mycelium in leaf primordia infects the subsequent developing leaves and the flower is usually converted to a leafy structure (virescence). Very occasionally, ears may be only partly diseased. The disease is fairly prevalent on local landraces of pearl millet in many areas. In the Gambia for example, 10-15 percent incidence occurs regularly on late planted millet and records suggest that this has been the usual level of disease each year since the 1930s.

The pathogen survives the dry season as dormant oospores in the soil. These germinate to cause primary infection of young seedlings soon after the first crop of early millet is planted. Diseased seedlings begin to produce sporangia 10-20 days later which are dispersed to cause secondary infection (Singh & Williams, 1980). Late sown millet can be infected either by oospores in the soil or by sporangia from the early millet crop. Oospores are produced in infected tissue and are returned to the soil in necrotic tissue. *S. graminicola* has been shown to be heterothallic, oospores mainly being produced in plants infected by a mixed sporangial inoculum containing at least two sets of compatibility types (Michelmore *et al.*, 1982), thus the pathogen is outbreeding as well as the host.

Multilocational trials coordinated by ICRISAT in India and Africa have shown that variability of disease incidence between locations is large and there is evidence of differences in cultivar susceptibility between Africa and Asia. It is necessary to understand whether this is due to differences in pathogenicity or to environmental effects. Quantifying these differences also present problems with systemic diseases because of the binomial distribution of disease incidence and the consequent difficulty of getting statistically valid comparisons of susceptibility with relatively small populations of plants.

METHODS

A project was established by the U.K. Overseas Development Administration at Reading University in 1979 to determine whether isolates from Africa and Asia showed significant difference in pathogenicity by direct comparison of their behavior under uniform conditions on a range of pearl millet varieties which were being used in the international trials. Inoculum consisted of oospore samples in leaf debris which can remain viable for several years; however, oospore infection of seedlings proved to be low and erratic. A system similar to that used at ICRISAT (Williams *et al.*, 1981) was employed. Oospore infection of the universal suscept 7042 was established and diseased plants of this variety were used as sources of sporangial inoculum in polythene tunnels with controlled minimum temperatures of 18-20°C. Test varieties were planted in pots between the rows of spreader plants (five seedlings/ pot and 20 pots/plot). There were two replicates of eight varieties x one inoculum source in each tunnel. Initial infection of seedlings occurs in the first few days and symptom expression is complete by four weeks when final records of disease incidence and symptom type are taken. Tunnels are sterilised between treatments to remove any trace of previous inoculum.

RESULTS

Five isolates (two from Kamboinse in Upper Volta, two from ICRISAT and one from Jamnagar in India) have been tested on eight potential differential cultivars. Because the data of disease incidence are binomially distributed, they cannot be analysed efficiently by conventional methods and a program devised at Rothamsted and Reading (the general linear interactive model - GLIM) was used on logit transformed data (Ball & Pike, in press). There were

significant differences between isolates and between cultivars and
slight difference in rank order of the cultivars with different inoculum sources. Not only were there differences in pathogenicity
between the Indian and African isolates, but substantial differences
occurred between isolates from the same area but from different host
sources.

The type of host reaction to infection could be assigned to two
categories, (a) much leaf distortion, stunting and chlorosis with relatively little sporulation, (b) little effect on leaf morphology
but heavy sporulation. The frequency with which each category occurred
in a cultivar was constant and did not vary with the different pathogen isolates; it therefore appears to be a stable characteristic of
the cultivar. Detailed results from these experiments are being published elsewhere (Ball, in press).

DISCUSSION

Although the data provides no definitive evidence for a vertical
pathosystem, it is doubtful whether the differential interaction is a
valid test of this where cultivars and isolates are genotypically
heterogeneous. Due to their outbreeding nature, pearl millet cultivars
behave more as multilines or variety mixtures and although the disease
has a qualitative effect at the level of the individual, at the population level disease is a quantitative character (i.e. the frequency
of individuals contracting the disease). Thus, despite the resemblance
of the isolate behavior to that of horizontal pathotypes, vertical
resistance may still operate, as is suggested by the lack of durability
of downy mildew resistance in many cultivars, but with many of its
characteristics obscured in a multiline population.

With systemic diseases, it is important to determine whether un-
infected plants are chance escapes which have not encountered a matching pathogen genotype, or whether they are resistant to all the
virulence types in the pathogen population. Where both host and pathogen are outbreeding the low frequency of matching encounters would play
play an important role in limiting disease incidence. The occasional
later infection of secondary tillers is indicative of a late matching
encounter. Furthermore, there is evidence for increasing disease incidence with increasing inoculum, thereby increasing the probability
of matching encounters. Nevertheless, the fact that some pearl millet
cultivars have remained resistant in many areas over several seasons
(ICRISAT, 1978-80) suggests that they have a genetic component which
confers stability of resistance to the pathogen population as a whole.
The relative importance of chance escape or true resistance can be
resolved by challenging clones with sporangial inoculum produced
from the same plant.

An interesting feature of this and other destructive systemic
diseases is that because plants infected at an early stage produce

no seed or pollen, susceptible genotypes would be eliminated fairly
rapidly from the population if the pathosystem was of the more
conventional vertical type. This does not happen in landraces,
where the level of disease remains fairly constant (but perhaps at
an agriculturally unacceptable high level) over many generations.
Robinson (1980) explained this by postulating an evolutionary stable
strategy involving many vertical genes, where maximum spatial and
temporal discontinuity of susceptible host tissue results in relatively rare matching alloinfection of each generation. With an
outbreeding pathogen, oospores produced in a diseased plant would
be genotypically different from those which infected it. Thus the
pathosystem can be viewed as an ever-changing mosaic of pathogen
genotypes interacting with an ever-changing mosaic of host genotypes.
Any shift of host genotypes can be matched by a similar shift in
pathogen genotype spectrum and any reduction in the frequency or
number of resistance factors in the outbreeding host population
would increase the frequency at which susceptible genotypes would
occur.

       In a variable outbreeding pathosystem only a proportion of the
total pathogenic variability within the pathogen population would be
sampled at any one location or time, or by any host or group of hosts.
Isolates sampled from the landraces would be representative of the
'wild' pathogen population but samples from field trials would
necessarily represent portions of this population selected out by the
cultivar genotype. Thus, changes in the type of cultivars grown over
a period would account for changes in the virulence of the sampled
inoculum. This could account for the differences in pathogenicity
of the Kamboinse inoculum from different cultivar sources and the
apparent change in pathogenicity with time at ICRISAT.

       One way of breaking this apparent vicious circle of continually
adapting pathogen aggressiveness is to restrict free gene transfer
in the pathogen population. The stunted reaction type which restricts
asexual sporulation and oospore production would provide a suitable
method of achieving this. Not only is this character apparently
stable with different inoculum sources on a particular variety, it
has the important epidemiological effect of restricting secondary
infection and inoculum carry-over, besides reducing gene flow in
the pathogen population.

REFERENCES

Ball, Sarah, L., in press, Pathogenic variability of Downy Mildew
       (Sclerospora graminicola (Sacc.) Schroet.) on Pearl Millet
       (Pennisetum americanum (L.)), Ann. appl. Biol., 101.
Ball, Sarah, L., and Pike, P.D., in press, Use of a statistical
       model to investigate the pathogenic variation of Sclerospora
       graminicola downy mildew on pearl millet, Ann. appl. Biol., 101.

ICRISAT, 1978-80, Reports of the 1978, 1979 and 1980 International
     Pearl Millet Downy Mildew Nursery.
Michelmore, R.W., Pawar, M.N., and Williams, R.J., 1982, Hetero-
     thallism in *Sclerospora graminicola*, *Phytopathology* (in press).
Robinson, R.A., 1980, New concepts in breeding for disease resistance,
     Ann. Rev. Phytopath., 18:189-210.
Singh, S.D., and Williams, R.J., 1980, The role of sporangia in the
     epidemiology of pearl millet downy mildew, *Phytopathology*, 70:
     1187-1190.
Williams, R.J., Singh, S.D., and Pawar, M.N., 1981, An improved field
     screening technique for downy mildew resistance in pearl millet,
     Plant Disease, 65:239-241.

PARTICIPANTS

A. ALLAVENA
Ist. Sperimentale per l'Orticoltura, Sezione di
20075 MONTANASO LOMBARDO (MI) Italy

L.J. ANTUNES DEL DUCA
EMBRAPA Passo Fundo, Caixa Postal 351
99.100 PASSO FUNDO (RS) Brazil

N. BANGURA
WARDA, ADRAO, E.J. Roye Memorial Bldg., P.O. Box 1019
MONROVIA Liberia

RITA BASILE
Ist. Sperimentale per la Patologia Vegetale
Via Casal de' Pazzi 250 - 00100 ROMA Italy

M.A. BEEK
UNDP/SF Project 381 (BRZ/69/535) Caixa Postal 351
99.100 PASSO FUNDO (RS) Brazil

C. BERNIER
Dept. Plant Sciences, The University of Manitoba
WINNIPEG (Manitoba R3T 2N2) Canada

O.S. BINDRA
FAO Host-Plant Resistance Expert, c/o UNDP, P.O.Box 913
KHARTOUM Sudan

A. BOTTALICO
Centro di Studi per le Tossine del C.N.R.
Via G. Amendola, 165/A - 70126 BARI Italy

ANGELA CAPUSSO-TOSI
Istituto di Nematologia Agraria del C.N.R.
Via G. Amendola, 165/A - 70126 BARI Italy

D. CASULLI
Ist. di Patologia Vegetale, Fac. Agraria
Via G. Amendola, 165/A - 70126 BARI Italy

F. CERVONE
Ist. Orto Botanico, Cattedra di Fisiologia Vegetale
Largo Cristina di Savoia, 24 - 00165 ROMA Italy

L. CHIARAPPA
Plant Protection Service FAO,
Via delle Terme di Caracalla - 00100 ROMA Italy

A. CICCARONE
Ist. Patologia Vegetale, Facoltà di Agraria
Via G. Amendola, 165/A - 70126 BARI Italy

M. CIRULLI
Ist. Patologia Vegetale, Facoltà di Agraria
Via G. Amendola, 165/A - 70126 BARI Italy

MARIA I. COIRO
Istituto di Nematologia Agraria del C.N.R.
Via G. Amendola, 165/A - 70126 BARI Italy

LUCIANA CORAZZA
Ist. Sperimentale per la Patologia Vegetale
Via Casal de' Pazzi, 250 - 00156 ROMA Italy

T.M. CROSBIE
Dept. Agronomy, Iowa State University of Science and
Technology
AMES (Iowa 50011) USA

F. DE CICCO
Ist. Patologia Vegetale, Facoltà di Agraria
Via G. Amendola, 165/A - 70126 BARI Italy

W.A.J. DE MILLIANO
Dept. Plant Pathology, Agricultural University
Binnenhaven 9 - 6709-PD WAGENINGEN The Netherlands

A. DINOOR
The Hebrew University of Jerusalem,
The Levi Eshkol School of Agriculture
P.O. Box 12 - REHOVOT 76-100 Israel

M. DI VITO
Istituto di Nematologia Agraria del C.N.R.
Via G. Amendola, 165/A - 70126 BARI Italy

F. ELIA
Istituto di Nematologia Agraria del C.N.R.
Via G. Amendola, 165/A - 70126 BARI Italy

G.L ERCOLANI
Ist. di Patologia Vegetale, Facoltà di Agraria
Via G. Amendola, 165/A - 70126 BARI

G. FANIZZA
Ist. Miglioramento Genetico, Facoltà di Agraria
Via G. Amendola, 165/A - 70126 BARI

F. FIUME
Ist. Sperimentale per l'Orticoltura
Via Piacenza, 21 -   84098 PONTECAGNANO (SA)   Italy

R.C. GARBER
Dept. Plant Pathology, Cornell University
334 Plant Sciences Bldg - ITHACA (N.Y. 14853)   USA

A. GRANITI
Ist. Patologia Vegetale, Facoltà di Agraria
Via G. Amendola, 165/A - 70126 BARI Italy

N. GRECO
Istituto di Nematologia Agraria del C.N.R.
Via G. Amendola, 165/A - 70126 BARI Italy

R.M. JIMENEZ-DIAZ
Escuela Tecnica Superior de Ingenieros Agronomos,
Finca "Alameda del Obispo", Apdo de Correos N. 246
CORDOBA   Spain

H.J. JØRGENSEN
Risø National Laboratory, Postbox 49
DK-4000 ROSKILDE Denmark

H. KHALIFA
Ministry of Agriculture, Agric. Research Corp.,
Cotton Breeding Section - WAD MEDANI Sudan

C. LAVIOLA
Ist. Patologia Vegetale, Università degli Studi
Viale delle Scienze, 2 - 90128 PALERMO Italy

N. LUISI
Ist. Patologia Vegetale, Facoltà di Agraria
Via G. Amendola, 165/A - 70126 BARI Italy

G.P. MARTELLI
Ist. Patologia Vegetale, Facoltà di Agraria
Via G. Amendola, 165/A - 70126 BARI Italy

J.M. McDERMOTT
Pestology Centre, Biological Sciences Dept.
Simon Fraser University
BURNABY (B.C. V5A 1S6) Canada

J. MEUNIER
IRHO, Departement Selection, GERDAT
B.P. 5035 - 34032 MONTPELLIER CEDEX France

A. MICKE
Plant Breeding & Genetics Section
Wagramerstrasse 5 Box 100
A-1400 WIEN Austria

L. MITTENPERGHER
Centro di Studio per la Patologia delle Specie Legnose
e Montane
Piazzale delle Cascine, 28 - 50144 FIRENZE Italy

LISA MUNK
Dept. Plant Pathology, The Royal Veterinary & Agric.
University
Agrovey 10, Højbakkegard - DK-2630 TÄSTRUP Denmark

HANNE ØSTERGARD
Risø National Laboratory
Postbox - DK-4000 ROSKILDE Denmark

O.T. PAGE
The International Potato Center
Apartado 5969 - LIMA Peru

M. PALMAS
Orto Botanico
Viale Fra Ignazio, 13 - 09100 CAGLIARI Italy

FRANCOISE PERSON-DEDRYVER
INRA - Lab. de Recherches de la Chaire de Zoologie
Domaine de la Motte-au-Vicomte, B.P. 29
35650 LE RHEU France

F. PIRO
Ist. Sperimentale per il Tabacco
Via P. Vitiello, 66 - 84014 SCAFATI (SA) Italy

S. PORCELLI
Ist. Sperimentale per l'Orticoltura
Via Piacenza, 21 - 84098 PONTECAGNANO (SA) Italy

P. PRESTE
Institute of Plant Pathology, University
P.O. Box 744 - S-750 07 UPPSALA Sweden

A. QUACQUARELLI
Ist. Patologia Vegetale, Facoltà di Agraria
Via G. Amendola, 165/A - 70126 BARI Italy

P. RADDI
Centro di Studio per la Patologia delle Specie Legnose
e Montane
Piazzale delle Cascine, 28 - 50144 FIRENZE Italy

J.-L. RENARD
I.R.H.O.
Boite Postale 8 - DABOU Ivory Coast

M. SALERNO
Ist. Patologia Vegetale, Facoltà di Agraria
Via G. Amendola, 165/A - 70126 BARI Italy

MARIA CORINNA SANGUINETI
Ist. Agronomia Generale e Coltivazioni Erbacee
Via Filippo Re, 8 - 40126 BOLOGNA Italy

F.T. SCARASCIA-MUGNOZZA
Facoltà di Agraria dell'Università della Tuscia
01100 VITERBO Italy

M. SCHIAVI
Istituto Sperimentale per l'Orticoltura
20075 MONTANASO LOMBARDO (MI) Italy

G. SIDHU
The University of Nebraska-Lincoln,
Institute of Agriculture & Natural Resources
LINCOLN (Nebraska 68583) USA

ANNA MARIA SIMEONE
Ist. Sperimentale per la Frutticoltura
Via Fioranello, 52
00040 CIAMPINO AEROPORTO (ROMA) Italy

A. SINISCALCO
Ist. Patologia Vegetale, Facoltà di Agraria
Via G. Amendola, 165/A - 70126 BARI Italy

D. SIPPEL
c/o UNDP,   P.O. Box 913   -   KHARTOUM   Sudan

D. SISTO
Ist. Patologia Vegetale, Facoltã di Agraria
Via G. Amendola, 165/A   -   70126 BARI   Italy

G.P. SORESSI
Ist. Sperimentale per l'Orticoltura
20075 MONTANASO LOMBARDO (MI)   Italy

E. TORRES
Centro Internacional Mejoramiento de Maiz y Trigo
LONDRES 40   Mexico 6 D.F.

H. Van der BEEK
IPHR (FAO)   c/o PUND,   CASIER-ONU
RABAT   Morocco

C.H.   Van SILFHOUT
Research Institute for Plant Protection
Binnenhaven 12, P.O. Box 42
6700-AA   WAGENINGEN   The Netherlands

J.M. WALLER
Commonwealth Mycological Institute,
Ferry Lane,
KEW,   RICHMOND   (Surrey TW9 3AF)   U.K.

F. WILLIAMS
Simon Fraser University,   Pestology Centre,
Biological Sciences Department
BURNABY   (B.C. V5A 1S6)   Canada

G. ZACCHEO
Istituto di Nematologia Agraria del C.N.R.
Via G. Amendola, 165/A   -   70126 BARI   Italy

GIUSEPPINA ZITELLI
Ist. Sperimentale per la Cerealicoltura
Via Cassia, 176   -   00191 ROMA   Italy

LECTURERS

J.A. BROWNING
Dept. Plant Sciences, Texas A&M University,
College of Agriculture
COLLEGE STATION (Texas 77843) USA

I.W. BUDDENHAGEN
Dept. Agronomy & Plant Pathology, University of California
DAVIS (Calif. 95616) USA

O. DE PONTI
Institute for Horticultural Plant Breeding
Postbus 16 - WAGENINGEN The Netherlands

A.B. ESKES
FAO Instituto Agronomico, Secao Genetica
Caixa Postal 28 - 13100 CAMPINAS (S.P.) Brazil

R. JOHNSON
Plant Breeding Institute
Maris Lane, Trumpington
CAMBRIDGE (CB2 2LQ) U.K.

J. KRANZ
Abt. Phytopathologie und Angew. Entomologie
"Justus Liebig-Universitat Giessen"
Tropen Institute
Schottstrasse 2-4
6300 GIESSEN Fed. Rep. Germany

F. LAMBERTI
Istituto di Nematologia Agraria del C.N.R.
Via G. Amendola, 165/A - 70126 BARI Italy

J.E. PARLEVLIET
Institute of Plant Breeding, Agricultural University,
166 Lawickse Allee
WAGENINGEN   The Netherlands

C.B. PERSON
Dept. of Botany, University of British Columbia
VANCOUVER   (B.C.)   Canada

C.A.J. PUTTER
Pathosystems Research Unit, University of Natal
P.O. Box 375
PIETERMARITZBURG 3200   Rep. South Africa

R.A. ROBINSON
Simon Fraser University, Pestology Centre,
Biological Sciences Dept.
BURNABY   (B.C.   V5A 1S6)   Canada

E.V. SHARP
Agric. Experiment Station, College of Agriculture
Montana State University
BOZEMAN   (Montana 59717)   USA

C.E. TAYLOR
Scottish Crop Research Institute
Invergowrie,   DUNDEE   DD2 5DA, Scotland,   U.K.

N.A. Van der GRAAFF
Plant Protection Service, FAO
Via delle Terme di Caracalla   -   00100   ROMA   Italy

M.S. WOLFE
Plant Breeding Institute
Maris Lane, Trumpington
CAMBRIDGE   (CB2 2LQ)   U.K.

J.C. ZADOKS
Dept. Phytopathology, Agricultural University
Binnenhaven 9   -   6709-PD WAGENINGEN   The Netherlands